Technician mathematics 2
Third edition

Series Editor:

D. R. Browning, B.Sc., F.R.S.C., C.Chem., A.R.T.C.S.
Formerly Principal Lecturer and Head of Chemistry, Bristol
Polytechnic

Technician mathematics 2

Third edition

J. O. Bird
B.Sc.(Hons), C.Math., F.I.M.A., C.Eng., M.I.E.E., F.Coll.P., F.I.E.I.E

A. J. C. May
B.A., C.Eng., M.I.Mech.E., F.I.E.I.E., M.B.I.M.

Longman
Scientific &
Technical

Longman Scientific & Technical
Longman Group UK Limited
Longman House, Burnt Mill, Harlow
Essex CM20 2JE, England
and Associated Companies throughout the world

First published 1978
Second edition 1982
Third edition 1994

British Library Cataloguing in Publication Data
A catalogue entry for this title is available from the
British Library.

ISBN 0-582-23427-1

Set by Keyword in 10/12 pt Times
Printed in Malaysia by VVP

Contents

Preface

Technician mathematics 2 is the second in a series which deal simply and carefully with the fundamental mathematics essential in the development of technicians and engineers.

The material for the third edition of this successful tectbook has been updated and expanded to cover (together with *Technician mathematics 3*) the main areas of the Business and Technology Education Council's 'Mathematics for Engineers' module for National Certificate and Diploma courses likely to be studied in a 'Mathematics' lesson. Four-figure tables are assumed to be redundant; the electronic calculator is now an accepted essential tool in technican and engineering studies. The modifications in the third edition reflect this change.

The definition and solution of engineering problems relies on the ability to represent systems and their behaviour in mathematical terms, which in turn, depends on the use of various mathematical tools. This textbook deals with some of those mathematical skills and concepts relevant to effective performance in engineering employment at technican level.

The aim of this book is to model simple engineering systems, to generate numerical values for system parameters, to manipulate data to determine system response in defined conditions, to evaluate the effects on systems of changes in variables and to communicate ideas mathematically.

Each topic considered in the text is presented in a way that assumes in the reader only the knowledge attained in *Technician mathematics 1*. This practical mathematics book contains over 250 detailed worked problems, followed by some 800 further problems with answers. Although specifically written for the early stages of BTEC National Certificate and Diploma Mathematics, the book will also be suitable for NVQ and gNVQ studies and for students studying A level Mathematics.

I would like to express my appreciation for the friendly cooperation and helpful advice given by the publishers, and to Mrs. Elaine Woolley for the excellent typing of the manuscript.

Finally, I would like to pay tribute to my colleague, friend and co-author Tony May who sadly died in 1991 after a short illness.

John O. Bird
Highbury College, Portsmouth

Chapter 1

Evaluation and transposition of formulae

1.1   Introduction

The statement $R = \dfrac{\rho l}{a}$ is called a **formula** for R in terms of ρ, l and a. The single term on the left-hand side is called the **subject of the formula**. If the value of three of the four symbols are given then the fourth may be determined. The most common method used for evaluating formulae is by using an electronic calculator (see *Technician Mathematics 1*).

1.2   Evaluation of formulae

Sequence of steps

A logical sequence of steps is necessary to evaluate certain expressions. For example, if $y = ax^b$ then it is important to realise that the power b refers only to the x term. If the power 'b' were intended to refer to the 'a' term as well, then a bracket would have to be employed, i.e. $y = (ax)^b$. If $a = 2$, $b = 4$ and $x = 3$, then to calculate:

(i) $y = ax^b$, x^b is calculated first, i.e. $3^4 = 81$, and then this value is multiplied by 'a'. Hence $y = 2 \times 81 = \mathbf{162}$.

(ii) $y = (ax)^b$, 'a' is firstly multiplied by x, i.e. $2 \times 3 = 6$, and then this value is raised to the power 'b'. Hence $y = (6)^4 = \mathbf{1\,296}$.

(iii) $y = (a + x)^b$, the value of the expression within the brackets must be determined first, i.e. $(a + x) = (2 + 3) = 5$, and then this value is raised to the power 'b'. Hence $y = (5)^4 = \mathbf{625}$.

(iv) $y = (x + b)(b - a)$, the values of the separate bracketed terms are determined first, i.e. $(x + b) = (3 + 4) = 7$ and $(b - a) = (4 - 2) = 2$. Hence $y = (7)(2) = \mathbf{14}$.

In more complicated expressions, such as $V = (\pi r^2 h) + \left(\dfrac{2}{3}\pi r^3\right)$, each bracketed term is calculated separately and then added together. Similarly, if $A = \sqrt{[s(s - a)(s - b)(s - c)]}$, then the sequence of steps is firstly to evaluate each of the bracketed terms $(s - a)$, $(s - b)$ and $(s - c)$, then to determine the product of s, $(s - a)$, $(s - b)$ and $(s - c)$, and then to take the square root of the result.

When using a calculator for such multiple step functions as shown above use is made of either

(a)   a **'pencil and paper' method**, whereby the result of each separate step of the sequence is noted on paper, or

(b)   the calculator **memory facility**.

(See Problems 1 to 9)

Tables of values

When plotting graphs it is often necessary to draw up a table of values by carrying out repeated calculations from a given equation or formula for different values of the variables. For example, if a table of integer values from $x = -2$ to $x = +3$ is required for the formula $y = 3.2x^2 - 2.6$ then it may be achieved as follows:

For each value of x the sequence of steps is (i) square x, (ii) multiply by 3.2, and (iii) subtract 2.6. A table is produced as shown below.

x	-2	-1	0	1	2	3
x^2	4	1	0	1	4	9
$3.2x^2$ -2.6	12.8 -2.6	3.2 -2.6	0 -2.6	3.2 -2.6	12.8 -2.6	28.8 -2.6
$y = 3.2x^2 - 2.6$	10.2	0.6	-2.6	0.6	10.2	26.2

This type of calculation can be performed quickly with a calculator and, as long as the correct sequence of steps is adhered to, intermediate values need not be noted in a table. In this case the table of values would simply be:

x	-2	-1	0	1	2	3
y	10.2	0.6	-2.6	0.6	10.2	26.2

(see Problem 10).

Errors, accuracy and approximate values

In order to find out how long it takes a person to walk 1 kilometre at constant speed, several methods may be adopted. A reasonable walking speed is 5 or 6 kilometres per hour: using this data, a result of 10 to 12 minutes can be calculated. So in integer values it will take, say, 11 minutes to walk

1 kilometre. For a more accurate result, a distance of 1 kilometre can be marked, using the recording meter on a car as a measuring device. A person can be timed by a wrist watch walking this distance and the resultobtained may be accurate correct to one decimal place. By using a stop watch and carefully measuring the distance using surveying equipment a result which is correct to two decimal places may be achieved. Ultimately, using a device calibrated at the National Physics Laboratory and an atomic clock which only varies from the correct time by about 5 seconds in 700 years, a result which is correct to several decimal places can be achieved.

In all these cases the result is correct to a stated accuracy only and is not the exact time taken by a person to walk 1 kilometre. In all problems in which the measurement of distance, time, weight, or other quantities occurs, an exact answer cannot be given, only an answer which is correct to a stated degree of accuracy. To take account of this an **error due to measurement** is said to exist.

When numbers are expressed as decimal fractions, an error frequently exists. The proper fraction $\frac{9}{17}$ is equal to 0.529 411 8 ... but will be expressed as, say, 0.529 correct to 3 significant figures or 0.529 4 correct to 4 decimal places. The error introduced in this case is called a **rounding off error** (or a truncation error). Care must be taken not to round off values too soon. For example $(1.325\,82)^4 = 3.089\,86$, correct to 6 significant figures, or 3.10 correct to 3 significant figures. However, if 1.325 82 had been rounded off to 1.33 before being raised to the power 4 then $(1.33)^4 = 3.129\,01$, correct to 6 significant figures or 3.13 correct to 3 significant figures.

Errors can occur in calculations. When evaluating an expression errors can be made when performing arithmetic operations such as multiplication, division, addition and subtraction. For example, if when multiplying 27.63 by 62.7 the answer obtained is 1 731.401 then an error has been made since $27.63 \times 62.7 = 1\,732.401$. This type of error is known as a **blunder**.

Another kind of error may occur due to the incorrect positioning of the decimal point after a calculation has been completed, and this is called an **order of magnitude error**. If, when determining the value of $\dfrac{274.3}{0.067\,1}$ the result is found to be 408.79 instead of 4 087.9, correct to 5 significant figures, then an order of magnitude error will have been made.

To take account of measurement errors it is usual to limit answers so that the result given **is not more than one significant figure greater than the least accurate number given in the data**. For example, it is required to find the resistance of a resistor in an electrical circuit when a voltage of 1.34 volts results in a current flow of 7.3 amperes. The relationship is $R = \dfrac{V}{I}$. Thus

$$R = \frac{1.34}{7.3} = 0.183\,561\,6 \ldots \text{ohms}$$

The least accurate number in the data is 7.3 A, since its value is only given correct to two significant figures. Thus the answer should not be expressed to more than a three significant figure accuracy. Hence

$$R = \frac{1.34}{7.3} = 0.184 \text{ ohms}$$

would be the normal way of writing this answer.

Blunders and order of magnitude errors can be reduced by determining approximate values of calculations. There is no hard and fast rule for doing this, but generally the problem is reduced to a mental arithmetic type problem by expressing numbers correct to one or two significant figures only and reducing the calculation to standard form if necessary. Then the approximate value of

$$\frac{289.63 \times 0.047\,1}{73.824} \text{ could be } \frac{3 \times 10^2 \times 5 \times 10^{-2}}{7 \times 10}$$

Using the laws of indices this becomes $\frac{15}{7} \times 10^{-1} \approx 0.2$. That is, an answer of about 0.2 will be expected. On performing the calculation, if an answer of 0.184 8 is obtained, it will indicate that the calculation has probably been done correctly. An answer of 0.482 7 will suggest that a blunder has occurred or an answer of 0.018 48 that an order of magnitude error has occurred. If either of these last two answers are obtained, the calculation should be checked and repeated if necessary.

Answers which do not seem feasible must be checked and repeated as necessary. If as a result of a calculation the area of a small triangle is found to be thousands of square kilometres, or the height of man tens of metres, then it is likely that an error has occurred. For example, if 5 kg of sugar cost 125p and it is required to find out how much 2 kg cost, then an answer of $125 \times \frac{5}{2}$ or 312.5p is obviously incorrect since it must cost less to buy 2 kg than 5 kg.

When determining an answer to a calculation the points listed below should be considered to ascertain whether the result obtained is likely to be correct and expressed to a reasonable degree of accuracy:

(a) the answer should not be more accurate than the data on which it is based;

(b) errors do exist in all problems based on measurements;

(c) errors do exist in all problems where rounding off has occurred;

(d) blunders and order of magnitude errors should be eliminated as far as possible by determining the approximate value of a calculation; and

(e) a check should be made to see whether the answer obtained seems feasible.

(See Problems 11 and 12)

Evaluation of polynomials

Two methods are available for the evaluation of polynomials:

(*a*)   **taking terms in sequence**

Consider a polynomial $y = ax^4 + bx^3 + cx^2 + dx + e$
Each term has to be determined separately and the individual terms
are then added together. When using a calculator a memory facility is
required or a 'paper and pencil' exercise must be used.

(*b*)   **nested multiplication**

$y = ax^4 + bx^3 + cx^2 + dx + e$ can be rewritten as

$$y = \{[(ax + b)x + c]x + d\}x + e$$

We can now evaluate the innermost bracket $(ax + b)$, multiply this by
x, add c, multiply $[(ax + b)x + c]$ by x etc. In this way multiplication
can be carried out using a calculator **continuously** without the use of
a memory facility. This technique is known as **nested multiplication**.

(See Problem 13)

In most practical branches of science and engineering the evaluation of
formulae is essential. Below are some worked problems typical of the types
of formulae met in day-to-day calculations.

Worked problems on evaluation of formulae

Problem 1. The surface area, A, of a hollow cone is given by the
formula $A = \pi rl$. Find the surface area in square centimetres,
when $r = 4.0$ cm and $l = 9.0$ cm.

Substituting given values for symbols in the formula $A = \pi rl$

$A = (\pi)(4.0)(9.0)$
  $= 113.097\,335\ldots\ \text{cm}^2$

Hence the surface area, $A = \textbf{113 cm}^2$, correct to 3 significant figures.

Problem 2. In an electrical circuit the voltage V volts is given by
$V = IR$. Find the voltage, when $I = 7.240$ amperes and
$R = 12.57$ ohms, correct to 4 significant figures.

$$V = IR$$

Therefore $V = (7.240)(12.57)$ volts

$$= 91.0068$$

Hence the voltage, $V = $ **91.01 volts**, correct to 4 significant figures.

Problem 3. A formula used for calculating velocity v m/s is given by $v = u + at$. If $u = 12.47$ m/s, $a = 5.46$ m/s^2 and $t = 4.92$ s, find v correct to 2 decimal places.

$$v = u + at$$

Therefore $v = 12.47 + (5.46)(4.92)$

$$= 12.47 + 26.8632$$

$$= 39.3332 \text{ m/s}$$

Hence the velocity, $v = $ **39.33 ms̀**, correct to 2 decimal places.

Problem 4. The area A m^2 of a circle is given by $A = \pi r^2$. Find the area correct to 2 decimal places given $r = 4.156$ m.

$$A = \pi r^2$$

Therefore $A = (\pi)(4.156)^2$ m^2

$$= 54.2626438 \ldots$$

Hence the area, $A = $ **54.26 m^2**, correct to 2 decimal places.

Problem 5. Power P watts in an electrical circuit may be expressed by the formula $P = \dfrac{V^2}{R}$. Evaluate the power correct to 2 decimal places, given that $V = 24.62$ volts and $R = 45.21$ ohms.

$$P = \frac{V^2}{R}$$

Therefore $P = \dfrac{(24.62)^2}{45.21}$ watts

$$= \frac{606.1444}{45.21} = 13.407308$$

Hence the power, $P = $ **13.41 watts**, correct to 2 decimal places.

Problem 6. The volume V cm^3 of a right circular cone is given by the formula $V = \dfrac{1}{3}\pi r^2 h$. Given $r = 5.637$ cm and $h = 16.41$ cm, find the volume in standard form correct to 3 significant figures.

$$V = \frac{1}{3}\pi r^2 h$$

Therefore $V = \dfrac{1}{3}(\pi)(5.637)^2(16.41)$

$$= 546.051\,077\ldots\text{cm}^3$$

Hence the volume $V = \mathbf{5.46 \times 10^2}$ **cm**^{3}, correct to 3 significant figures.

Problem 7. If force F newtons is given by

$$F = \frac{G m_1 m_2}{d^2}$$

where m_1 and m_2 are masses, d their distance apart and G a constant, find the force given $G = 6.67 \times 10^{-11}$ Nm^2 kg^{-2}, $m_1 = 8.43$ kg, $m_2 = 17.2$ kg and $d = 24.2$ m. Express the answer in standard form to 3 significant figures.

$$F = \frac{G m_1 m_2}{d^2}$$

$$= \frac{(6.67)(10^{-11})\ \text{Nm}^2\ \text{kg}^{-2}\ (8.43)\ \text{kg}(17.2)\ \text{kg}}{(24.2)^2\ \text{m}^2}$$

$$= \frac{(6.67)(8.43)(17.2)(10^{-11})\ \text{N}}{585.6}$$

$$= \frac{967.123\,32}{585.6} \times 10^{-11}\ \text{N}$$

$$= 1.651\,508\,4 \times 10^{-11}\ \text{N}$$

$$= 1.65 \times 10^{-11}\ \text{N, correct to 3 significant figures.}$$

Hence the force, $F = \mathbf{1.65 \times 10^{-11}}$ **N**.

Problem 8. The time t seconds of swing of a simple pendulum is given by the formula

$$t = 2\pi \sqrt{\left(\frac{l}{g}\right)}$$

Find the time, correct to 3 decimal places, given $l = 13.0$ m and $g = 9.81$ m s^{-2}.

$$t = 2\pi \sqrt{\left(\frac{l}{g}\right)}$$

$$t = (2)(\pi)\sqrt{\left(\frac{13.0 \text{ m}}{9.81 \text{ m s}^{-2}}\right)}$$

$$= (2)(\pi)\sqrt{(1.325\,178\,39 \text{ s}^2)}$$

$$= (2)(\pi)(1.151\,163\,93) \text{ s}$$

$$= 7.232\,976\,28\ldots$$

Hence the time $t = \mathbf{7.233}$ **s**

Problem 9. A formula for resistance variation with temperature is $R = R_0(1 + \alpha t)$. Given $R_0 = 15.42$ ohms, $\alpha = 0.002\,70$ and $t = 78.4°$C evaluate R, correct to 2 decimal places.

$$R = R_0(1 + \alpha t)$$

$$= 15.42[1 + (0.002\,7)(78.4)] \text{ ohms}$$

$$= 15.42[1 + 0.211\,68]$$

$$= 15.42(1.211\,68)$$

$$= 18.684\,106 \text{ ohms}$$

Hence $R = \mathbf{18.68}$ **ohms**, correct to 2 decimal places.

Problem 10. The vapour pressure p of triethylamine at various temperatures T is given by $\lg p = 7.60 - \dfrac{770}{T - 3.12}$. Draw up a table of values for p when T has values of 100 K, 120 K, 140 K and 160 K. Give each value of p correct to 3 decimal places.

The sequence of steps is as follows:

(i)   When $T = 100$ K, $T - 3.12 = 100 - 3.12 = 96.88$ K

(ii) $\dfrac{770}{T - 3.12} = \dfrac{770}{96.88} = 7.948\,0$

(iii) $7.60 - \dfrac{770}{T - 3.12} = 7.60 - 7.948\,0 = -0.348\,0$

(iv) $p = \text{antilog}\left(7.60 - \dfrac{770}{T - 3.12}\right) = \text{antilog}(-0.348\,0)$

$$\equiv 10^{-0.348\,0} = 0.448\,7$$

$$= \mathbf{0.449}, \text{ correct to 3 decimal places.}$$

A similar sequence of steps is used for the remaining values of T and a table of values is produced as shown below.

T	100	120	140	160
$T - 3.12$	96.88	116.88	136.88	156.88
$\dfrac{770}{T - 3.12}$	7.9480	6.5880	5.6254	4.9082
$7.60 - \dfrac{770}{T - 3.12}$	−0.3480	1.0120	1.9746	2.6918
$p = \mathrm{antilog}\left(7.60 - \dfrac{770}{T - 3.12}\right)$	0.449	10.280	94.319	491.813

Problem 11. The area of a rectangle $A = bh$. The base, b, of the rectangle when measured was found to be 4.63 centimetres and the height, h, 9.4 centimetres.
Determine the area of the rectangle.

Area $= bh = 9.4 \times 4.63 = 43.522$ cm^2

The approximate value is $9 \times 5 = 45$ cm^2, so there are no obvious blunder or order of magnitude errors. However, it is not usual in a measurement type question to state the answer to an accuracy greater than one significant figure more than the least accurate number in the data; this is 9.4 cm, so the result should not have more than 3 significant figures.
Thus

Area = **43.5 cm^2**

would be the normal way of expressing the result.

Problem 12. State which type of error or errors have been made in the following problems:

(a) $\dfrac{37.4}{16.8 \times 0.071} = 1.335$ correct to 3 decimal places;

(b) $16.814 \times 0.0038 = 0.03689$ correct to 4 significant figures;

(c) In the formula $C = 2\pi r$, r was measured as 1.72 cm, giving a value for C of 10.8071 correct to 4 decimal places; and

(d) $127.8 \times 1.632 = 208.6$

(a) $\dfrac{37.4}{16.8 \times 0.071}$ is approximately equal to $\dfrac{4 \times 10}{2 \times 10 \times 7 \times 10^{-2}}$

or $\frac{4}{14} \times 10^2$, that is about 30. **Thus two errors have occurred, both a blunder and an order of magnitude error.**

(b) $16.814 \times 0.003\,8$ is approximately equal to $1.7 \times 10 \times 4 \times 10^{-3}$, that is about 7×10^{-2} or 0.07. **Thus a blunder has been made in the calculation.**

(c) $C = 2\pi r = 2 \times \pi \times 1.72$. The answer should be expressed to 4 significant figures only (one more than the least accurate data figure, 1.72). **Hence the error is that of introducing a greater accuracy than exists in the data.**

(d) $127.8 \times 1.632 = 208.569\,6$, **hence a rounding off error has occurred.** This answer should have stated $127.8 \times 1.632 = 208.6$ correct to 4 significant figures.

Problem 13. Evaluate $3x^3 - 2x^2 + 4x - 8$ when $x = 4.5$ using (a) terms in sequence, (b) nested multiplication.

Let $y = 3x^3 - 2x^2 + 4x - 8$
Hence when $x = 4.5$, $y = 3(4.5)^3 - 2(4.5)^2 + 4(4.5) - 8$

(a) $y = 273.375 - 40.5 + 18 - 8$
      $= \mathbf{242.875}$

(b) $y = 3x^3 - 2x^2 + 4x - 8 = [(3x - 2)x + 4]x - 8$
      $y = [(3 \times 4.5 - 2)4.5 + 4]4.5 - 8$
      $= \mathbf{242.875}$

Further problems on evaluation of formulae may be found in Section 1.4 (Problems 1–32), page 17.

1.3   Transposition of formulae

Formulae play a very important part in mathematics, science and engineering, as by them it is possible to give a concise, accurate and generalised statement of laws.

From Section 1.2 it can be seen that some formulae contain several symbols. Usually, one of the symbols is isolated on one side of the equation and is called the subject of the formula. Sometimes a symbol other than the subject is required to be calculated. In such circumstances it is usually easiest to rearrange the formula to make a new subject before numbers are substituted for symbols. This rearranging process is called **transposing the formula** or simply **transposition**. Transposition should be treated carefully because an error can result in a statement that is not true, with possibly serious effects in certain branches of industry.

Simple cases of transposition of formulae are introduced in *Technician mathematics 1*. More difficult examples are contained in the following worked problems.

Worked problems on transposition of formulae

Formulae involving roots and powers

Problem 1. Transpose the formula $K = \frac{1}{2}mv^2$ to make v the subject.

$\frac{1}{2}mv^2 = K$

Whenever the prospective new subject is a squared term the object must be to isolate that term on the left-hand side (L.H.S.) and then take the square root of both sides of the equation. Multiplying both sides by 2 gives

$$2(\tfrac{1}{2}mv^2) = 2K$$

Therefore $mv^2 = 2K$

Dividing both sides by m gives

$$\frac{mv^2}{m} = \frac{2K}{m}$$

Therefore $v^2 = \dfrac{2K}{m}$

Taking the square root of both sides gives

$$\sqrt{v^2} = \sqrt{\left[\frac{2K}{m}\right]}$$

Therefore $v = \sqrt{\left[\dfrac{2K}{m}\right]}$

Problem 2. In a right-angled triangle of sides a, b and c, a mathematical expression for Pythagoras's theorem is $a^2 = b^2 + c^2$. Transpose the formula to find b.

$b^2 + c^2 = a^2$

and $b^2 = a^2 - c^2$

Taking the square root of both sides gives

$$\sqrt{b^2} = \sqrt{(a^2 - c^2)}$$

Therefore $b = \sqrt{(a^2 - c^2)}$

Problem 3. If $t = 2\pi \sqrt{\left(\dfrac{l}{g}\right)}$ find l in terms of t, π and g.

Whenever the prospective new subject is involved within a square root sign it is advisable to isolate that term on the L.H.S. and then to square both sides of the equation.

$$2\pi \sqrt{\left(\frac{l}{g}\right)} = t$$

Dividing both sides by 2π gives

$$\frac{2\pi \sqrt{\left(\dfrac{l}{g}\right)}}{2\pi} = \frac{t}{2\pi}$$

Therefore $\sqrt{\left(\dfrac{l}{g}\right)} = \dfrac{t}{2\pi}$

Squaring both sides gives

$$\left\{\sqrt{\left(\frac{l}{g}\right)}\right\}^2 = \left(\frac{t}{2\pi}\right)^2$$

Therefore $\dfrac{l}{g} = \left(\dfrac{t}{2\pi}\right)^2$

Multiplying both sides by g gives

$$g\left(\frac{l}{g}\right) = g\left(\frac{t}{2\pi}\right)^2$$

Therefore $\qquad l = g\left(\dfrac{t}{2\pi}\right)^2$

or $\qquad l = \dfrac{gt^2}{4\pi^2}$

Both of the above answers for l are correct, the former being the neater version.

Problem 4. If $Z = \sqrt{(R^2 + X^2)}$, make R the subject of the equation.

$$\sqrt{(R^2 + X^2)} = Z$$

Squaring both sides gives

$$\{\sqrt{(R^2 + X^2)}\}^2 = Z^2$$

Therefore $\qquad R^2 + X^2 = Z^2$

Subtracting X^2 from both sides gives

$R^2 = Z^2 - X^2$

Taking the square root of both sides gives

$$\sqrt{R^2} = \sqrt{(Z^2 - X^2)}$$

Therefore $\quad R = \sqrt{(Z^2 - X^2)}$

Problem 5. The volume v of a sphere is given by the formula $v = \dfrac{4}{3}\pi r^3$. Express r in terms of v.

$\dfrac{4}{3}\pi r^3 = v$

Multiplying both sides by 3 gives

$4\pi r^3 = 3v$

Dividing both sides by 4π gives

$r^3 = \dfrac{3v}{4\pi}$

In this case, to find r, the cube root of both sides needs to be taken. Thus

$$\sqrt[3]{r^3} = \sqrt[3]{\left(\frac{3v}{4\pi}\right)}$$

Therefore $\quad r = \sqrt[3]{\left(\dfrac{3v}{4\pi}\right)}$

Formulae where factorisation is necessary

Problem 6. Transpose the formula $A = \dfrac{b^2 x + b^2 y}{c}$ to make b the subject.

$\dfrac{b^2 x + b^2 y}{c} = A$

Multiplying both sides by c gives

$b^2 x + b^2 y = Ac$

Factorising the L.H.S. gives

$$b^2(x + y) = Ac$$

Dividing both sides by $(x + y)$ gives

$$b^2 = \frac{Ac}{(x + y)}$$

Taking the square root of both sides gives

$$b = \sqrt{\left[\frac{Ac}{(x + y)}\right]}$$

Problem 7. If $p = \dfrac{b - d}{\sqrt{(xa + xc)}}$, make x the subject of the formula.

$$\frac{b - d}{\sqrt{(xa + xc)}} = p$$

Multiplying both sides of the equation by $\sqrt{(xa + xc)}$ gives

$$b - d = p\sqrt{(xa + xc)}$$

or

$$p\sqrt{(xa + xc)} = b - d$$

Dividing both sides by p gives

$$\sqrt{(xa + xc)} = \frac{b - d}{p}$$

Squaring both sides gives

$$(xa + xc) = \left[\frac{b - d}{p}\right]^2$$

Factorising the L.H.S. gives

$$x(a + c) = \left[\frac{b - d}{p}\right]^2$$

Dividing both sides by $(a + c)$ gives

$$x = \frac{\left[\dfrac{b - d}{p}\right]^2}{(a + c)}$$

Therefore $x = \dfrac{(b - d)^2}{p^2(a + c)}$

Both of the above answers for x are correct, the latter being the neater version.

Formulae in which the new subject occurs in more than one term

---

Problem 8. If $ab = 2b + c - xb$, express b in terms of a, c and x.

If $ab = 2b + c - xb$, then:

$ab - 2b + xb = c$

Factorising the L.H.S. gives

$b(a - 2 + x) = c$

Dividing both sides by $(a - 2 + x)$ gives

$$b = \frac{c}{a - 2 + x}$$

Problem 9. If $x = \dfrac{y}{1 + y}$ make y the subject of the formula.

$$\frac{y}{1 + y} = x$$

Multiplying both sides by $(1 + y)$ gives

$$y = x(1 + y)$$

Therefore $y = x + xy$

The prospective new subject, y, is now on both sides of the equation. Hence all terms in y should now be isolated on the L.H.S. Thus:

$y - xy = x$

Factorising the L.H.S. gives

$y(1 - x) = x$

Dividing both sides by $(1 - x)$ gives

$$y = \frac{x}{1 - x}$$

Problem 10. Transpose the formula $I = \dfrac{iR}{R + r}$ to make R the subject.

$$\frac{iR}{R + r} = I$$

Multiplying both sides by $(R + r)$ gives

$$iR = I(R + r)$$

Therefore $iR = IR + Ir$

Isolating terms in R on the L.H.S. gives

$$ir - IR = Ir$$

Factorising the L.H.S. gives

$$R(i - I) = Ir$$

Dividing both sides by $(i - I)$ gives

$$R = \frac{Ir}{i - I}$$

Problem 11. Transpose the formula $x = \dfrac{ab^2}{b^2 + d} - c$ to make b the subject.

$$\frac{ab^2}{b^2 + d} - c = x$$

Hence
$$\frac{ab^2}{b^2 + d} = x + c$$

and
$$ab^2 = (x + c)(b^2 + d)$$

Therefore
$$ab^2 = xb^2 + xd + cb^2 + cd$$

and
$$ab^2 - xb^2 - cb^2 = xd + cd$$

Factorising gives: $b^2(a - x - c) = d(x + c)$

and
$$b^2 = \frac{d(x + c)}{(a - x - c)}$$

Hence
$$b = \sqrt{\left[\frac{d(x + c)}{(a - x - c)}\right]}$$

Problem 12. If $\dfrac{e}{f} = \sqrt{\left[\dfrac{x + y}{x - y}\right]}$ express y in terms of e, f and x.

$$\sqrt{\left[\frac{x + y}{x - y}\right]} = \frac{e}{f}$$

Hence $$\frac{x+y}{x-y} = \frac{e^2}{f^2}$$

and $$f^2(x+y) = e^2(x-y)$$

Therefore $$f^2x + f^2y = e^2x - e^2y$$

and $$f^2y + e^2y = e^2x - f^2x$$

Factorising gives: $y(f^2 + e^2) = x(e^2 - f^2)$

Hence $$y = \frac{x(e^2 - f^2)}{(e^2 + f^2)}$$

Further problems on transposition of formulae may be found in the following Section (1.4) (Problems 33–82), page 20.

1.4   Further problems

Evaluation of formulae

1. The area A of a rectangle is given by the formula $A = lb$. Evaluate the area when l is 14.21 cm and b is 7.46 cm
   $[A = 106.0 \text{ cm}^2]$

2. The circumference C of a circle is given by the formula $C = 2\pi r$. Find the circumference given $r = 6.20$ cm $[C = 38.96 \text{ cm}]$

3. The area A of a triangle may be evaluated by using the formula $A = \frac{1}{2}bh$. Find the area when $b = 7.0$ mm and $h = 12.5$ mm
   $[A = 43.75 \text{ mm}^2]$

4. A formula used when dealing with gas laws is $R = \dfrac{PV}{T}$.
   Evaluate R when P is $2\,000 \dfrac{\text{kN}}{\text{m}^2}$, V is 5 m³ and T is 200 K
   $\left[\dfrac{50 \text{ kN m}}{\text{K}} = \dfrac{50 \text{ kJ}}{\text{K}} \right]$

5. In the following formula evaluate X when $p = 751$, $a = 5$
   and $t = 21$: $X = p\left[1 + \dfrac{at}{100}\right]$ $[1\,540]$

6. The voltage drop, V, in an electrical circuit is given by the formula $V = E - IR$. Evaluate the voltage drop when $E = 4.81$ volts, $I = 0.680$ amperes and $R = 5.37$ ohms
   $[V = 1.158 \text{ volts}]$

7. $a = \dfrac{uv}{u+v}$. Evaluate a when $u = 10.93$ and $v = 7.380$
   $[a = 4.405]$

8. When a number of cells are connected together the current I amperes of the combination is given by the formula

   $I = \dfrac{nE}{R + nr}$. Evaluate the current when $n = 40$, $E = 2.3$

   volts, $R = 2.9$ ohms and $r = 0.60$ ohms
   $[I = 3.42$ amperes$]$

9. The power P watts of an electrical circuit is given by the

   formula $P = \dfrac{V^2}{R}$. Find the power when V is 14.8 volts and R

   is 19.5 ohms $[P = 11.23$ watts$]$

10. If $W = 8.20$, $v = 10.0$ and $g = 9.81$ evaluate B given that

    $B = \dfrac{Wv^2}{2g}$ $[B = 41.79]$

11. If $F = \frac{1}{2}m(v^2 - u^2)$ find F when $m = 17.0$, $v = 14.8$ and
    $u = 9.24$ $[F = 1\,136]$

12. The time t seconds of oscillation for a simple pendulum is

    given by $t = 2\pi\sqrt{\dfrac{l}{g}}$. Evaluate the time of oscillation when

    $l = 61.42$ m and $g = 9.81\ \dfrac{\text{m}}{\text{s}^2}$ $[t = 15.72$ s$]$

Evaluate the subject of each of the formulae shown in
Problems 13 to 23.

13. $S = 2\pi r^2 + 2\pi rh$ when $\pi = 3.14$, $r = 2.42$ and $h = 3.47$
    $[89.51]$

14. $E = \frac{1}{2}LI^2$ when $L = 3.55$ and $I = 0.490$ $[0.426]$

15. $A = \dfrac{V}{100}\left[Q - \dfrac{mV^2}{g}\right]$ when $Q = 50.28$, $m = 17$, $V = 5.0$ and

    $g = 9.81$ $[0.348]$

16. $I = \dfrac{V}{\sqrt{(R^2 + X^2)}}$ when $V = 240$, $R = 10.0$ and $X = 15.0$

    $[13.31]$

17. $a = \dfrac{T(1 - S)}{S(T + 1)}$ when $T = 2.31$ and $S = 2.62$ $[-0.431\,5]$

18. $X = T\left(1 + \dfrac{b}{10}\right)^n$ when $T = b = n = \dfrac{3}{4}$ $[0.791\,8]$

19. $A = \sqrt{[s(s - a)(s - b)(s - c)]}$ where $s = \dfrac{a + b + c}{2}$ and

    $a = 2.0$, $b = 3.0$ and $c = 4.0$ $[2.90]$

20. $y = \sqrt{\left(\dfrac{ak}{p} - \dfrac{b}{q}\right)}$ when $a = 0.72$, $k = 0.620$, $p = 0.003\,00$,

    $b = 41.36$ and $q = 0.480$ [7.914]

21. $Z = \sqrt{\left[R^2 + \left(\omega L - \dfrac{1}{\omega C}\right)^2\right]}$ when $R = 24$, $L = 0.30$, $w = 352$

    and $C = 6.5 \times 10^{-5}$ [66.4]

22. $S = ut + \frac{1}{2}at^2$ when $u = 10.1$, $t = 5.20$ and $a = -3.80$
    [1.144]

23. $a = \dfrac{T}{\sqrt{(ku^2 - lv^2)}}$ when $T = 17.42$, $k = 0.073\,0$, $u = 5.46$,

    $l = 0.089\,0$ and $v = 3.21$ [± 15.52]

24. Experimental values of load W newtons and distance

    l metres are related by a law of the form $W = \dfrac{-2.05}{l} + 36.0$.

    Draw up a table of values as l increases from 0.1 m to 0.7 m
    in 0.1 m steps.

l	0.1	0.2	0.3	0.4	0.5	0.6	0.7
W	15.50	25.75	29.17	30.88	31.90	32.58	33.07

25. The relationship between the solubility S of potassium chlorate
    and temperature $t°C$ is $S = 3 + 0.10t + 0.004t^2$. Draw up a
    table of values as t increases from $20°C$ to $100°C$ in $10°$ steps.

t	20	30	40	50	60	70	80	90	100
S	6.6	9.6	13.4	18.0	23.4	29.6	36.6	44.4	53.0

In Problems 26 to 32, state the type of error or errors which have
been made, if, indeed, there is an error present.

26. $73 \times 68.247 = 4\,982.03$
    [Accuracy error, add correct to 2 decimal places]

27. $18 \times 0.08 \times 6 = 86.4$ [Order of magnitude error]

28. $\dfrac{47.3}{9 \times 0.071} = 5\,301.2$ correct to 1 decimal place   [Blunder]

29. $\dfrac{16.7 \times 0.02}{41} = 8.15 \times 10^{-3}$ correct to 3 significant figures

    [No error]

30. The force, $P = mf$, where m is the mass and f the acceleration. The value of P was found to be 1 775.61 N when m was 181 kg and f was 9.81 m s^{-2}.
    [Measured values, hence $P = 1\,776$ N]

31. $6.73 \div 0.006 = 4\,211.6$ [Blunder]

32. $\dfrac{3.1 \times 0.008}{37.6 \times 0.347} = 9.100\,7 \times 10^{-4}$ [Blunder]

Transposition of formulae

Make the symbol indicated in round brackets the subject of each of the formulae shown in Problems 33 to 65 and express each in its simplest form.

33. $R = PQ^2V$ (Q)

$$\left[Q = \sqrt{\frac{R}{PV}} \right]$$

34. $X = 3Mn^2L^2$ (L)

$$\left[L = \sqrt{\frac{X}{3Mn^2}} \right]$$

35. $y^2 = 4a\left(x + \dfrac{c^2}{4a} \right)$ (x)

$$\left[x = \frac{y^2 - c^2}{4a} \right]$$

36. $a^2 = b^2 + c^2$ (c)
    $[c = \sqrt{(a^2 - b^2)}]$

37. $\dfrac{x^2}{P^2} + \dfrac{y^2}{q^2} = 1$ (P)

$$\left[P = \frac{xq}{\sqrt{(q^2 - y^2)}} \right]$$

38. $w = a\sqrt{c}$ (c)

$$\left[c = \left(\frac{w}{a} \right)^2 \right]$$

39. $t = 2\pi \sqrt{\dfrac{l}{g}}$ (g)

$$\left[g = \frac{4\pi^2 l}{t^2} \right]$$

40. $V = \sqrt{(2gh)}$ (h)

$$\left[h = \frac{V^2}{2g} \right]$$

41. $E = \dfrac{V^2 t}{R}$ (V)

$$\left[V = \sqrt{\frac{ER}{t}} \right]$$

42. $v^2 = u^2 + 2as$ (u)
    $[u = \sqrt{(v^2 - 2as)}]$

43. $s = v + \frac{1}{2}at^2$ (t)

$$\left[t = \sqrt{\frac{2(s - v)}{a}} \right]$$

44. $A = \dfrac{\pi r^2 \theta}{360}$ (r)

$$\left[r = \sqrt{\frac{360A}{\pi\theta}} \right]$$

45. $v = w\sqrt{(a^2 - y^2)}$ (y)

$$\left[y = \frac{\sqrt{(a^2 w^2 - v^2)}}{w} \right]$$

46. $M = \sqrt{\dfrac{a+x}{y}}$ (x)

$[x = M^2 y - a]$

47. $S = \dfrac{n}{2}[2a + (n-1)d]$ (d)

$\left[d = \dfrac{2(S-an)}{n(n-1)}\right]$

48. $A = b\left(\dfrac{1}{1-d} - r\right)$ (d)

$\left[d = \dfrac{A + b(r-1)}{A + br}\right]$

49. $S = \sqrt{\left(\dfrac{b^2}{4}\right) + a^2}$ (b)

$[b = 2\sqrt{(s^2 - a^2)}]$

50. $Z = \sqrt{[r^2 + (2\pi f L)^2]}$ (f)

$\left[f = \dfrac{\sqrt{(Z^2 - r^2)}}{2\pi L}\right]$

51. $Z = \sqrt{[R^2 + (X_1 - X_2)^2]}$ (X_2)

$[X_2 = X_1 - \sqrt{(Z^2 - R^2)}]$

52. $P = \dfrac{e^2 m - e^2 n}{S}$ (e)

$\left[e = \sqrt{\dfrac{PS}{m-n}}\right]$

53. $l = \dfrac{\pi h}{4d^2}(D_2^2 - D_1^2)$ (D_2)

$\left[D_2 = \sqrt{\left(\dfrac{4ld^2}{\pi h} + D_1^2\right)}\right]$

54. $M = \dfrac{bd^3}{12} + \dfrac{bd^3}{3}$ (d)

$\left[d = \sqrt[3]{\dfrac{12M}{5b}}\right]$

55. $M = \pi(R^4 - r^4)$ (r)

$\left[r = \sqrt[4]{\dfrac{\pi R^4 - M}{\pi}}\right]$

56. $m = \dfrac{2(T - an)}{n(n-l)}$ (a)

$\left[a = \dfrac{2T - mn(n-l)}{2n}\right]$

57. $V = \dfrac{Er}{R+r}$ (r)

$\left[r = \dfrac{VR}{E-V}\right]$

58. $a + x = \dfrac{y}{2+y}$ (y)

$\left[y = \dfrac{2(a+x)}{1-x-a}\right]$

59. $V = \dfrac{Kab}{b-a}$ (a)

$\left[a = \dfrac{Vb}{Kb+V}\right]$

60. $A = \dfrac{p-q}{2+pq}$ (q)

$\left[q = \dfrac{p-2A}{pA+1}\right]$

61. $m = \dfrac{uL}{L+rcR}$ (L)

$\left[L = \dfrac{mrCR}{u-m}\right]$

62. $\dfrac{1}{x} = \dfrac{4K-2b}{3c+5K}$ (K)

$\left[K = \dfrac{3c+2bx}{4x-5}\right]$

63. $x^2 = \dfrac{y^2 - r^2}{y^2}$ (y)

$$\left[y = \dfrac{r}{\sqrt{(1 - x^2)}} \right]$$

64. $\dfrac{e}{f} = \dfrac{1 + r^2}{1 - r^2}$ (r)

$$\left[r = \sqrt{\dfrac{e - f}{e + f}} \right]$$

65. $\dfrac{D}{d} = \sqrt{\dfrac{f + p}{f - p}}$ (p)

$$\left[p = \dfrac{f(D^2 - d^2)}{(D^2 + d^2)} \right]$$

66. The passage of sound waves through walls is governed by the equation $v = \sqrt{\dfrac{(K + 4G/3)}{\rho}}$ where K is the bulk modulus and G the shear modulus. Make G the subject of the formula

$$\left[G = \dfrac{3}{4}(v^2\rho - K) \right]$$

67. If P is the safe load which may be carried by a steel plate weakened by rivet holes then $P = f(b - nd)t$. Make f, the safe working stress in the steel and then n, the number of rivet holes, the subject of the formula

$$\left[f = \dfrac{P}{(b - nd)t}; \quad n = \dfrac{fbt - P}{fdt} \right]$$

68. The safe working stress of timber, f, is related to the moment of resistance, M, of a beam of rectangular cross section of depth and breadth, b and d, by the equation $f = \dfrac{6M}{bd^2}$. Make d the subject of the formula $\left[d = \sqrt{\dfrac{6M}{bf}} \right]$

69. The modulus of elasticity of a structural material (E) is given by the formula $E = \dfrac{Wl}{Ax}$. Make x the subject of the formula

$$\left[x = \dfrac{Wl}{EA} \right]$$

70. The outside resistance R_t of a cold store is given by the equation $\dfrac{R_t}{R} = \dfrac{\theta - \theta_o}{\theta - \theta_d}$ where R is the resistance of the surface and θ, θ_o, θ_d are the store temperature, the minimum outside temperature and the dew point temperature respectively.

Make θ the subject of the formula $\left[\theta = \dfrac{R_t\theta_d - R\theta_o}{R_t - R} \right]$

71. Van der Waals' equation for the pressure of a real gas (p) is
$\left(p + \dfrac{a}{v^2}\right)(V - b) = RT$. Make p the subject of the formula

$$\left[p = \frac{RT}{V - b} - \frac{a}{v^2} \right]$$

72. The observed growth yield, Y, of an organism is given by
$\dfrac{1}{Y} = \dfrac{m}{\mu} + \dfrac{1}{Y_G}$. Make Y_G, the true growth yield, the subject of
the equation $\left[Y_G = \dfrac{\mu Y}{\mu - mY} \right]$

73. The viscosity coefficient of a liquid (η) is given by the
equation $\eta = \dfrac{\pi Pr^4 t}{8vl}$. Make v the subject of the formula

$$\left[v = \frac{\pi Pr^4 t}{8\eta l} \right]$$

74. The velocity v of an electron is given by $v = \sqrt{\dfrac{2Ve}{m}}$. Make m
the subject of the formula $\left[m = \dfrac{2Ve}{v^2} \right]$

75. Make λ, the wavelength of X-rays, the subject of the
following formula $\dfrac{\mu}{\rho} = \dfrac{CZ^4 \lambda^{5/2} N}{A}$ $\left[\lambda = \sqrt[5]{\left(\dfrac{A\mu}{\rho CZ^4 N}\right)^2} \right]$

76. Given that $28t(p - d) = \dfrac{23\pi d^2}{4}$, find p when $t = 0.500$ and
$d = 1.20\sqrt{t}$ [1.778 or 0.080 5]

77. Given $I = \dfrac{nE}{R + nr}$ find n when $I = 2$, $E = 1.8$, $R = 2.4$ and
$r = 0.5$ [6]

78. The velocity, V, of water in a pipe occurs in the following
formula $h = \dfrac{0.03LV^2}{2dg}$. Express V as the subject of the
formula and find its value when $h = 0.614$, $L = 168$,
$d = 0.250$ and $g = 9.81$ $\left[V = \sqrt{\dfrac{2hdg}{0.03L}}, 0.773\,0 \right]$

79. Transpose the formula $T = 4\pi\sqrt{\dfrac{(M + 3m)l}{3(M + 2m)g}}$ to make m the
subject $\left[m = \dfrac{M(16\pi^2 l - 3gT^2)}{6(gT^2 - 8\pi^2 l)} \right]$

80. Transpose $p = \dfrac{y+k}{\dfrac{y}{p_1}+\dfrac{k}{p_2}}$ to make y the subject. Evaluate y to 2

decimal places given that $p = 6.21$, $k = 32.8$, $p_1 = 5.22$ and

$p_2 = 7.31$ $\left[y = \dfrac{kp_1(p_2 - p)}{p_2(p - p_1)},\ 26.02 \right]$

81. In an electrical alternating current circuit the impedance Z is

given by the formula $Z = \sqrt{\left[R^2 + \left(\omega L - \dfrac{1}{\omega C} \right)^2 \right]}$. Make L

the subject of the formula and evaluate when $Z = 50$,
$R = 30$, $\omega = 314$ and $C = 4 \times 10^{-6}$

$\left[L = \dfrac{1}{\omega}\left(\dfrac{1}{\omega C} + \sqrt{(Z^2 - R^2)} \right),\ L = 2.66 \right]$

82. The sag S at the centre of a wire of length l supported at
two points distance d apart is given by the formula

$S = \sqrt{\dfrac{3d(l - d)}{8}}$. Transpose the formula to make l the

subject. Find the length of the wire to the nearest millimetre
given that d is 1.62 m and S is 82.0 cm

$\left[l = \dfrac{8S^2}{3d} + d,\ l = 2.727\ \text{m} \right]$

Chapter 2

Indices and logarithms

2.1 Laws of indices

The laws of indices, with an example of each, are summarised below:

1. $a^m \times a^n = a^{m+n}$; $2^3 \times 2^4 = 2^{3+4} = 2^7 = \textbf{128}$

2. $\dfrac{a^m}{a^n} = a^{m-n}$; $\dfrac{3^4}{3^2} = 3^{4-2} = 3^2 = \textbf{9}$

3. $(a^m)^n = a^{mn}$; $(2^3)^2 = 2^{3 \times 2} = 2^6 = \textbf{64}$

4. $a^{\frac{m}{n}} = \sqrt[n]{a^m}$; $8^{\frac{4}{3}} = \sqrt[3]{8^4} = 2^4 = \textbf{16}$

5. $a^{-n} = \dfrac{1}{a^n}$; $5^{-2} = \dfrac{1}{5^2} = \dfrac{1}{\textbf{25}}$

6. $a^0 = 1$; $7^0 = 1$

Worked problems on laws of indices

Problem 1. Evaluate (a) $\dfrac{2^2 \times 2^3 \times 2}{2^4}$ (b) $16^{\frac{3}{4}}$ (c) $(5)(3x^0)$

(a) $\dfrac{2^2 \times 2^3 \times 2}{2^4} = \dfrac{2^{2+3+1}}{2^4}$ (by law 1) $= \dfrac{2^6}{2^4} = 2^{6-4}$ (by law 2)

$$= 2^2 = \textbf{4}$$

(b) $16^{\frac{3}{4}} = \sqrt[4]{16^3}$ (by law 4) $= 2^3 = \textbf{8}$

     (Note, it does not matter whether the 4th root of 16 is found first, or whether 16 cubed is found first – the same answer will result.)

(c) $(5)(3x^0) = (15)(3)$, since $x^0 = 1$ from law 6

$$= \textbf{15}$$

Problem 2. Evaluate (a) $25^{-\frac{1}{2}}$ (b) $8^{-\frac{2}{3}}$ (c) $\dfrac{(2^2)^3(3^3)^2}{(2\times3)^4}$

(a) $25^{-\frac{1}{2}} = \dfrac{1}{25^{\frac{1}{2}}}$ (from law 5) $= \dfrac{1}{\sqrt{25}}$ (from law 4)

$$= \dfrac{1}{\pm5} = \pm\dfrac{1}{5}$$

(b) $8^{-\frac{2}{3}} = \dfrac{1}{8^{\frac{2}{3}}}$ (from law 5) $= \dfrac{1}{\sqrt[3]{8^2}}$ (from law 4)

$$= \dfrac{1}{2^2} = \dfrac{1}{4}$$

(c) $\dfrac{(2^2)^3(3^3)^2}{(2\times3)^4} = \dfrac{(2^{2\times3})(3^{3\times2})}{(2^4)(3^4)}$ (from law 3) $= \dfrac{2^6\times3^6}{2^4\times3^4}$

$$= 2^{6-4}\times3^{6-4} \text{ (from law 2)}$$

$$= 2^2\times3^2 = 4\times9$$

$$= \mathbf{36}$$

Further problems on laws of indices may be found in Section 2.5 (Problems 1–10), page 33.

2.2   Definition of a logarithm

If a number y can be written in the form b^x then the index x is called the logarithm of y to base b, i.e.

if $y = b^x$ then $x = \log_b y$ (1)

Thus, since $1\,000 = 10^3$ then $3 = \log_{10} 1000$ **(Note that $\log_{10} y$ is usually written as lg y)**

Similarly, since $81 = 3^4$ then $4 = \log_3 81$

To evaluate x when $x = \log_2 64$, the expression is initially written in its equivalent index form, i.e. $64 = 2^x$. However, $64 = 2^6$, hence $2^6 = 2^x$, from which $x = 6$. Thus $\log_2 64 = 6$

Worked problems on the definition of a logarithm

Problem 1. Evaluate (a) $\log_2 16$ (b) $\log_{10} 10000$ (c) $\log_5 125$

(a) Let $x = \log_2 16$, then $2^x = 16$ from the definition of a logarithm. However, $16 = 2^4$, thus $2^x = 2^4$, from which $x = 4$.
Hence $\mathbf{\log_2 16 = 4}$

(b) Let $x = \log_{10} 10\,000$ then $10^x = 10\,000$ from the definition of a logarithm (equation (1)).

However, $10\,000 = 10^4$, thus $10^x = 10^4$, from which $x = 4$.

Hence $\log_{10} \mathbf{10\,000 = 4}$

(c) Let $x = \log_5 125$ then $5^x = 125$, from the definition of a logarithm. However, $125 = 5^3$, thus $5^x = 5^3$, from which $x = 3$.

Hence $\log_5 \mathbf{125 = 3}$

Problem 2. Evaluate (a) $\lg 0.01$ (b) $\log_4 8$ (c) $\log_3 \dfrac{1}{27}$

(a) Let $x = \lg 0.01$ then $10^x = 0.01$ from the definition of a logarithm. However, $0.01 = 10^{-2}$, thus $10^x = 10^{-2}$, from which $x = -2$.

Hence $\lg \mathbf{0.01 = -2}$

(b) Let $x = \log_4 8$ then $4^x = 8$ from the definition of a logarithm. However, $8 = 2^3$ and $4 = 2^2$, thus $(2^2)^x = 2^3$, i.e. $2^{2x} = 2^3$, from which $2x = 3$ and $x = \dfrac{3}{2}$

Hence $\log_4 \mathbf{8 = 1\dfrac{1}{2}}$

(c) Let $x = \log_3 \dfrac{1}{27}$ then $3^x = \dfrac{1}{27}$ by the definition of a logarithm. However, $\dfrac{1}{27} = \dfrac{1}{3^3} = 3^{-3}$, thus $3^x = 3^{-3}$, from which $x = -3$.

Hence $\log_3 \dfrac{1}{27} = \mathbf{-3}$

Problem 3. Solve the following equations for x:

(a) $\lg x = 5$ (b) $\log_3 x = 2$ (c) $\log_4 x = -2\dfrac{1}{2}$

(a) If $\lg x = 5$ then $x = 10^5$ from the definition of a logarithm, i.e. $x = \mathbf{100\,000}$

(b) If $\log_3 x = 2$ then $x = 3^2 = \mathbf{9}$

(c)   If $\log_4 x = -2\frac{1}{2}$ then $x = 4^{-2\frac{1}{2}} = 4^{-\frac{5}{2}} = \frac{1}{4^{\frac{5}{2}}} = \frac{1}{\sqrt{(4)^5}} = \frac{1}{\pm(2)^5}$

$$= \pm\frac{1}{32}$$

Further problems on the definition of a logarithm may be found in Section 2.5 (Problems 11–25), page 34.

2.3   Laws of logarithms

There are three laws of logarithms which may each be deduced from the definition of a logarithm and the laws of indices.

Law 1

Let $\log_b M = x$ and $\log_b N = y$.

Then, by the definition of a logarithm, $M = b^x$ and $N = b^y$.

Therefore $MN = b^x b^y = b^{x+y}$ from the first law of indices.

This equation may be written in its equivalent logarithmic form, i.e.

$$\log_b MN = x + y$$

Hence $\boxed{\log_b MN = \log_b M + \log_b N}$ \hfill (1)

For example, $\log_{10} 6 = \log_{10}(2 \times 3) = \log_{10} 2 + \log_{10} 3$ (which may be checked).

By similar reasoning to that above, law 1 may be extended to apply to any number of functions.

For example, $\log_b (MNP) = \log_b M + \log_b N + \log_b P$.

Law 2

As with law 1, let $\log_b M = x$ and $\log_b N = y$,

from which $M = b^x$ and $N = b^y$.

Therefore $\dfrac{M}{N} = \dfrac{b^x}{b^y} = b^{x-y}$ from the second law of indices.

This equation may be written in its equivalent logarithmic form, i.e.

$$\log_b \frac{M}{N} = x - y$$

Hence $\boxed{\log_b \dfrac{M}{N} = \log_b M - \log_b N}$ \hfill (2)

For example, $\log_{10} \dfrac{3}{2} = \log_{10} 3 - \log_{10} 2$ (which may be checked).

Law 3

Let $\log_b N = x$, then $N = b^x$.

Raising each side by the power a gives: $N^a = (b^x)^a = b^{ax}$, from the third law of indices.

This equation may be written in its equivalent logarithmic form, i.e.

$$\log_b = N^a = ax$$

Hence $\boxed{\log_b N^a = a \log_b N}$ (3)

For example, $\log_{10} 2^5 = \log_{10} 2$ (which may be checked).

Law 3 is true when a is positive, negative or fractional.

For example, $\log_{10} \sqrt[3]{5^2} = \log_{10} 5^{\frac{2}{3}} = \dfrac{2}{3} \log_{10} 5$

The three laws of logarithms may be combined to simplify certain expressions.

For example, $\log\left(\dfrac{5 \times \sqrt{3}}{2^4}\right) = \log 5 + \log \sqrt{3} - \log 2^4$ (by laws 1 and 2)

$$= \log 5 + \log 3^{\frac{1}{2}} - \log 2^4$$

$$= \log 5 + \dfrac{1}{2} \log 3 - 4 \log 2 \text{ (by law 3)}.$$

In this example the base of the logarithm is not stated; in fact, the result is true whatever base is chosen. In practice, logarithms to a base of 10, written as 'lg', or logarithms to a base of e, written as 'ln', are used, where e is equal to 2.718 3 correct to 4 decimal places (see Chapter 8).

Indicial equations

The laws of logarithms may be used to solve certain equations involving powers – called indicial equations. For example, to solve, say, $2^x = 8$, logarithms to a base of 10 are taken of both sides, i.e.

$$\log_{10} 2^x \qquad = \log_{10} 8$$

and $x \log_{10} 2 \qquad = \log_{10} 8$ from the third law of logarithms.

Rearranging gives $x = \dfrac{\log_{10} 8}{\log_{10} 2} = \dfrac{0.903\,1}{0.301\,0} = 3$ which may readily be checked.

[Note, $\dfrac{\log_{10} 8}{\log_{10} 2}$ is not equal to $\log_{10}(8/2)$]

Worked problems on laws of logarithms and indicial equations

Problem 1. Write $\log\left(\dfrac{16 \times \sqrt[4]{5}}{27}\right)$ in terms of log 2, log 3 and log 5 to any base.

$\log\left(\dfrac{16 \times \sqrt[4]{5}}{27}\right) = \log 16 + \log \sqrt[4]{5} - \log 27$, by the first and second laws of logarithms

$\qquad = \log 2^4 + \log 5^{\frac{1}{4}} - \log 3^3$, by the laws of indices

$\qquad = \mathbf{4 \log 2} + \dfrac{1}{4}\mathbf{\log 5} - \mathbf{3 \log 3}$, by the third law of logarithms.

Problem 2. Write $\log_b 300$ in terms of $\log_b 2$, $\log_b 3$ and $\log_b 5$.

$300 = 2 \times 2 \times 3 \times 5 \times 5 = 2^2 \times 3 \times 5^2$

Hence $\log_b 300 = \log_b(2^2 \times 3 \times 5^2) = \log_b 2^2 + \log_b 3 + \log_b 5^2$, by the first law of logarithms

$\qquad = \mathbf{2 \log_b 2} + \mathbf{\log_b 3} + \mathbf{2 \log_b 5}$, by the third law of logarithms

Problem 3. Simplify $\log 27 - \log 9 + \log 81$.

$27 = 3^3$, $9 = 3^2$ and $81 = 3^4$

Hence $\log 27 - \log 9 + \log 81 = \log 3^3 - \log 3^2 + \log 3^4$

$\qquad = 3 \log 3 - 2 \log 3 + 4 \log 3$, by the third law of logarithms

$\qquad = (3 - 2 + 4) \log 3 = \mathbf{5 \log 3}$

Problem 4. Evaluate $\dfrac{\log_x 32 - \log_x 4 + \log_x 8}{\log_x 256}$

$$\frac{\log_x 32 - \log_x 4 + \log_x 8}{\log_x 256} = \frac{\log_x 2^5 - \log_x 2^2 + \log_x 2^3}{\log_x 2^8},$$

when expressed as powers of 2

$$= \frac{5 \log_x 2 - 2 \log_x 2 + 3 \log_x 2}{8 \log_x 2},$$

by the third law of logarithms

$$= \frac{6 \log_x 2}{8 \log_x 2} = \frac{6}{8} = \frac{3}{4}$$

Problem 5. Solve the equation: $\log x^4 - \log x^3 = \log 3x - \log 2x$.

Since $\log x^4 - \log x^3 = \log 3x - \log 2x$

then $\log\left(\dfrac{x^4}{x^3}\right) = \log\left(\dfrac{3x}{2x}\right)$, by the second law of logarithms,

i.e. $\log x = \log\left(\dfrac{3}{2}\right)$

Taking antilogarithms of both sides gives $x = \dfrac{3}{2} = 1\dfrac{1}{2}$

Problem 6. Solve the equation $2^x = 5$, correct to 4 significant figures.

Taking logarithms to base 10 of both sides of $2^x = 5$ gives

$$\log_{10} 2^x = \log_{10} 5$$

i.e. $x \log_{10} 2 = \log_{10} 5$,

from the third law of logarithms

Rearranging gives $x = \dfrac{\log_{10} 5}{\log_{10} 2} = \dfrac{0.698\,97}{0.301\,02} = \mathbf{2.322}$, correct to 4 significant figures.

Problem 7. Solve the equation $3^{x+1} = 2^{2x-3}$, correct to 2 decimal places.

Taking logarithms to base 10 of both sides of $3^{x+1} = 2^{2x-3}$ gives:

$$\log_{10} 3^{x+1} = \log_{10} 2^{2x-3}$$

i.e. $(x + 1) \log_{10} 3 = (2x - 3) \log_{10} 2$

$x \log_{10} 3 + \log_{10} 3 = 2x \log_{10} 2 - 3 \log_{10} 2$

$x(0.477\,1) + 0.477\,1 = 2x(0.301\,0) - 3(0.301\,0)$

i.e. $0.4771x + 0.4771 = 0.6020x - 0.9030$

$0.4771 + 0.9030 = 0.6020x - 0.4771x$

$1.3801 = 0.1249x$

from which, $x = \dfrac{1.3801}{0.1249} = \mathbf{11.05}$,

correct to 2 decimal places.

Problem 8. Solve the equation $x^{2.5} = 37.5$, correct to 4 signifcant figures.

Taking logarithms to base 10 of both sides of $x^{2.5} = 37.5$ gives

$\log_{10} x^{2.5} = \log_{10} 37.5$

i.e. $2.5 \log_{10} x = \log_{10} 37.5$

Hence $\log_{10} x = \dfrac{\log_{10} 37.5}{2.5} = \dfrac{1.5740}{2.5} = 0.6296$

Thus $x = $ antilog $0.6296 \ (= 10^{0.6296}) = \mathbf{4.262}$, correct to 4 significant figures.

Further problems on laws of logarithms and indicial equations may be found in Section 2.5 (Problems 26–44), page 34.

2.4   Some properties of logarithms

(i) **$\log_b 1 = 0$**

Let $\log_b 1 = x$ then $b^x = 1$ from the definition of a logarithm.

If $b^x = 1$ then $x = 0$, from the sixth law of indices.

Hence $\log_b 1 = 0$.

($\log_{10} 1 = 0$, for example, has been met before.)

(ii) **$\log_b b = 1$**

Let $\log_b b = x$ then $b^x = b$ from the definition of a logarithm.

If $b^x = b$ then $x = 1$.

Hence $\log_b b = 1$.

($\log_{10} 10 = 1$, for example, has been met before.)

(iii) **$\log_b 0 \rightarrow -\infty$**

Let $\log_b 0 = x$ then $b^x = 0$ from the definition of a logarithm.

If $b^x = 0$, and if b is a positive real number, then x must approach minus infinity.

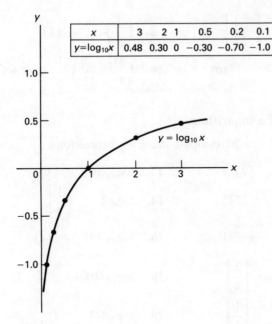

x	3	2	1	0.5	0.2	0.1
$y=\log_{10}x$	0.48	0.30	0	-0.30	-0.70	-1.0

Fig. 2.1

(For example, $2^{-2} = 0.25$, $2^{-20} = 9.54 \times 10^{-7}$, $2^{-200} = 6.22 \times 10^{-61}$ and so on.)

Hence $\log_b 0 \to -\infty$

Figure 2.1 shows a graph of $y = \log_{10} x$ which shows that as $x \to 0$, $\log_{10} x \to -\infty$.

(Although the logarithm to the base 10 of x is shown in Fig. 2.1, the same general shape occurs for a logarithm to any base.)

2.5 Further problems

Laws of indices
In Problems 1 to 10, evaluate the given expressions.

1. $\dfrac{3^5}{3^2 \times 3}$ [9]

2. $5p^0$ [5]

3. $\left(\dfrac{27}{8}\right)^{\frac{2}{3}}$ $\left[2\dfrac{1}{4}\right]$

4. $16^{-\frac{1}{4}}$ $\left[\dfrac{1}{2}\right]$

5. $\left(\dfrac{16}{81}\right)^{\frac{3}{4}}$ $\left[\dfrac{8}{27}\right]$

6. $\dfrac{(2^3)^3(5^2)^3}{5^5(2^2 \times 5)^2}$ $\left[6\dfrac{2}{5}\right]$

7. $49^{-\frac{1}{2}}$ $\left[\pm\dfrac{1}{7}\right]$

8. $\left(\dfrac{1}{2^3}\right)^{-1}$ [8]

9. $(8^{\frac{1}{2}} \times 5^{\frac{3}{2}})^{\frac{2}{3}}$ [10]

10. $\left(\dfrac{81^{\frac{1}{4}} \times 9^{\frac{1}{2}}}{3^2 \times 27^{\frac{2}{3}}}\right)^{-1}$ $[\pm 9]$

Definition of a logarithm

In Problems 11 to 20, evaluate the given expressions.

11. $\log_3 9$ [2]

12. $\log_{10} 10$ [1]

13. $\log_5 25$ [2]

14. $\log_8 2$ $\left[\dfrac{1}{3}\right]$

15. $\log_2 \dfrac{1}{8}$ $[-3]$

16. $\log_{11} 121$ [2]

17. $\log_{16} 8$ $\left[\dfrac{3}{4}\right]$

18. $\log_{10} 0.001$ $[-3]$

19. $\log_{27} 3$ $\left[\dfrac{1}{3}\right]$

20. $\log_7 343$ [3]

Solve the equations in Problems 21 to 25.

21. $\log_{10} x = 3$ [1 000]

22. $\log_3 x = -2$ $\left[\dfrac{1}{9}\right]$

23. $\log_2 x = 4$ [16]

24. $\lg x = -2$ [0.01]

25. $\log_8 x = -\dfrac{4}{3}$ $\left[\dfrac{1}{16}\right]$

Laws of logarithms and indicial equations

Write the expressions given in Problems 26 to 28 in terms of log 2, log 3 and log 5 to any base.

26. log 30   [log 2 + log 3 + log 5]
27. log 450   [log 2 + 2 log 3 + 2 log 5]

28. $\log\left(\dfrac{125 \times \sqrt[3]{8}}{\sqrt[4]{81^3}}\right)$ [log 2 − 3 log 3 + 3 log 5]

29. Simplify: log 64 − log 128 + log 32   [4 log 2]
30. Simplify: log 125 + log 25 − log 625   [log 5]

31. Evaluate $\dfrac{\frac{1}{2}\log 16 - \frac{1}{3}\log 8}{\log 4}$ $\left[\dfrac{1}{2}\right]$

32. Evaluate $\dfrac{\log 25 - \log 125 + \frac{1}{2}\log 625}{3 \log 5}$ $\left[\dfrac{1}{3}\right]$

33. Solve the equation:

$$\log(x - 1) + \log(x + 1) = 2 \log(x + 2) \qquad \left[-1\frac{1}{4}\right]$$

34. Solve the equation: $\log 2x^3 - \log x = \log 16 - \log x$ [2]

35. Solve the equation:

$$2 \log x^2 - 3 \log x = \log 8x - \log 4x \qquad [2]$$

In Problems 36 to 44, solve the indicial equations for x, each correct to 4 significant figures.

36. $3^x = 7.42$ [1.824]
37. $3.5^{2x} = 15.4$ [1.091]
38. $78.62 = 5.1^x$ [2.679]
39. $0.035^x = 2.74$ [$-0.300\,1$]
40. $2^{x+1} = 3^{2x-1}$ [1.191]
41. $5.2^{2x-1} = 3.6^{x+2}$ [2.088]
42. $x^{1.2} = 17.3$ [10.76]
43. $3x^{3.5} = 51.6$ [2.254]
44. $x^{-0.45} = 0.521$ [4.258]

Chapter 3

Continued fractions

3.1 Determining quotients and convergents

Any given fraction of less than unity may be expressed in the form shown below for the fraction $\dfrac{5}{8}$.

$$\frac{5}{8} = \cfrac{1}{\cfrac{8}{5}} = \cfrac{1}{1+\cfrac{3}{5}} = \cfrac{1}{1+\cfrac{1}{\cfrac{5}{3}}} = \cfrac{1}{1+\cfrac{1}{1+\cfrac{2}{3}}} = \cfrac{1}{1+\cfrac{1}{1+\cfrac{1}{\cfrac{3}{2}}}} = \cfrac{1}{1+\cfrac{1}{1+\cfrac{1}{1+\cfrac{1}{2}}}}$$

This form can be expressed generally as

$$\cfrac{1}{A + \cfrac{K}{B + \cfrac{L}{C + \cfrac{M}{D + N}}}} \text{ etc.}$$

Comparing the given example with the general form shows A, B, C and D are 1, 1, 1 and 2 respectively. A fraction written in the general form is called a **continued fraction** and the integers A, B, C and D are called the **quotients** of the continued fraction. The quotients may be used to obtain closer and closer approximations to the original fraction, called **convergents**. These approximations to fractions are used to obtain practical ratios for gearwheels or for a dividing head (used to give a required angular displacement), the practical ratios usually being very close to the desired ratios. (See Problems 4 and 5)

A tabular method is used to determine the convergents of a fraction. The values in Table 3.1 are obtained as follows.

Table 3.1

		1	2	3	4	5
a			1	1	1	2
b $\begin{cases} bn \\ \\ bd \end{cases}$		$\dfrac{0}{1}$	$\dfrac{1}{1}$	$\dfrac{1}{2}$	$\dfrac{2}{3}$	$\dfrac{5}{8}$

(i) The quotients are written in cells $a2$, $a3$, $a4$ and $a5$, leaving the first cell empty.

(ii) The fraction $\dfrac{0}{1}$ is always written in cell $b1$.

(iii) The reciprocal of the quotient in cell $a2$ is always written in cell $b2$.

(iv) The fraction in cell $b3$ is given by $\dfrac{a3 \times b2n + b1n}{a3 \times b2d + b1d}$

where the suffix n signifies a number in the numerator and suffix d signifies a number in the denominator. Thus $b2n$ signifies the numerator of the fraction in cell $b2$ and $b1d$ signifies the denominator of the fraction in $b1$. Thus, $b3 = \dfrac{1 \times 1 + 0}{1 \times 1 + 1} = \dfrac{1}{2}$

(v) The fraction in cell $b4$ is given by $\dfrac{a4 \times b3n + b2n}{a4 \times b3d + b2d}$

i.e. $b4 = \dfrac{1 \times 1 + 1}{1 \times 2 + 1} = \dfrac{2}{3}$

(vi) The fraction in cell $b5$ is given by $\dfrac{a5 \times b4n + b3n}{a5 \times b4d + b3d}$

i.e. $b5 = \dfrac{2 \times 2 + 1}{2 \times 3 + 2} = \dfrac{5}{8}$

For a fraction having a larger number of quotients, the general rule for obtaining the N'th fraction, where $N = 3, 4, 5, 6, \ldots$ is given by:

$$bN = \frac{N\text{'th quotient} \times (N-1)^{\text{th}} \text{ numerator} + (N-2)^{\text{th}} \text{ numerator}}{N\text{'th quotient} \times (N-1)^{\text{th}} \text{ denominator} + (N-2)^{\text{th}} \text{ denominator}}$$

The values of the fractions in row b, i.e. the convergents, are

$\dfrac{1}{1}, \dfrac{1}{2}, \dfrac{2}{3}$ and $\dfrac{5}{8}$ and their values are getting closer and closer to $\dfrac{5}{8}$ when viewed from left to right. The values are also alternately greater than, then less than, $\dfrac{5}{8}$.

A simple way of obtaining the quotients of $\dfrac{5}{8}$ is as shown.

First Step: divide numerator into the denominator and obtain a remainder.

$$\text{(numerator)} \quad 5 \enclose{longdiv}{8} \, (\, 1 \quad \text{(First quotient)}$$
$$\underline{5}$$
$$3 \quad \text{(remainder)}$$

Second Step: divide remainder of step 1 into numerator of step 1.

$$\text{(numerator)} \quad 3 \enclose{longdiv}{5} \, (\, 1 \quad \text{(Second quotient)}$$
$$\underline{3}$$
$$2 \quad \text{(remainder)}$$

Third Step: divide remainder of step 2 into numerator of step 2.

$$\text{(numerator)} \quad 2 \enclose{longdiv}{3} \, (\, 1 \quad \text{(Third quotient)}$$
$$\underline{2}$$
$$1 \quad \text{(remainder)}$$

This procedure is continued until a zero remainder is obtained. The complete method of obtaining the quotients for $\frac{5}{8}$ is as shown.

$$5 \enclose{longdiv}{8} \, (\, 1 \qquad\qquad\qquad \text{1st quotient}$$
$$\underline{5}$$
$$3 \enclose{longdiv}{5} \, (\, 1 \qquad\qquad \text{2nd quotient}$$
$$\underline{3}$$
$$2 \enclose{longdiv}{3} \, (\, 1 \qquad \text{3rd quotient}$$
$$\underline{2}$$
$$1 \enclose{longdiv}{2} \, (\, 2 \quad \text{4th quotient}$$
$$\underline{2}$$
$$0$$

Procedure for finding quotients and convergents

To find the quotients and convergents of, say, $\frac{12}{19}$,

(i) the quotients are obtained as shown above.

$$
\begin{array}{r}
12 \overline{)\ 19\ (\ 1} \\
\underline{12} \\
7 \overline{)\ 12\ (\ 1} \\
\underline{7} \\
5 \overline{)\ 7\ (\ 1} \\
\underline{5} \\
2 \overline{)\ 5\ (\ 2} \\
\underline{4} \\
1 \overline{)\ 2\ (\ 2} \\
\underline{2} \\
0
\end{array}
$$

i.e. the quotients are 1, 1, 1, 2 and 2.

(ii) A tabular method is now used to determine the convergents.

(a) The quotients are listed in the first row, leaving the first cell empty.

	1	1	1	2	2

(b) Put $\dfrac{0}{1}$ in the first cell of the second row and the reciprocal of the first quotient in the second cell.

	1	1	1	2	2
$\dfrac{0}{1}$	$\dfrac{1}{1}$				

(c) For each subsequent cell apply the relationship

$\dfrac{t}{u} = \dfrac{R \times r + p}{R \times s + q}$, where these values are as shown.

P	Q	R
$\dfrac{p}{q}$	$\dfrac{r}{s}$	$\dfrac{t}{u}$

This gives:

		1	1	1	2	2
$\dfrac{0}{1}$	$\dfrac{1}{1}$	$\dfrac{1}{2}$	$\dfrac{2}{3}$	$\dfrac{5}{8}$	$\dfrac{12}{19}$

i.e. the convergents of $\dfrac{12}{19}$ are $\dfrac{1}{1}, \dfrac{1}{2}, \dfrac{2}{3}, \dfrac{5}{8}$ and $\dfrac{12}{19}$

The principal uses of convergents are in mechanical and production engineering. As stated previously, they may be used (i) to obtain the best approximation within the limited number of gear wheels available, to a required gear ratio, and (ii) to obtain the best approximation to a required angular displacement when using a dividing head. Worked Problems 4 and 5 illustrate these uses.

Worked problems on continued fractions

Problem 1. Determine the quotients and convergents of $\dfrac{26}{55}$

Using the procedure given above:

(i)   The quotients are obtained by repeated division as shown.

$$
\begin{array}{r}
26 \overline{)\ 55\ }(\ 2 \\
52 \\
\hline
3 \overline{)\ 26\ }(\ 8 \\
24 \\
\hline
2 \overline{)\ 3\ }(\ 1 \\
2 \\
\hline
1 \overline{)\ 2\ }(\ 2 \\
2 \\
\hline
0
\end{array}
$$

i.e. **the quotients are 2, 8, 1 and 2**.

(ii)   Using the tabular approach to determine the convergents gives:

	2	8	1	2
$\dfrac{0}{1}$	$\dfrac{1}{2}$	$\dfrac{8}{17}$	$\dfrac{9}{19}$	$\dfrac{26}{55}$

These values are obtained as follows:

(a) the quotients are written into the first row, leaving the first cell empty,

(b) the first cell of the second row is $\dfrac{0}{1}$ (always), and the second cell is the reciprocal of the first quotient,

(c) applying the relationship given on page 39 to obtain the remaining convergents gives the values shown,

i.e. the second convergent is $\dfrac{8 \times 1 + 0}{8 \times 2 + 1} = \dfrac{8}{17}$

the third convergent is $\dfrac{1 \times 8 + 1}{1 \times 17 + 2} = \dfrac{9}{19}$

the fourth convergent is $\dfrac{2 \times 9 + 8}{2 \times 19 + 17} = \dfrac{26}{55}$

The last convergent is always the value of the original fraction. If this is not so, then an arithmetic error has been made either in determining the values of the quotients or in determining the values of the convergents.

Hence, the convergents are $\dfrac{1}{2}, \dfrac{8}{17}, \dfrac{9}{19}$ and $\dfrac{26}{55}$

Problem 2. If the third convergent of $\dfrac{123}{267}$ is used as an approximation to this fraction, determine, correct to 3 significant figures, the difference and the percentage difference between this convergent and the fraction.

Using the procedure given previously to obtain the first three quotients, gives:

$$123 \quad \overline{)\ 267\ }\ (\ 2$$
$$\underline{246}$$
$$21 \quad \overline{)\ 123\ }\ (\ 5$$
$$\underline{105}$$
$$18 \quad \overline{)\ 21\ }\ (\ 1$$
$$\underline{18}$$
$$3$$

i.e. the first three convergents are 2, 5 and 1

Using a tabular approach to find the first three convergents gives

		2	5	1
0	1	5	6
1	2	11	13

i.e. the first three convergents are $\dfrac{1}{2}$, $\dfrac{5}{11}$ and $\dfrac{6}{13}$

To find the difference and percentage difference, the fraction and the convergent are expressed as decimal fractions. Working correct to, say, seven significant figure accuracy, gives:

$$\frac{123}{267} = 0.460\,674\,2$$

$$\frac{6}{13} = 0.461\,538\,5$$

The difference between the third convergent and the fraction is +0.000 864, correct to 3 significant figures.

The percentage difference is given by $\dfrac{\text{difference} \times 100}{\text{true value of fraction}}$

i.e. $\dfrac{0.000\,864 \times 100}{0.460\,674\,2} = \mathbf{0.188\%}$, correct to 3 significant figures.

Problem 3. Determine the convergents of the fraction $\dfrac{143}{267}$ and list the error and percentage error resulting from taking each convergent as an approximation of the fraction. What conclusions can you draw from these results?

The quotients are obtained by repeated division.

$$143 \overline{)\ 267\ } (\ 1$$
$$143$$
$$124 \overline{)\ 143\ } (\ 1$$
$$124$$
$$19 \overline{)\ 124\ } (\ 6$$
$$114$$
$$10 \overline{)\ 19\ } (\ 1$$
$$10$$
$$9 \overline{)\ 10\ } (\ 1$$
$$9$$
$$1 \overline{)\ 9\ } (\ 9$$
$$9$$
$$\overline{0}$$

i.e. the quotients are 1, 1, 6, 1, 1 and 9

The convergents are obtained using a tabular method (see Problem 1).

		1	1	6	1	1	9
$\dfrac{0}{1}$	$\dfrac{1}{1}$	$\dfrac{1}{2}$	$\dfrac{7}{13}$	$\dfrac{8}{15}$	$\dfrac{15}{28}$	$\dfrac{143}{267}$

i.e. the convergents are $\dfrac{1}{1}, \dfrac{1}{2}, \dfrac{7}{13}, \dfrac{8}{15}, \dfrac{15}{28}$ and $\dfrac{143}{267}$

The convergents are expressed as decimal fractions, the error is obtained by subtracting the values of the convergents from the value of the fraction, and the percentage error is determined using:

$$\frac{\text{difference} \times 100}{\text{true value of fraction}}$$

The results are as shown on p. 44, working to 5 decimal place accuracy for decimal fractions and 2 decimal place accuracy for the percentage error.

Convergent	Decimal equivalent of convergent	Fraction	Error	% Error
$\dfrac{1}{1}$	1.000 00	0.535 58	+0.464 42	+86.71
$\dfrac{1}{2}$	0.500 00	0.535 58	−0.035 58	−6.64
$\dfrac{7}{13}$	0.538 46	0.535 58	+0.002 88	+0.54
$\dfrac{8}{15}$	0.533 33	0.535 58	−0.002 25	−0.42
$\dfrac{15}{28}$	0.535 71	0.535 58	+0.000 13	+0.02
$\dfrac{143}{267}$	0.535 58	0.535 58	0.000 00	0

From these results, it can be seen that:

(i) the errors and hence percentage errors are alternately greater than and then less than the true value,

(ii) each convergent is nearer in value to the true value than the previous one,

(iii) the percentage error decreases very rapidly to a value of less than 1% after two convergents have been considered.
(NB: this will not always be the case, but values do converge very rapidly on the final value.)

Problem 4. A gear train is to provide a ratio of $\dfrac{125}{233}$. If the accuracy is to be within $\pm 0.000\,15$, determine a suitable gearwheel ratio using gearwheels having fewer than 50 teeth.

The convergents of $\dfrac{125}{233}$ are obtained as shown previously.

Quotients:

$$125 \overline{) 233} \ (\ 1$$
$$\underline{125}$$
$$108 \overline{) 125} \ (\ 1$$
$$\underline{108}$$
$$17 \overline{) 108} \ (\ 6$$
$$\underline{102}$$
$$6 \overline{) 17} \ (\ 2$$
$$\underline{12}$$
$$5 \overline{) 6} \ (\ 1$$
$$\underline{5}$$
$$1 \overline{) 5} \ (\ 5$$
$$\underline{5}$$
$$0$$

i.e. the quotients are 1, 1, 6, 2, 1 and 5
The convergents are as shown

	1	1	6	2	1	5
$\dfrac{0}{1}$	$\dfrac{1}{1}$	$\dfrac{1}{2}$	$\dfrac{7}{13}$	$\dfrac{15}{28}$	$\dfrac{22}{41}$	$\dfrac{125}{233}$

The accuracy of each convergent is determined, as shown in Problem 3, working to 6 decimal place accuracy.

Convergent	Decimal equivalent of convergent	Fraction	Error
$\dfrac{1}{1}$	1.000 000	0.536 481	+0.463 519
$\dfrac{1}{2}$	0.500 000	0.536 481	−0.036 481
$\dfrac{7}{13}$	0.538 462	0.536 481	+0.001 981
$\dfrac{15}{28}$	0.535 714	0.536 481	−0.000 767
$\dfrac{22}{41}$	0.536 585	0.536 481	+0.000 104

Hence, the gear train $\dfrac{22}{41}$ gives the required ratio within an

accuracy of $\pm 0.000\,15$

Problem 5. A dividing head used to give angular rotation requires
one turn of the handle to give a rotation of 9°. The index plate of
the dividing head has every hole circles from 30 holes to 60 holes
to accurately locate the handle for parts of a turn. It is required
to produce an angular rotation of 7.37° in a workpiece. Determine
the nearest approximation to this value which can be obtained by
the dividing head.

[An 'every hole circle' of, say, 30 holes, has 30 holes
symmetrically spaced round the circumference of the circle, called
the index plate, the centre of each hole being

$$\frac{360°}{30 \text{ holes}},$$

i.e. 12° displaced from the next hole. The handle associated with the
index plate of the dividing head can be engaged, using a pin, with
any of the holes on the index plate. An index plate having an
'every hole circle' of, say, 40 to 60 holes, contains 20 circles of
holes, the first having 40 symmetrically spaced holes, the second
41 holes ... and the last 60 holes. If the rotation of a workpiece
of, say, 13° 30' is required, this is achieved by rotating the handle
one complete turn to give 9° 0' of rotation to the workpiece and then turning the handle half a turn to give
4° 30' of rotation to the workpiece. The workpiece is locked in
this position by engaging the handle in the 20th hole of a 40-hole
circle or the 21st hole of a 42-hole circle ..., or the 30th hole of a
60-hole circle.]

$$\text{Number of turns of dividing head} = \frac{\text{angle required}}{9°}$$

$$= \frac{7.37}{9}$$

$$= \frac{737}{900}$$

The quotients of this fraction are:

$$
\begin{array}{r}
737 \overline{)\ 900\ }(\ 1 \\
737 \\
\hline
163 \overline{)\ 737\ }(\ 4 \\
652 \\
\hline
85 \overline{)\ 163\ }(\ 1 \\
85 \\
\hline
78 \overline{)\ 85\ }(\ 1 \\
78 \\
\hline
7 \overline{)\ 78\ }(\ 11 \\
77 \\
\hline
1 \overline{)\ 7\ }(\ 7 \\
7 \\
\hline
0
\end{array}
$$

that is, 1, 4, 1, 1, 11 and 7.

The convergents are:

	1	4	1	1	11	7
$\dfrac{0}{1}$	$\dfrac{1}{1}$	$\dfrac{4}{5}$	$\dfrac{5}{6}$	$\dfrac{9}{11}$	$\dfrac{104}{127}$	$\dfrac{737}{900}$

The convergent nearest to the fraction, within the limitation of an every hole circle of 60, is $\dfrac{9}{11}$, since 104 and 127 hole circles are not available. Since the index plate of the dividing head contains symmetrical circular patterns of holes from 30 holes to 60 holes only, it is necessary to express $\dfrac{9}{11}$ as a fraction $\dfrac{a}{b}$, where a and b are numbers between 30 and 60. This is achieved by multiplying both numerator and denominator of $\dfrac{9}{11}$ by 4, i.e. $\dfrac{9}{11} \times \dfrac{4}{4} = \dfrac{36}{44}$,

that is, **36 holes on the 44-hole circle**.

[Check: this gives an angular displacement of $\dfrac{36 \times 9}{44}$, i.e. 7.36°, an error of $\dfrac{0.01 \times 100}{7.37}$ or 0.136%]

3.2   Further problems

In Problems 1 to 5, determine the convergents of the fractions given.

1. $\dfrac{43}{83}$ $\left[\dfrac{1}{1}, \dfrac{1}{2}, \dfrac{14}{27}, \dfrac{43}{83}\right]$

2. $\dfrac{68}{89}$ $\left[\dfrac{1}{1}, \dfrac{3}{4}, \dfrac{13}{17}, \dfrac{68}{89}\right]$

3. $\dfrac{113}{355}$ $\left[\dfrac{1}{3}, \dfrac{7}{22}, \dfrac{113}{355}\right]$

4. $\dfrac{24}{91}$ $\left[\dfrac{1}{3}, \dfrac{1}{4}, \dfrac{4}{15}, \dfrac{5}{19}, \dfrac{24}{91}\right]$

5. $\dfrac{27}{59}$ $\left[\dfrac{1}{2}, \dfrac{5}{11}, \dfrac{11}{24}, \dfrac{27}{59}\right]$

6. Determine the error and percentage error if the 5th convergent of $\dfrac{67}{114}$ is used as an approximation to this value. $[-0.000\,516, -0.088\%]$

7. The 4th convergent of $\dfrac{146}{257}$ is used as an approximation to this fraction. Determine the difference and percentage difference between the 4th and final convergents. $[-0.000\,526, -0.093\%]$

8. Show that the convergents of $\dfrac{17}{27}$ are alternately larger than then smaller than the fraction and determine the error and percentage error between each convergent and the fraction. $\begin{bmatrix} 0.370\,370, 58.52\% \\ -0.139\,630, -20.59\% \\ 0.037\,037, 5.88\% \\ -0.004\,630, -0.74\% \end{bmatrix}$

9. Use the method of continued fractions to determine a gearwheel ratio close to $\dfrac{37}{47}$, if only wheels having a maximum of 35 teeth and a minimum of 20 teeth are available. $\left[\dfrac{22}{28}\right]$

10. A gearwheel ratio of $\frac{28}{95}$ is required to provide a particular thread when using a centre lathe. If only gearwheels having between 18 and 80 teeth are available, use continued fractions to find a close approximation to the required ratio. $\left[\dfrac{20}{68}\right]$

11. A dividing head gives a 9° rotation per turn of the handle and has an every hole index plate from 30 to 60 holes. Calculate the best setting to give an angular rotation of 22.96°.

    [2 turns and a setting of 32 holes on a 58-hole circle]

12. Determine the best setting to give an angular rotation of 50° 44′ 29″ using a 9° per turn dividing head and an every hole index plate from 30 to 60 holes.

    [5 turns and a setting of 37 holes on a 58-hole circle]

Chapter 4

Analytical solution of quadratic equations

4.1 The roots of a quadratic equation

An equation of the general form $ax^2 + bx + c = 0$, where a, b and c are constants, is called a **quadratic equation**. (A **quadratic expression** in x is an expression of the form $ax^2 + bx + c$.) When solving a quadratic equation, there can be two values of x which satisfy the equation and these solutions are called the **roots** of the equation.

If the roots of a quadratic equation are known, then it is possible to form the equation. Let the roots be, say, $x = \alpha$ and $x = \beta$. Then $(x - \alpha) = 0$ and/or $(x - \beta) = 0$. Since one or both of these equations is (are) equal to 0, their product will be equal to 0. Hence

$$(x - \alpha)(x - \beta) = 0$$

Removing the brackets by multiplication gives:

$$x^2 - (\alpha + \beta)x + \alpha\beta = 0 \tag{1}$$

The general form of a quadratic equation is $ax^2 + bx + c = 0$ and on dividing each term by a, this can be written as

$$x^2 + \frac{b}{a}x + \frac{c}{a} = 0 \tag{2}$$

Comparing the coefficients of equations (1) and (2):

$$x^2 \boxed{-(\alpha + \beta)}\ x \boxed{+\alpha\beta} = 0$$
$$\text{and}\quad x^2 \boxed{+\ \frac{b}{a}}\ x \boxed{+\frac{c}{a}} = 0$$

This shows that $-(\alpha + \beta) = \dfrac{b}{a}$ and $\alpha\beta = \dfrac{c}{a}$

Thus, the sum of the roots, $(\alpha + \beta)$, is equal to $-\dfrac{b}{a}$ and the product of the roots, $(\alpha\beta)$, is equal to $\dfrac{c}{a}$

For example, to form the quadratic equation whose roots are 2 and 3: substituting $\alpha = 2$ and $\beta = 3$ in the equation

$$x^2 - (\alpha + \beta)x + \alpha\beta = 0 \text{ gives}$$

$$x^2 - (2 + 3)x + 2(3) = 0$$

That is, $x^2 - 5x + 6 = 0$

Three analytical methods for solving quadratic equations will now be considered:

(a) Factorisation (where possible) – see Section 4.2.
(b) Completing the square – see Section 4.3.
(c) Formula – see Section 4.4.

(A graphical solution of quadratic equations is discussed in Chapter 6.)

4.2   Solution of quadratic equations by factorisation

Multiplying out $(2x - 1)(x + 3)$ gives $2x^2 + 6x - x - 3$, i.e. $2x^2 + 5x - 3$. The reverse process of moving from $2x^2 + 5x - 3$ to $(2x - 1)(x + 3)$ is called **factorising**.

A procedure to factorise a quadratic **expression** such as $(2x^2 + 5x - 3)$ is as follows:

(i)   The factors of the first term, $2x^2$, are $2x$ and x and these are placed in brackets thus: $(2x \quad)(x \quad)$.
(i)   The factors of the last term, -3, are $+1$ and -3, or -1 and $+3$.
(iii) The only combination of (i) and (ii) to give a middle term of $+5x$ is -1 and $+3$.
      Hence $(2x^2 + 5x - 3) = (2x - 1)(x + 3)$

Note that the product of the two inner terms added to the product of the two outer terms must equal the middle term, $+5x$ in this case.

Similarly, factorising $(a^2 + 2ab + b^2)$ gives $(a + b)(a + b)$, i.e. $(a + b)^2$, and factorising $(a^2 - 2ab + b^2)$ gives $(a - b)(a - b)$, i.e. $(a - b)^2$. Both $(a + b)^2$ and $(a - b)^2$ are referred to as **'perfect squares'**.

$(a^2 - b^2)$ is referred to as **'the difference of two squares'** and factorises as $(a + b)(a - b)$. Hence, for example,

$$(23^2 - 12^2) = (23 + 12)(23 - 12)$$
$$= (35)(11) = 385$$

Not all quadratic expressions will factorise. $(a^2 + b^2)$ is an example of a simple quadratic expression which will not factorise.

When a quadratic equation can be factorised, this gives the easiest method of solving it. If the general form of the quadratic equation $ax^2 + bx + c = 0$ can be expressed as the factors $(x - \alpha)(x - \beta) = 0$, then $x = \alpha$ and $x = \beta$ are the roots of the equation. For the expression $ax^2 + bx + c$ to be split into its factors, say $(px + q)(rx + s)$, three criteria must be met:

(i) pr must be equal to a;
(ii) qs must be equal to c; and
(iii) $ps + qr$ must be equal to b.

To factorise $6x^2 + 13x + 6$, pr can either be 6(1) or 3(2). Also qs can be 6(1) or 3(2). In order to meet the third criterion, a solution is $p = 3$, $q = 2$, $r = 2$ and $s = 3$.

Hence $6x^2 + 13x + 6 = (3x + 2)(2x + 3)$.
Thus to solve the equation $6x^2 + 13x + 6 = 0$:
factorising gives $(3x + 2)(2x + 3) = 0$
and hence $(3x + 2) = 0$ and/or $(2x + 3) = 0$.

When $(3x + 2) = 0$, $x = -\dfrac{2}{3}$ and

when $(2x + 3) = 0$, $x = -\dfrac{3}{2}$

Checking these solutions:

When $x = -\dfrac{2}{3}$, L.H.S. $= 6\left[-\dfrac{2}{3}\right]^2 + 13\left[-\dfrac{2}{3}\right] + 6 = 0 = $ R.H.S.

When $x = -\dfrac{3}{2}$, L.H.S. $= 6\left[-\dfrac{3}{2}\right]^2 + 13\left[-\dfrac{3}{2}\right] + 6 = 0 = $ R.H.S.

Hence $x = -\dfrac{2}{3}$ and $x = -\dfrac{3}{2}$ are the roots of the equation $6x^2 + 13x + 6 = 0$.

4.3   Solution of quadratic equations by 'completing the square'

An expression such as x^2 or $(x + 3)^2$ or $(x + a)^2$ is called a **perfect square**.
   If $x^2 = 5$ then $x = \pm\sqrt{5}$
If $(x + 3)^2 = 7$ then $(x + 3) = \pm\sqrt{7}$
      i.e. $x = -3 \pm \sqrt{7}$
If $(x + a)^2 = 8$ then $(x + a) = \pm\sqrt{8}$
      i.e. $x = -a \pm \sqrt{8}$

Hence, if a quadratic equation can be rearranged so that one side of the equation is a perfect square and the other side of the equation is a number, then the solution of the equation is readily obtained by taking the square root of each side of the equation as in the examples above.

The process of rearranging one side of a quadratic equation into a perfect square before solving is called **'completing the square'**.

Now $(x + a)^2 = x^2 + 2ax + a^2$

Thus, in order to make the quadratic expression $x^2 + 2ax$ into a perfect square it is necessary to add (half the coefficient of x)^2

i.e. $\left[\dfrac{2a}{2}\right]^2$ or a^2

For example, $x^2 + 5x$ becomes a perfect square by adding

$\left[\dfrac{5}{2}\right]^2$, i.e. $x^2 + 5x + \left[\dfrac{5}{2}\right]^2 = \left(x + \dfrac{5}{2}\right)^2$

To solve the quadratic equation $2x = 1 - 8x^2$ by completing the square

Procedure

1. Rearrange the equation so that all terms are on the same side of the equals sign and the coefficient of the x^2 term is positive.

    Hence, if $2x = 1 - 8x^2$
    then $8x^2 + 2x - 1 = 0$.

2. Make the coefficient of the x^2 term unity.
    In this case this is achieved by dividing throughout by 8.

    Hence $\dfrac{8x^2}{8} + \dfrac{2x}{8} - \dfrac{1}{8} = \dfrac{0}{8}$

    i.e. $x^2 + \dfrac{1}{4}x - \dfrac{1}{8} = 0$

3. Rearrange the equation so that the x^2 and x terms are on one side of the equals sign and the constant is on the other side of the equals sign.

    Hence, if $x^2 + \dfrac{1}{4}x - \dfrac{1}{8} = 0$

    then $x^2 + \dfrac{1}{4}x = \dfrac{1}{8}$

4. Add to both sides of the equation (half the coefficient of x)^2.

    In this case the coefficient of x is $\dfrac{1}{4}$

    Half the coefficient of x is $\dfrac{1}{8}$

(Half the coefficient of x)^2 is $\left[\dfrac{1}{8}\right]^2$

Thus $x^2 + \dfrac{1}{4}x + \left[\dfrac{1}{8}\right]^2 = \dfrac{1}{8} + \left[\dfrac{1}{8}\right]^2$

The left-hand side is now a perfect square.

$$\left(x + \dfrac{1}{8}\right)^2 = \dfrac{1}{8} + \left(\dfrac{1}{8}\right)^2$$

5. Evaluate the side of the equation that does not contain the terms in x (in this case, the right-hand side).

$$\left(x + \dfrac{1}{8}\right)^2 = \dfrac{1}{8} + \dfrac{1}{64} = \dfrac{9}{64}$$

6. Take the square root of both sides of the equation (remembering that the square root of a number gives a $\pm$ answer).

$$\sqrt{\left(x + \dfrac{1}{8}\right)^2} = \sqrt{\dfrac{9}{64}}$$

i.e. $\left(x + \dfrac{1}{8}\right) = \pm\dfrac{3}{8}$

7. Solve the simple equations:

$$x = -\dfrac{1}{8} \pm \dfrac{3}{8}$$

Hence $x = -\dfrac{1}{8} + \dfrac{3}{8} = \dfrac{1}{4}$

or $x = -\dfrac{1}{8} - \dfrac{3}{8} = -\dfrac{1}{2}$

Hence $x = \dfrac{1}{4}$ or $x = -\dfrac{1}{2}$ the root of the quadratic equation

$2x = 1 - 8x^2$

4.4 Solution of quadratic equations by formula

By solving the general form of the quadratic equation, it is possible to express the variable in terms of the constants a, b and c. The procedure shown can be used to solve any quadratic equation but it is usually easier just to apply the formula obtained. The method of solution shown involves making the left-hand side of an equation into a perfect square as in the method shown in Section 4.3 above.

The general form of a quadratic equation is:

$$ax^2 + bx + c = 0$$

Dividing throughout by a and putting the constant term on the right-hand side of the equation gives:

$$x^2 + \frac{b}{a}x = -\frac{c}{a}$$

In order to make the left-hand side a perfect square, the square of one-half of the coefficient of the term in x is added to each side of the equation. This gives:

$$x^2 + \frac{b}{a}x + \left[\frac{b}{2a}\right]^2 = \left[\frac{b}{2a}\right]^2 - \frac{c}{a}$$

or

$$\left[x + \frac{b}{2a}\right]^2 = \frac{b^2}{4a^2} - \frac{c}{a} = \frac{b^2 - 4ac}{4a^2}$$

Taking the square root of each side of the equation gives:

$$x + \frac{b}{2a} = \frac{\pm\sqrt{(b^2 - 4ac)}}{2a}, \text{ since } \sqrt{(4a^2)} = \pm 2a$$

Hence

$$\boxed{x = \frac{-b \pm \sqrt{(b^2 - 4ac)}}{2a}}$$

The term $b^2 - 4ac$ is called the **discriminant** and the values of a, b and c in the discriminant determine the number and type of roots obtained, when solving a quadratic equation. When b^2 is greater than $4ac$, two unequal roots are obtained. When b^2 is equal to $4ac$, only one root is obtained (this can be considered as two equal roots). When b^2 is less than $4ac$, the roots obtained can only be expressed in complex-number form and are beyond the scope of this text. (See *Technician mathematics 3*).

For example, to solve the equation $3x^2 + 8x - 7 = 0$ correct to 2 decimal places using the formula, the equation must be compared with the general equation. Then

$$ax^2 + bx + c = 0$$

and

$$3x^2 + 8x - 7 = 0$$

By comparing coefficients, $a = 3$, $b = 8$ and $c = -7$.

Applying the formula, $x = \dfrac{-b \pm \sqrt{(b^2 - 4ac)}}{2a}$, gives

$$x = \frac{-8 \pm \sqrt{[8^2 - (4)(3)(-7)]}}{2(3)}$$

$$= \frac{-8 \pm \sqrt{(64 + 84)}}{6} = \frac{-8 \pm \sqrt{148}}{6}$$

$$= \frac{-8 \pm 12.166}{6}$$

That is, $x = -3.36$ and/or 0.69 correct to 2 decimal places.

Checking:
When $x = -3.36$: $3(-3.36)^2 + 8(-3.36) - 7 = -0.01$, that is, very nearly equal to the right-hand side value.

When $x = 0.69$: $3(0.69)^2 + 8(0.69) - 7 = -0.05$, that is, very nearly equal to the right-hand side value.

Hence $x = -3.36$ and $x = 0.69$ are the roots of the equation $3x^2 + 8x - 7 = 0$ correct to two decimal places.

Worked problems on quadratic equations

Problem 1. The roots of a quadratic equation are $-\dfrac{2}{5}$ and $\dfrac{1}{3}$. Find the equation.

If the roots of a quadratic equation are α and β, then $(x - \alpha)(x - \beta) = 0$. Hence, if $\alpha = -\dfrac{2}{5}$ and $\beta = \dfrac{1}{3}$, then

$$\left[x - \left(-\frac{2}{5}\right)\right]\left[x - \frac{1}{3}\right] = 0$$

i.e. $\left(x + \dfrac{2}{5}\right)\left(x - \dfrac{1}{3}\right) = 0$

$$x^2 + \frac{2}{5}x - \frac{1}{3}x - \frac{2}{15} = 0$$

Multiplying throughout by 15 gives:
$$15x^2 + 6x - 5x - 2 = 0$$
i.e. $\mathbf{15x^2 + x - 2 = 0}$

$$\left[\text{Alternatively, from Section 4.1, } x^2 - \left(-\frac{2}{5} + \frac{1}{3}\right)x + \left(-\frac{2}{5} \cdot \frac{1}{3}\right) = 0\right.$$

$$x^2 + \frac{1}{15}x - \frac{2}{15} = 0$$

$$\left.15x^2 + x - 2 = 0\right]$$

Problem 2. Find the equations whose roots are (*a*) 3 and -3, (*b*) 1.6 and -0.3.

(*a*)   Since 3 and -3 are the roots of a quadratic equation then
$$(x - 3)(x + 3) = 0$$
i.e. $x^2 - 3x + 3x - 9 = 0$
i.e. $\mathbf{x^2 - 9 = 0}$

(b)   Since 1.6 and -0.3 are the roots of a quadratic equation then

$$(x - 1.6)(x + 0.3) = 0$$
i.e. $x^2 - 1.6x + 0.3x - 0.48 = 0$
i.e. $x^2 - 1.3x - 0.48 = 0$

Problem 3. Solve the equation $k^2 + 5k - 14 = 0$ by factorisation.

The only possible integer coefficients to make k^2 are 1×1, hence the factors will be of the form:

$(k \quad)(k \quad)$

The term (-14) can be obtained by integers $14 \times (-1)$, $(-14) \times 1$, $7 \times (-2)$ or $(-7) \times 2$. The only combination to give a middle term of $+5$ is $1 \times 7 + 1 \times (-2)$, the 1's being the coefficients of k and the 7 and (-2) the factors to make (-14).

Hence the factors are $(k + 7)(k - 2)$

The equation is $(k + 7)(k - 2) = 0$

For this equation to be true, $(k + 7) = 0$, or $(k - 2) = 0$, or both $(k + 7)$ and $(k - 2)$ are equal to 0.

If $(k + 7) = 0$, then $k = -7$
If $(k - 2) = 0$, then $k = 2$

Checking:
When $k = -7$, $k^2 + 5k - 14 = (-7)^2 + 5(-7) - 14 = 0$, which is equal to the right-hand side.
   When $k = 2$, $k^2 + 5k - 14 = (2)^2 + 5(2) - 14 = 0$, which is equal to the right-hand side.
   Hence **the roots of the equation $k^2 + 5k - 14 = 0$ are $k = -7$ and $k = 2$.**

Problem 4. Determine the roots of the equation $2r^2 + 10r + 8 = 0$ by: (a) factorisation; (b) completing the square; and (c) formula.

(a) *By factorisation*
The only possible integer coefficients to make $2r^2$ are 2 and 1, hence the factors will be:

$(2r \quad)(r \quad)$

The number term, 8, can be obtained by integers 8×1 or 4×2. There are, in fact, two other possibilities, i.e. -8×-1 or -4×-2. However, the middle term of the quadratic equation (i.e. $10r$) is positive, which makes the former two suggestions the only possible ones. The only combinations to give a middle term of $+10$ is $(2 \times 4) + (1 \times 2)$, the first 2 and the 1 being the coefficients of r and the 4 and last 2 the factors to make $+8$.

Hence the factors are $(2r + 2)(r + 4)$.

The equation is $(2r + 2)(r + 4) = 0$
Hence either $2r + 2 = 0$, i.e. $r = -1$
    or $r + 4 = 0$, i.e. $r = -4$

Checking: $2r^2 + 10r + 8 = 0$.

When $r = -1$, L.H.S. $= 2(-1)^2 + 10(-1) + 8 = 0 =$ R.H.S.
When $r = -4$, L.H.S. $= 2(-4)^2 + 10(-4) + 8 = 0 =$ R.H.S.

(b) By 'completing the square'

$2r^2 + 10r + 8 = 0$

Making the coefficient of r^2 unity gives:

$r^2 + 5r + 4 = 0$

Rearranging gives:

$r^2 + 5r = -4$

Adding to both sides (half the coefficient of r)^2 gives:

$$r^2 + 5r + \left[\frac{5}{2}\right]^2 = \left[\frac{5}{2}\right]^2 - 4$$

The left-hand side is now a perfect square.

$$\text{Hence } \left(r + \frac{5}{2}\right)^2 = \frac{25}{4} - 4 = \frac{9}{4}$$

Taking the square root of both sides gives:

$$r + \frac{5}{2} = \sqrt{\frac{9}{4}} = \pm\frac{3}{2}$$

$$\text{Hence } r = -\frac{5}{2} \pm \frac{3}{2}$$

$$\text{i.e. } r = -\frac{5}{2} + \frac{3}{2} = -1$$

$$\text{or } r = -\frac{5}{2} - \frac{3}{2} = -4 \text{ as above.}$$

(c) By formula

$2r^2 + 10r + 8 = 0$

The general equation of a quadratic is $ax^2 + bx + c = 0$.

By comparison therefore $a = 2$, $b = 10$ and $c = 8$

$$r = \frac{-b \pm \sqrt{(b^2 - 4ac)}}{2a}$$

$$= \frac{-10 \pm \sqrt{[10^2 - (4)(2)(8)]}}{2(2)}$$

$$= \frac{-10 \pm \sqrt{(100 - 64)}}{4} = \frac{-10 \pm 6}{4}$$

$$= \frac{-16}{4} \text{ or } \frac{-4}{4}$$

i.e. $r = -4$ or -1 as above.

Hence **the roots of the equation $2r^2 + 10r + 8 = 0$ are $r = -1$ and $r = -4$.**

Problem 5. Solve the following quadratic equations by factorising:

(a) $x^2 - 4x + 4 = 0$ (b) $4x^2 - 9 = 0$

(a)   Since $x^2 - 4x + 4 = 0$
then $(x - 2)(x - 2) = 0$
i.e. $(x - 2)^2 = 0$ (the left-hand side is a **perfect square**).
Hence $x = 2$ is the only root of the equation $x^2 - 4x + 4 = 0$
(Check: $(2)^2 - 4(2) + 4 = 0$)

(b)   Since $4x^2 - 9 = 0$, the left-hand side is the **difference of two squares**, $(2x)^2$ and $(3)^2$
Hence $(2x + 3)(2x - 3) = 0$.

Thus, either $(2x + 3) = 0$, i.e. $x = -\frac{3}{2}$,

and/or $(2x - 3) = 0$, i.e. $x = \frac{3}{2}$

Hence $x = \pm\frac{3}{2}$ are the solutions of the equation

$4x^2 - 9 = 0$

$\left(\text{Check: When } x = +\frac{3}{2}, \ 4\left(+\frac{3}{2}\right)^2 - 9 = 4\left(\frac{9}{4}\right) - 9 = 0;\right.$

$\left.\text{When } x = -\frac{3}{2}, \ 4\left(-\frac{3}{2}\right)^2 - 9 = 4\left(\frac{9}{4}\right) - 9 = 0\right)$

Problem 6. Solve the equation $\dfrac{y + 2}{4} + \dfrac{3}{y - 1} = 5$ by completing the square, giving the answers correct to 4 significant figures.

Multiplying through by $4(y - 1)$ gives:

$$4(y - 1)\left(\frac{y + 2}{4}\right) + 4(y - 1)\left(\frac{3}{y - 1}\right) = 4(y - 1)(5)$$

i.e. $(y - 1)(y + 2) + 12 = 20(y - 1)$

$$y^2 + y - 2 + 12 = 20y - 20$$

$$y^2 - 19y = -30$$

Adding to each side (half the coefficient of $y)^2$, i.e. $\left(-\dfrac{19}{2}\right)^2$, gives:

$$y^2 - 19y + \left[-\frac{19}{2}\right]^2 = -30 + \left[-\frac{19}{2}\right]^2$$

The left-hand side is now a perfect square.

Hence $\left(y - \dfrac{19}{2}\right)^2 = -30 + 90.25 = 60.25$

Taking the square root of both sides of the equation gives:

$$y - \frac{19}{2} = \sqrt{60.25}$$

i.e. $y - 9.5 = \pm 7.762$

$$y = 9.5 \pm 7.762$$

Hence $y = 17.26$ or 1.738 correct to 4 significant figures.

Checking:
When $y = 17.26$, L.H.S. $= \dfrac{17.26 + 2}{4} + \dfrac{3}{17.26 - 1}$

$$= \frac{19.26}{4} + \frac{3}{16.26}$$

$$= 4.815\,0 + 0.184\,5$$

$$= 4.999\,5$$

$$= 5.000 \text{ correct to 4 significant figures}$$

$$= \text{R.H.S.}$$

When $y = 1.738$, L.H.S. $= \dfrac{1.738 + 2}{4} + \dfrac{3}{1.738 - 1}$

$$= \frac{3.738}{4} + \frac{3}{0.738}$$

$$= 0.934\,5 + 4.065\,0$$

$$= 4.999\,5$$

$$= 5.000 \text{ correct to 4 significant figures}$$

$$= \text{R.H.S.}$$

Problem 7. Solve the equation $3y^2 + 6y + 2 = 0$ by formula, giving the roots correct to 2 decimal places.

Comparing the equation given with the general equation $ax^2 + bx + c = 0$ shows that $a = 3$, $b = 6$ and $c = 2$.

The general solution is $y = \dfrac{-b \pm \sqrt{(b^2 - 4ac)}}{2a}$

hence $y = \dfrac{-6 \pm \sqrt{[6^2 - 4(3)(2)]}}{2(3)}$

or $y = \dfrac{-6 \pm \sqrt{12}}{6}$

i.e. $y = \dfrac{-6 + 3.464}{6}$ and/or $\dfrac{-6 - 3.464}{6}$

Then $y = -0.42$ and/or -1.58 correct to 2 decimal places.

Checking:
When $y = -0.42$, $3y^2 + 6y + 2 = 3(-0.42)^2 + 6(-0.42) + 2$ which is equal to 0.009, that is, very nearly equal to the right-hand side. Also when $y = -1.58$, $3y^2 + 6y + 2 = 3(-1.58)^2 + 6(-1.58) + 2$ which is equal to 0.009, that is, very nearly equal to the right-hand side. Thus **the roots of the equation $3y^2 + 6y + 2 = 0$ are $y = -0.42$ and $y = -1.58$ correct to 2 decimal places.**

Further problems on quadratic equations may be found in Section 4.7 (Problems 1–31), page 68.

4.5 Practical problems involving quadratic equations

There are many practical problems where a quadratic equation has first to be formed, from given information, before it is solved. Some typical examples are shown in the following worked problems.

Worked practical problems involving quadratic equations

Problem 1. The area of a rectangle is 18.3 cm^2 and its width is 2.7 cm shorter than its length. Calculate the dimensions of the rectangle, correct to 3 significant figures.

Let the length of the rectangle by x cm, say. Then its width will be $(x - 2.7)$ cm.

The area = length × width = $x(x - 2.7) = 18.3$ cm^2

i.e. $x(x - 2.7) = 18.3$

or $x^2 - 2.7x - 18.3 = 0$

Solving this quadratic equation by formula gives:

$$x = \frac{-(-2.7) \pm \sqrt{[(-2.7)^2 - 4(1)(-18.3)]}}{2}$$

$$= \frac{2.7 \pm \sqrt{(7.29 + 73.2)}}{2}$$

$$= \frac{2.7 \pm 8.972}{2} = 5.836 \text{ or } -3.136$$

The second solution is not possible as negative length does not exist, hence **the dimensions are: length 5.84 cm and width (5.84 − 2.7), i.e. 3.14 cm, correct to 3 significant figures.**

Checking:

Area = 5.84 × 3.14 = 18.3 cm^2, correct to 3 significant figures.

Problem 2. The height S metres of a mass thrown vertically upward at time t seconds is given by $S = 40t - 13t^2$. Find out how long the mass will take after being thrown to reach a height of 25 metres: (*a*) on the ascent and (*b*) on the descent.

Height $S = 40t - 13t^2$

When $S = 25$ m then $25 = 40t - 13t^2$

i.e. $13t^2 - 40t + 25 = 0$, which is a quadratic equation.

Using the formula $t = \dfrac{-b \pm \sqrt{[b^2 - 4ac]}}{2a}$

where $a = 13$, $b = -40$ and $c = 25$ gives:

$$t = \frac{-(-40) \pm \sqrt{[(40)^2 - (4)(13)(25)]}}{2(13)}$$

$$= \frac{40 \pm \sqrt{[1\,600 - 1\,300]}}{26} = \frac{40 \pm 17.32}{26}$$

$$= \frac{57.32}{26} \text{ or } \frac{22.68}{26}$$

$$= \textbf{2.205 s or 0.872 s}$$

Checking:

$S = 40t - 13t^2$

When $t = 0.872$ s, R.H.S. $= 40(0.872) - 13(0.872)^2$

$$= 34.88 - 9.885$$

$$= 24.995 = 25.0 \text{ correct to 3 significant figures}$$

$$= \text{height of mass.}$$

When $t = 2.205$ s, R.H.S. $= 40(2.205) - 13(2.205)^2$

$$= 88.20 - 63.206$$

$$= 24.994 = 25.0 \text{ correct to 3 significant figures}$$

$$= \text{height of mass.}$$

Hence **the mass will reach a height of 25 m after 0.872 s on the ascent and after 2.205 s on the descent.**

Problem 3. The total surface area of a closed cylindrical can is 125 cm^2. Calculate the radius of the cylinder if its height is 5.24 cm.

For a closed cylinder having radius r and height h the total surface area is given by $2\pi rh + 2\pi r^2$

i.e. $125 = 2\pi r(5.24) + 2\pi r^2$

$$(2\pi)r^2 + (10.48\pi)r - 125 = 0$$

$$r^2 + \left(\frac{10.48\pi}{2\pi}\right)r - \frac{125}{2\pi} = 0$$

Hence $r^2 + 5.24r - 19.89 = 0$

Using the quadratic formula:

$$r = \frac{-5.24 \pm \sqrt{[(5.24)^2 - (4)(1)(-19.89)]}}{2(1)}$$

$$= \frac{-5.24 \pm \sqrt{[27.46 + 79.56]}}{2}$$

$$= \frac{-5.24 \pm 10.345}{2} = \frac{5.105}{2} \text{ or } \frac{-15.585}{2}$$

$$= 2.553 \text{ cm or } -7.793 \text{ cm}$$

Thus **the radius of the cylinder is 2.553 cm.** (The negative value is meaningless and is neglected.)

[*Check:*

Total surface area, $125 = 2\pi rh + 2\pi r^2$

When $r = 2.553$ and $h = 5.24$ then

R.H.S. $= 2\pi(2.553)(5.24) + 2\pi(2.553)^2 = 125.0 =$ L.H.S.]

Problem 4. A rectangular building is 20.0 m long and 12.0 m wide. A concrete path of constant width is laid all the way around the building. If the area of the path is 75.0 m² calculate its width to the nearest millimetre.

Figure 4.1 shows the rectangular building with its surrounding path of width d m.

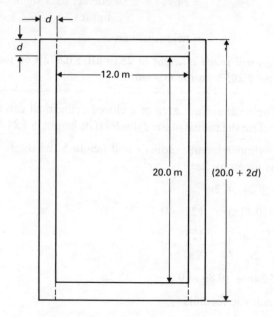

Fig. 4.1

Area of path $= 2(12.0 \times d) + 2d(20.0 + 2d)$

i.e. $75.0 = 24.0d + 40.0d + 4d^2$

or $4d^2 + 64.0d - 75.0 = 0$

$$d = \frac{-64.0 \pm \sqrt{[(64.0)^2 - 4(4)(-75.0)]}}{2(4)}$$

$$= \frac{-64.0 \pm \sqrt{[4\,096 + 1\,200]}}{8} = \frac{-64.0 \pm 72.774}{8}$$

$$= \frac{-136.774}{8} \text{ or } \frac{8.774}{8}$$

$$= -17.096\ 8 \text{ m or } 1.096\ 8 \text{ m}$$

Checking:

$4d^2 + 64.0d - 75.0 = 0$

When $d = 1.096\ 8$, L.H.S. $= 4(1.096\ 8)^2 + 64.0(1.096\ 8) - 75.0$

$$= 4.812 + 70.195 - 75.0 = 0.007$$

which is very nearly equal to 0.

Thus the width of the path to the nearest millimetre is 1 m 97 mm (neglecting the meaningless negative result).

Problem 5. Figure 4.2 shows a rectangular metal plate. A circle of radius R has been cut out of the centre and the quadrants of circles of radius r from the corners as shown. The remaining area forms a template of area 200 cm². Calculate the diameter of the central circular area if $R = 2r$.

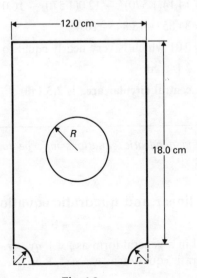

Fig. 4.2

$$\text{Area of template} = [(12.0)(18.0 + r)] - \left[\pi R^2 + 2\left(\frac{\pi r^2}{4}\right) \right]$$

$$\text{i.e. } 200 = [216.0 + 12.0r] - \left[\pi(2r)^2 + \frac{\pi r^2}{2} \right]$$

$$200 = 216.0 + 12.0r - 4\pi r^2 - \frac{\pi r^2}{2}$$

$$200 = 216.0 + 12.0r - \frac{9}{2}\pi r^2$$

$$\frac{9}{2}\pi r^2 - 12.0r - 16.0 = 0$$

$$14.14r^2 - 12.0r - 16.0 = 0$$

Hence $r = \dfrac{-(-12.0) \pm \sqrt{[(-12.0)^2 - (4)(14.14)(-16.0)]}}{2(14.14)}$

$$= \frac{12.0 \pm \sqrt{[144.0 + 905.0]}}{28.28}$$

$$= \frac{12.0 \pm 32.39}{28.28} = 1.570 \text{ or } -0.721$$

Thus, neglecting the meaningless negative result, $r = 1.570$ cm.

Checking:

$14.14r^2 - 12.0r - 16.0 = 0$

When $r = 1.570$, L.H.S. $= 14.14(1.570)^2 - 12.0(1.570) - 16.0$

$$= 34.85 - 18.84 - 16.0$$

$$= 0.01 \text{ which is very nearly equal to } 0$$

$R = 2r = 2(1.570) = 3.140$ cm

Hence **the diameter of the central circular area is 2(3.140),** i.e. **6.280 cm.**

Further practical problems involving quadratic equations may be found in Section 4.7 (Problems 32–49), page 70.

4.6   Algebraic solution of linear and quadratic equations simultaneously

A linear equation can be expressed in a general form as $px + qy = r$, where p, q and r are constants, and a quadratic equation as $y = ax^2 + bx + c$, where a, b and c are constants.

Both these equations in general form have an infinite number of solutions. For each value of x selected, there is a corresponding value of y. However, when considering the equations simultaneously, there is a maximum of two values of x and y which will satisfy both equations.

For these two equations to have values of x and corresponding values of y which simultaneously satisfy both equations, the value of x and y in each equation must be the same, i.e. the y-value in the linear equation must be equal to the y-value in the quadratic equation. Making y the subject of the linear equation gives $y = \dfrac{r - px}{q}$ or $y = -\dfrac{p}{q}x + \dfrac{r}{q}$. By letting $-\dfrac{p}{q} = m$ and $\dfrac{r}{q} = k$, this equation can be written in the standard linear form of $y = mx + k$. For a simultaneous solution of both equations

$$y = mx + k = ax^2 + bx + c$$

$$\text{or } ax^2 + (b - m)x + (c - k) = 0$$

This is a quadratic equation and can be solved for x by the methods shown in Sections 4.2 to 4.4. For example, to find the values of x and y which simultaneously satisfy the equations

$$y = 2x^2 - 5x + 2 \tag{1}$$

$$\text{and } y = 2x - 3 \tag{2}$$

the right-hand sides of each equation are equated.

Then, $2x^2 - 5x + 2 = 2x - 3$

Hence $2x^2 - 7x + 5 = 0$

$$x = \frac{7 \pm \sqrt{(49 - 40)}}{4}$$

$$= 2.5 \text{ or } 1$$

By substituting these values of x in either equation (1) or equation (2) corresponding values of y can be found. When $x = 2.5$, using equation $y = 2(2.5) - 3 = 2$ and when $x = 1$, $y = 2(1) - 3 = -1$.

Checking in equation (1), when $x = 2.5$ and $y = 2$, the right-hand side $2x^2 - 5x + 2 = 2(2.5)^2 - 5(2.5) + 2 = 2$, i.e. the value of y. When $x = 1$ and $y = -1$, the right-hand side is $2x^2 - 5x + 2 = 2(1)^2 - 5(1) + 2 = -1$, the value of y. Hence $x = 2.5$, $y = 2$ and $x = 1$, $y = -1$ are the solutions the simultaneous equations $y = 2x^2 - 5x + 2$ and $y = 2x - 3$.

(A graphical solution to these equations is shown in Chapter 6, page 100

Worked problems on the algebraic solution of linear and quadratic equations simultaneously

Problem 1. Determine the values of x and y which simultaneously satisfy the equations

$$y = 2x^2 - 4x + 1$$

and $y = 4 - 3x$.

For a simultaneous solution, the values of y must be equal, hence the right-hand side of each equation can be equated. Thus

$$2x^2 - 4x + 1 = 4 - 3x$$

i.e. $2x^2 - x - 3 = 0$

Factorising, $(2x - 3)(x + 1) = 0$

Hence $x = \dfrac{3}{2}$ or -1

In the equation $y = 4 - 3x$: when $x = \dfrac{3}{2}$, $y = 4 - 3\left(\dfrac{3}{2}\right) = -\dfrac{1}{2}$ and

when $x = -1$, $y = 4 - 3(-1) = 7$

Checking in the equation $y = 2x^2 - 4x + 1$ gives:

when $x = \dfrac{3}{2}$, R.H.S. $= 2\left(\dfrac{3}{2}\right)^2 - 4\left(\dfrac{3}{2}\right) + 1 = -\dfrac{1}{2}$,

i.e. the calculated value of y

when $x = -1$, R.H.S. $= 2(-1)^2 - 4(-1) + 1 = 7$,

i.e. the calculated value of y

Hence the simultaneous solutions occur when

$$x = \dfrac{3}{2}, y = -\dfrac{1}{2} \text{ and when } x = -1, y = 7$$

Further problems on the algebraic solution of linear and quadratic equations simultaneously may be found in the following Section (4.7) (Problems 50–56), page 73.

4.7   Further problems

Quadratic equations

1. Form the quadratic equations in x whose roots are:
    (a) 4 and 3; (b) 3 and -2; and (c) -1 and -7.
    (a) $[x^2 - 7x + 12 = 0]$ (b) $[x^2 - x - 6 = 0]$
    (c) $[x^2 + 8x + 7 = 0]$

2. The roots of three quadratic equations are as shown.
    Determine the equations in terms of x.
    (a) $\dfrac{1}{2}$ and $\dfrac{1}{3}$; (b) $\dfrac{2}{3}$ and $-\dfrac{4}{5}$; and (c) $-\dfrac{7}{3}$ and $-\dfrac{9}{7}$
    (a) $[6x^2 - 5x + 1 = 0]$ (b) $[15x^2 + 2x - 8 = 0]$
    (c) $[21x^2 + 76x + 63 = 0]$

3. Determine the quadratic equations in x whose roots are:
    (*a*) 1.4 and 3.8; (*b*) -4.1 and 2.7; and (*c*) -16.3 and -18.9
    (*a*) $[x^2 - 5.2x + 5.32 = 0]$ (*b*) $[x^2 + 1.4x - 11.07 = 0]$
    (*c*) $[x^2 + 35.2x + 308.07 = 0]$
4. Solve the following equations:
    (*a*) $c^2 - 9 = 0$; (*b*) $x^2 - 27 = 0$; and (*c*) $(t + 1)^2 = 19$.
    (*a*) $[\pm 3]$ (*b*) $[\pm 5.196]$ (*c*) $[3.359 \text{ or } -5.359]$
5. Solve the equations shown, by factorisation.
    (*a*) $x^2 + 10x = -21$; (*b*) $a^2 = 6 + a$; (*c*) $4x^2 + 8x + 3 = 0$.
    (*a*) $[-3, -7]$ (*b*) $[3, -2]$ (*c*) $\left[-\dfrac{3}{2}, -\dfrac{1}{2}\right]$
6. Solve the equations:
    (*a*) $2r^2 = r + 3$, (*b*) $\dfrac{1}{6}y^2 - \dfrac{1}{3}y = 4$, (*c*) $x^2 - \dfrac{5}{6}x = -\dfrac{1}{6}$
    (*a*) $\left[-1, \dfrac{3}{2}\right]$ (*b*) $[6, -4]$ (*c*) $\left[\dfrac{1}{2}, \dfrac{1}{3}\right]$
7. Solve the quadratic equations shown, by factorisation.
    (*a*) $12z^2 = -z + 6$, (*b*) $x^2 + 6 = 5x$, (*c*) $2p^2 + 10p + 8 = 0$
    (*a*) $\left[\dfrac{2}{3}, -\dfrac{3}{4}\right]$ (*b*) $[3, 2]$ (*c*) $[-4, -1]$
8. (*a*) Complete the following equation:
    $x^2 + 6x + \square = (x + \square)^2$
    (*b*) Hence solve the equation $x^2 + 6x + 7 = 0$
    (*a*) $[x^2 + 6x + 9 = (x + 3)^2]$ (*b*) $[-4.414 \text{ or } -1.586]$
9. Solve the following equations by the method of completing
    the square. (*a*) $x^2 - 3x - 4 = 0$, (*b*) $2x^2 + 9x - 5 = 0$,
    (*c*) $6x^2 + 10x = 4$
    (*a*) $[-1 \text{ or } 4]$ (*b*) $\left[\dfrac{1}{2} \text{ or } -5\right]$ (*c*) $\left[\dfrac{1}{3} \text{ or } -2\right]$

Solve the equations in Problems 10–17 correct to 3 decimal
places using the method of completing the square.
10. $3r^2 + 4r - 5 = 0$ $[-2.120 \text{ or } 0.786]$
11. $3(b^2 - 1) = 7b$ $[-0.370 \text{ or } 2.703]$
12. $3(2a^2 - 1) = 4a$ $[-0.448 \text{ or } 1.115]$
13. $3.2 - 5p^2 = 4p$ $[-1.294 \text{ or } 0.494]$
14. $\dfrac{2}{x - 3} + 4 = \dfrac{x + 4}{3}$ $[2.000 \text{ or } 9.000]$
15. $ax^2 + bx + c = 0$ $\left[\dfrac{-b \pm \sqrt{(b^2 - 4ac)}}{2a}\right]$
16. $2(h + 2) + 4 = 3h(h - 1)$ $[-1.000 \text{ or } 2.667]$
17. $4.6t^2 + 3.5t - 1.75 = 0$ $[-1.105 \text{ or } 0.344]$

Solve the equations in Problems 18–22 correct to 2 decimal places using the formula for solving quadratic equations.

18. $b^2 - 1 = 4b$ $[4.24, -0.24]$
19. $y^2 - y - 1 = 0$ $[1.62, -0.62]$
20. $1.6t^2 = 6.2t + 2.7$ $[-0.39, 4.27]$
21. $2t^2 + 5t = 1$ $[-2.69, 0.19]$
22. $x^2 = 2x + 2$ $[2.73, -0.73]$

Solve the equations in Problems 23–31 correct to 3 decimal places using the formula.

23. $3 = x(7 - 3x)$ $[1.768, 0.566]$
24. $2t^2 + 5t = 4$ $[0.637, -3.137]$
25. $4 + 2a^2 = 7a$ $[2.781, 0.719]$
26. $g(g + 3) - 2g(g - 2) = 5$ $[6.193, 0.807]$
27. $\dfrac{x + 1}{5} + \dfrac{4}{x + 2} = 3$ $[12.633, -0.633]$
28. $\dfrac{y + 1}{y - 1} = y - 2$ $[3.732, 0.268]$
29. $\dfrac{3t + 1}{4t - 3} = \dfrac{3t - 2}{2t + 1}$ $[3.423, 0.243]$
30. $5.76r^2 + 2.86r - 1.35 = 0$ $[0.296, -0.792]$
31. $2.48d^2 - 3.93d - 1.41 = 0$ $[1.886, -0.301]$

Practical problems involving quadratic equations

32. The height S metres of a mass thrown vertically upwards at time t seconds is given by $S = 90t - 17t^2$. Determine how long the mass will take, after being thrown, to reach a height of 60 metres: (*a*) on the ascent and (*b*) on the descent.
(*a*) $[0.782 \text{ s}]$ (*b*) $[4.512 \text{ s}]$

33. Two moles of ethanoic acid react with 3 moles of ethanol to give x moles of ethyl ethanoate at equilibrium. Find x given that $\dfrac{x^2}{(3 - x)(2 - x)} = 4$ $[x = 1.57]$

34. The power (in watts) developed in an electrical circuit is given by $P = 20I - 16I^2$, where I is the current in amperes. Determine the current necessary to produce a power of 3 watts in the circuit.
$[0.174 \text{ A or } 1.076 \text{ A correct to 3 decimal places}]$

35. The area of a triangle is 23.1 cm² and its perpendicular height is 5.3 cm more than its base length. Find the length of the base correct to 3 decimal places. $[4.645 \text{ cm}]$

36. The sag l m in a cable stretched between two supports
    distance x m apart is given by $l = \dfrac{35}{3x} + x$
    Determine the distance between the supports when the sag is
    35 m.   [$\frac{1}{3}$ m or $34\frac{2}{3}$ m]

37. The bending moment in newton metres at a point in a beam
    is given by $M = \dfrac{2.5x(15 - x)}{2}$, where x metres is the distance
    from the point of support. Find the value of x when the
    bending moment is 63 N m.
    [5.081 m or 9.919 m, correct to 3 decimal places]

38. A cable is laid due south from P to Q and then due west
    from Q to R. PQ is 3 km longer than QR. The distance from
    P to R direct is 10 km. Calculate the distance from P to Q
    and also the total length of cable to be laid.
        [8.41 km, 13.82 km]

39. The current i through an electronic device in terms of the
    voltage v is given as $i = 0.003v^2 + 0.011v$. If $i = 4 \times 10^{-3}$
    calculate the values of v.   [$\frac{1}{3}$ or -4]

40. Two resistors, when connected in series, have a total
    resistance of 50 ohms. When connected in parallel their
    resistance is 10.5 ohms. If one of the resistors has a resistance
    of R_1 ohms: (a) show that $R_1{}^2 - 50R_1 + 525 = 0$; and
    (b) calculate the resistance of each.   [15 ohms, 35 ohms]

41. A cylindrical closed can has a total surface area of 468 cm^2.
    Calculate the radius of the cylinder if its height is 6.35 cm.
    [6.021 cm]

42. A tennis court measures 24 m by 11 m. In the layout of a
    number of courts an area of ground must be allowed for at
    the ends and at the side of each court. If a border of
    constant width is allowed around each court and the total
    area of the court and its border is 1 080 m^2, find the width of
    the border.   [8 m]

43. Calculate the diameter of a solid cylinder which has a height
    of 40.0 cm and a total surface area of 4.80 m^2.   [139.3 cm]

44. Figure 4.3 shows a container in the form of a cylindrical
    portion with a hemispherical end. If its total surface area is
    1.75 m^2, find the value of radius R.   [24.78 cm]

45. The total area of the metal template shown in Fig. 4.4 is
    425.6 cm^2. Find the value of r.   [1.994 cm]

46. Two circles have diameters which differ by 4.00 cm. If the
    sum of the areas of the circles is 146.0 cm^2, determine their
    radii.   [3.716 cm, 5.716 cm]

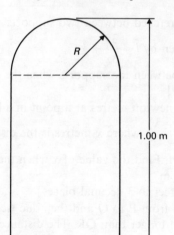

Fig. 4.3

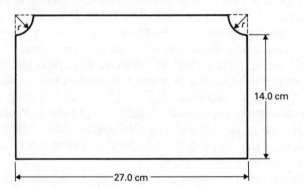

Fig. 4.4

47. A train travels 150 km at a constant speed. If it had travelled 20 km h⁻¹ faster the time for the journey would have been reduced by $37\frac{1}{2}$ minutes. Determine the original speed of the train.   [60 km h⁻¹]

48. If the total surface area of a solid cone is 526.4 cm² and its slant height is 17.6 cm, find its base diameter.
     [13.70 cm]

49. The solubility s in g per 100 cm³ of potassium nitrate at temperature $t\,°C$ is given by the equation
     $s = 13 + 0.613t + 0.017t^2$
     Determine the temperature at which the solubility is 100 g per 100 cm³.   [$t = 55.75°C$]

Algebraic solution of linear and quadratic equations simultaneously

50. Determine the values of x which simultaneously satisfy the equations $y = x^2 + 5x - 3$ and $y = 3x - 2$.
   $[-2.414$ or $0.414]$

51. Determine the solutions of the simultaneous equations $y = 5x + 4 - 2x^2$ and $y = 4 - x$.
   $[(0, 4)$ and $(3, 1)]$

52. Solve the simultaneous equations for x, correct to 3 decimal places: $y = 3.1x - 4.2$ and $y = 1.4x^2 - 2.3x - 5.1$.
   $[x = 4.017$ or $-0.160]$

Solve the simultaneous equations given in Problems 53 to 56 for x, giving the results correct to three decimal places.

53. $y = 3x^2 - 4x - 5$
   $y + 2 = 3x$ $[2.703, -0.370]$

54. $y + 5x + 3 = 3x^2$
   $y + 4x = 3$ $[1.591, -1.257]$

55. $y - \dfrac{6}{7} = x^2 + \dfrac{2}{5}x$

   $\dfrac{3}{4}x + \dfrac{4}{3}y = 5$ $[-2.249, 1.286]$

56. $9.3 + y = 4.8x^2 + 3.4x$
   $2.2x + 1.4y = -3.6$ $[0.774, -1.810]$

Chapter 5

Straight line graphs

5.1 The general equation of a straight line

A graph is a pictorial representation of information showing how one quantity varies with another related quantity. The most used method of showing a relationship between two sets of data is to use **cartesian** or **rectangular axes** as shown in Fig. 5.1. The points on a graph are called **coordinates**. Point P in Fig. 5.1 has coordinates $(2, 3)$, i.e. 2 units in the x-direction and 3 units in the y-direction. Similarly, point Q has coordinates $(-3, 4)$ and R has coordinates $(-2, -3)$. The origin has coordinates $(0, 0)$. Let a relationship between two variables x and y be $y = 5x + 3$.

When $x = 0$, $y = 5(0) + 3 = 3$

When $x = 1$, $y = 5(1) + 3 = 8$

When $x = 2$, $y = 5(2) + 3 = 13$

The coordinates $(0, 3)$, $(1, 8)$ and $(2, 13)$ have been produced from the equation by selecting arbitrary values of x and are shown plotted in Fig. 5.2. When the points are joined together a straight line graph results. The **gradient or slope** of the straight line is the ratio of the change in value of y to the change in the value of x between any two points on the line. If as x increases, y also increases then the gradient is positive.

In Fig. 5.2, gradient of AB $= \dfrac{\text{change in } y}{\text{change in } x} = \dfrac{\text{AC}}{\text{BC}} = \dfrac{23 - 8}{4 - 1} = \dfrac{15}{3} = 5$

If as x increases, y decreases then the gradient is negative.

An example is shown in Fig. 5.3(a), where

$$\text{gradient of DE} = \frac{\text{DF}}{\text{FE}} = \frac{3 - -3}{-1 - 2} = \frac{6}{-3} = -2$$

Figure 5.3(b) shows a straight line graph $y = 2$. Since the straight line is horizontal the gradient is zero.

The value of y when x is zero is called the **y-axis intercept value**. In Fig. 5.2, the y-axis intercept is 3 and in Fig. 5.2(a) it is 1. If the general equation of a graph is of the form $y = mx + c$, where m and c are constants, the graph will always be a straight line, m representing the gradient and c

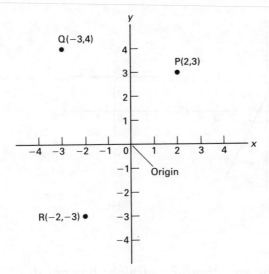

Fig. 5.1

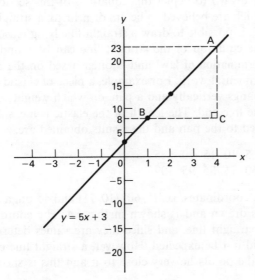

Fig. 5.2

the y-axis intercept. Thus $y = 5x - 2$ represents a straight line of gradient 5 and y-axis intercept -2. Similarly, $y = -2x + 7$ represents a straight line of gradient -2 and y-axis intercept 7.

Straight line graphs are discussed in Chapter 8 of *Technician Mathematics 1*.

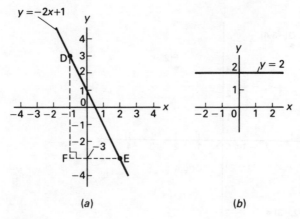

Fig. 5.3

5.2 Deducing the equation of a straight line graph

It has been shown in Section 5.1 that when the equation of a straight line is given, a graph can be drawn to depict this equation. In this section, the coordinates of points which are believed to lie on, or near to, a straight line are given and provided it is possible to draw a straight line lying reasonably near to these points, the equation of the straight line can be found. This technique is called **'determination of law'** and is often based on the results obtained by doing experimental work. For example, a piece of elastic is tied to a support so that it hangs vertically and a pan, on which weights can be placed, is attached to the free end. The length of the elastic is measured as various weights are added to the pan and the results obtained are:

Load, W (newtons)   5 10 15 20   25
Length, l (centimetres) 60 71 85 95 109

By using these values as coordinates, i.e. (5, 60) (10, 71) and so on, a graph of length against load is drawn and is shown in Fig. 5.4. The coordinates do not lie exactly on a straight line, and since they are values determined by experiment this would not be expected. However, a straight line can be chosen so that most of the points lie very close to it and this is shown by line PQ in Fig. 5.4.

By selecting any two points on line PQ, the gradient can be determined, giving the value of m in the general equation of a straight line $y = mx + c$. It is usual to select these points to give a horizontal whole number value, since gradient $= \dfrac{\text{vertical rise or fall}}{\text{horizontal distance}}$ and division by a whole number is easier than division by a decimal fraction. Also values corresponding to corners of squares of the graph paper can usually be most accurately determined from

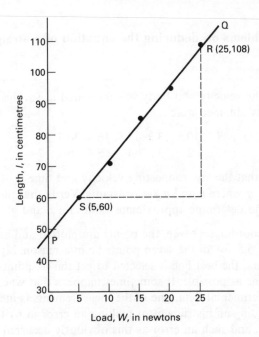

Fig. 5.4 Graph to determine the law relating length and load

the graph. Points R and S lying on line PQ have coordinates of (25, 108) and (5, 60), giving a gradient of

$$\frac{108 - 60}{25 - 5} \quad \text{or} \quad \frac{48}{20} \quad \text{or} \quad 2.4$$

The graph cuts the vertical axis when the load is zero at a length of 48 centimetres, thus giving the value of c in the straight line equation, $y = mx + c$. In this particular problem, the y-axis is equivalent to length l centimetres and the x-axis is equivalent to load W newtons. Hence the straight line equation $y = mx + c$ becomes $l = mW + c$. The law relating length and load is $l = 2.4W + 48$.

In general, when a set of coordinate values are given or are obtained experimentally and it is believed that they follow a law of the form $y = mx + c$, then, if a straight line can be drawn which is reasonably close to most of the coordinate values, this verifies that a law of the form $y = mx + c$ exists, relating the variables. The constants m and c can be calculated by drawing the graph, and determining the gradient to give the value of m and the y-axis intercept value to give the value of c.

Worked problems on deducing the equation of a straight line graph

---

Problem 1. The velocity of a body was measured at various times and the results obtained were:

Velocity (m s^{-1}) v 7.7 10.5 13.3 15.5 16.3 20.5 23
Time (s) t 1   2   3   4   5   6   7

It is believed that the law connecting velocity and time is of the form $v = u + at$ where u and a are constants. Verify that the law is as stated and determine approximate values for u and a.

Using the coordinates given, the points are plotted and are shown in Fig. 5.5. Six of the seven points lie in a reasonably straight line and the best line is selected to get the six points as close to the line as possible. It sometimes happens that when obtaining experimental data, one of the measurements is incorrect due to misreading an instrument or making an error in writing down a result, and such an error as this obviously occurred at a time of five seconds. Hence the point corresponding to $t = 5$ is disregarded. To determine the gradient, two points on the line are

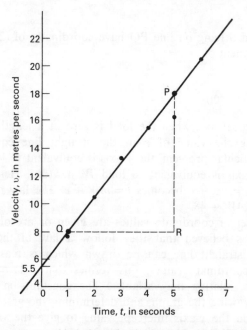

Fig. 5.5 Variation of velocity with time (Problem 1)

selected, these being point P, having coordinates (5, 18) and point Q, having coordinates (1, 8). The gradient is

$$\frac{\text{vertical rise}}{\text{horizontal distance}} = \frac{\text{PR}}{\text{QR}} = \frac{10}{4} = 2.5$$

The graph cuts the vertical axis when $t = 0$ at 5.5 metres per second. Thus, since a straight line graph can be drawn reasonably near to most of the coordinates, it is verified that a law of the form $v = u + at$ does relate the variables v and t. Also, since the gradient u is 2.5 and the vertical axis intercept value a is 5.5, **the law relating the data is $v = 5.5 + 2.5t$**

It can be seen that this law is written in the form $y = c + mx$ which is just the same as $y = mx + c$.

Problem 2. The crushing strength of mortar in tonnes varies with the percentage of water used in its preparation as shown below:

| Crushing strength (F) | 1.52 | 1.31 | 1.14 | 0.94 | 0.76 | 0.61 |
| % of water used (x) | 8.0 | 10.0 | 12.0 | 14.0 | 16.0 | 18.0 |

Prove that these values obey the law $F = ax + b$ where a and b are constants and determine approximate values for a and b.

Determine from the law of the graph (i) the crushing strength corresponding to 5% of water used, and (ii) the percentage of water used for a crushing strength of 1 tonne, assuming the graph remains linear outside the given range of values.

The coordinates given show that the crushing strength (F) decreases as the percentage of water used increases, that is, a graph of these coordinates has a negative gradient. When this happens, an allowance must be made on the vertical scale selected, so that the graph cuts the axis within the scale. One method of determining this is to do a sketch with two points roughly plotted and extrapolate to find roughly where the graph is likely to cut the vertical axis. In this problem such a sketch shows that the intercept value is about 2.2. The graph in Fig. 5.6 shows the points plotted and a straight line graph drawn through them. The gradient of the graph is obtained by selecting any two points on the graph, say P and Q. This gives a gradient of

$$\frac{1.5 - 0.6}{8 - 18} = \frac{0.9}{-10} = -0.09$$

(Since the crushing strength decreases as water content increases, the gradient is negative.) The vertical axis intercept value is

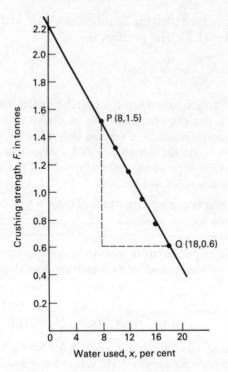

Fig. 5.6 Variation of crushing strength with water used
(Problem 2)

2.23 tonnes. Then since the coordinates given produce a straight line
graph, they do obey the law $F = ax + b$. The constants a and b
are such that **the law is $F = -0.09x + 2.23$**

(i) When the % water used is 5%, $x = 5$

Hence the crushing strength $F = -0.09(5) + 2.23 = \mathbf{1.78}$
tonnes.

(ii) Since $F = -0.09x + 2.23$ then $0.09x = 2.23 - F$

$$\text{and } x = \frac{2.23 - F}{0.09}$$

When the crushing strength, $F = 1$ tonne then the % water
used,

$$x = \frac{2.23 - 1}{0.09} = \mathbf{13.67\%}$$

Problem 3. The resistance R of a copper winding is measured at
various temperatures $t°C$ and the results are shown in the table:

R ohms		127.5	129.4	131.7	134.1	136
t degrees Celsius	50	54	59	64	68	

Show that the law relating resistance and temperature is of the form $R = at + b$. Determine the approximate value of the resistance at $61°C$ and the approximate temperature at which the resistance is 130 ohms. State the law relating resistance and temperature.

If the scales are selected with each one starting at zero, the coordinates will be bunched up in the top right-hand corner of the graph. Because of this, the scales have been selected as shown in Fig. 5.7, and from the graph which is drawn we can verify the law and determine the gradient. However, the vertical axis intercept value cannot be obtained since it is the value at $t = 0°C$ which is required. By selecting a point on the graph and knowing the gradient, the vertical axis intercept value can be calculated.

Since the graph is a straight line, the law relating R and t is of the form $R = at + b$. From the graph, when $t = 61°C$, R is equal to 132.7 ohms and when $R = 130$ ohms, $t = 55.4°C$. The gradient

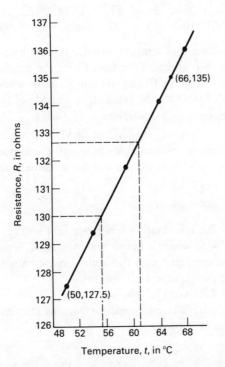

Fig. 5.7 Variation of resistance with temperature (Problem 3)

'a' is $\dfrac{135 - 127.5}{66 - 50}$ or $\dfrac{7.5}{16}$ or approximately equal to 0.47. To calculate b, any point on the graph, say (66, 135) is selected and substituting in the law: $R = at + b$, for R, a and t gives:

$135 = 0.47 \times 66 + b$

i.e. $b \approx 104$

Thus the law relating resistance and temperature is

$R = 0.47t + 104$

Further problems on deducing the equation of a straight line graph may be found in the following Section 5.3 (Problems 1–10).

5.3 Further problems

1. The volume of a gas, v, was measured at various temperatures t and the results were:

Volume	v (m³)	26.68	27.16	27.56	28.08	28.52	29.00
Temperature	t (°C)	10	20	30	40	50	60

   Prove that the volume and temperature are related by a law of the form $v = at + b$ and find the law. From the graph determine the volume at 35°C and the temperature when the volume is 30.0 m³, assuming the law is true outside of the range of values given. $\left[\begin{array}{l} v = 0.046t + 26.14 \\ \quad 27.75\ \text{m}^2, 83.9°\text{C} \end{array}\right]$

2. The following values of resistance R ohms and corresponding voltage V volts are obtained from a test on a filament lamp.

R ohms	74	103	122	136	147
V volts	48	71	86	98	106

   It is thought that the law relating resistance and voltage is of the form $R = aV + b$, where a and b are constants. Test to see if this law is obeyed and find the approximate values of a and b. Determine the voltage at 100 ohms.
   [$a = 1.25$, $b = 14$; 68.8 volts]

3. The strain, ε, in a wire was measured for various stresses, σ and the results are as shown:

Stress, σ N mm^{-2}	10.8	21.6	33.3	37.8	45.9
Strain, ε	0.000 12	0.000 24	0.000 37	0.000 42	0.000 51

Show that stress is related to strain by a law of the form
$\sigma = k\varepsilon$ where k is a constant, and determine the law for the
wire under test. $[\sigma = 90\,000\varepsilon]$

4. Tests carried out to determine the breaking stress, σ of rolled
   copper at various temperatures, t, gave the following results:

   Breaking stress,
   σ N cm^{-2} 8.90 8.45 8.20 7.70 7.50 7.10
   Temperature, $t\,°$C 60   210 300 410 500 600

   It is believed that the law relating breaking stress and
   temperature is of the form $\sigma = kt + l$ where k and l are
   constants. Verify that this law is true and find approximate
   values for k and l and hence state the law.

   $$\left[\sigma = -\frac{1}{300}t + 9.1\right]$$

5. The resistance R of a copper winding is measured at various
   temperatures $t°$C and the results are as shown below:

   R ohms   112.5 114.9 117.8 119.4 122.3
   t degrees Celsius   45   51   58   62   69

   Show that the law relating resistance and temperature is of
   the form $R = at + b$, where a and b are constants. State the
   law. Find also the value of resistance at $55°$C and the value of
   the temperature at 120 ohms. $[R = 0.41t + 94;\ 116.6$
   ohms; $63.4°$C]

6. The data show the force F newtons which, when applied to a
   lifting machine, overcomes a corresponding load of L newtons.

   F newtons 19   35   50   93 125 147
   L newtons 40 120 230 410 540 680

   Verify that the equation relating F and L is of the form
   $F = kL + c$ where k and c are constants and determine the
   law. Assuming the law holds true outside of the range given,
   find the force necessary to lift a load of 1 000 newtons.
   $[F = 0.2L + 11;\ 211$ newtons]

7. The velocity v of a body before a trial begins is given by

   Time t, s -110 -90 -70 -50 -30 -10
   Velocity,
   v, m s^{-1} 10.10   9.54   9.00   8.42   7.86   7.30

   (where $t = -30$ seconds means 30 seconds before the trial
   starts). Prove that the law is of the form $v = u + at$.
   Determine u and a and state the law. $[v = 7.02 - 0.028t]$

8. The mass, m, of steel joist varies with length l as follows:

Mass, m (kg)	90.0	95.0	100.0	110.0	120.0	130.0
Length, l (m)	3.29	3.48	3.67	4.03	4.40	4.78

Prove the law relating mass and length is of the form
$m = pl + q$ where p and q are constants and find the law.
$[m = 27.4l - 0.55]$

9. The speed n rev/min of a motor changes when the voltage V across the armature is varied. The results are shown in the table below:

n rev/min	560	720	900	1070	1240	1440
V volts	80	100	120	140	160	180

Verify that the readings are related by a law of the form
$V = pn + q$, where p and q are constants, and determine the approximate values of p and q. $[p = 0.118, q = 14.2]$

10. The variation of pressure p in a vessel with temperature T is believed to follow a law of the form $p = aT + b$. Verify that the variables are related by this law and determine the law.

p kilopascals	248	253	257	262	266	270
T kelvins	273	278	283	288	293	298

$[p = 0.87T + 11]$

Chapter 6

Graphical solution of equations

6.1 Graphical solution of linear simultaneous equations

A graph depicting the equations

$$x + y = 5 \tag{1}$$
$$x - y = 2 \tag{2}$$

over a range $x = 0$ to $x = 4$ is shown in Fig. 6.1.

The only point at which one value of x and one value of y simultaneously lie on each of these two straight lines is at point P, the point of intersection of the lines. The coordinates of P give the solution to the linear simultaneous equations, and are $x = 3\frac{1}{2}$, $y = 1\frac{1}{2}$.

Checking in both equations:

The left-hand side of equation (1) gives: $3\frac{1}{2} + 1\frac{1}{2} = 5$, which is equal to the right-hand side of equation (1).

The left-hand side of equation (2) gives: $3\frac{1}{2} - 1\frac{1}{2} = 2$, which is equal to the right-hand side of equation (2).

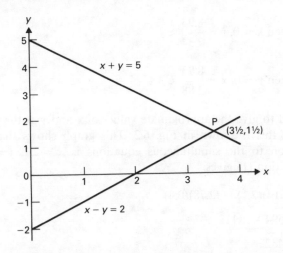

Fig. 6.1   Graph showing the equations $x + y = 5$ and $x - y = 2$

Hence, $x = 3\frac{1}{2}$, $y = 1\frac{1}{2}$ satisfies both equations simultaneously and is the required solution.

If a graphical result is required to a high degree of accuracy, the approximate value of the variables at the point of intersection can be found by drawing a graph. A second graph, having a much smaller range of values can then be drawn, which will magnify the point of intersection and hence give greater accuracy. For example, to solve the equations

$$3.14x - 2.78y = 5.71 \tag{1}$$

$$2.88x + 7.34y = 8.93 \tag{2}$$

correct to 2 decimal places by a graphical method, the procedure is as follows. To draw the first graph, values are selected which simplify the calculations needed to determine the coordinates. Obvious ones are when $x = 0$ and when $y = 0$.

For equation (1), when $x = 0$, $y = \dfrac{-5.71}{2.78} \approx -2.1$

and when $y = 0$, $x = \dfrac{5.71}{3.14} \approx 1.8$

A third coordinate can be calculated as a check point, although this is not necessary in this case as a final check in the original equations will be made to establish if any errors have occurred.

For equation (2), when $x = 0$, $y = \dfrac{8.93}{7.34} \approx 1.2$

and when $y = 0$, $x = \dfrac{8.93}{2.88} \approx 3.1$

A graph is produced to give the approximate values of x and y at the point of intersection, and this is shown in Fig. 6.2. This graph shows that the approximate solution to the simultaneous equations is $x = 2.1$, $y = 0.4$. Substituting in equation (1) gives:

L.H.S. $= (3.14)(2.1) - (2.78)(0.4)$

$= 6.594 - 1.112$

$= 5.482$

A second graph can now be drawn over the range, say $x = 2.0$ to $x = 2.3$, to magnify the scales on the axes and hence to ascertain the point of intersection of the two graphs more accurately.

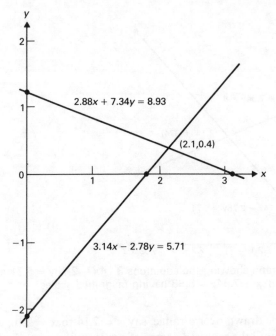

Fig. 6.2 Graph showing the equations $3.14x - 2.78y = 5.71$ and $2.88x + 7.34y = 8.93$

For equation (1), when $x = 2.0$, $y = \dfrac{-5.71 + 2(3.14)}{2.78}$

$$\text{or } y = 0.205$$

and when $x = 2.3$, $y = \dfrac{-5.71 + 2.3(3.14)}{2.78} = 0.544$

For equation (2), when $x = 2.0$, $y = \dfrac{8.93 - 2(2.88)}{7.34} = 0.432$

and when $x = 2.3$, $y = \dfrac{8.93 - 2.3(2.88)}{7.34} = 0.314$

This graph, having a magnified scale, is shown in Fig. 6.3. The point of intersection has coordinates (2.149, 0.374).

Checking in both equations:

The left-hand side of equation (1) is $(3.14 \times 2.149) - (2.78 \times 0.374) = 5.708$, which is very nearly equal to the right-hand side of equation (1). The left-hand side of equation (2) is $(2.88 \times 2.149) + (7.34 \times 0.374) = 8.937$, which is very nearly equal to the right-hand side of equation (2). As the equations have not been used in the calculations, it would have been satisfactory, in this case, to check using one equation only. If greater accuracy still is required,

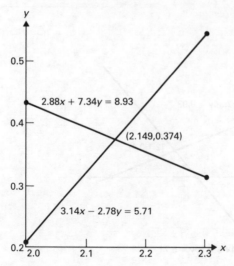

Fig. 6.3 Graph showing the equations $3.14x - 2.78y = 5.71$ and $2.88x + 7.34y = 8.93$ having magnified scales

a third graph can be drawn over a range, say $x = 2.14$ to $x = 2.16$, and this process continued until the required degree of accuracy is achieved, although generally it is quicker to achieve accurate results by calculation.

Worked problems on the graphical solution of simultaneous equations

Problem 1. Solve graphically the equations:

$$\frac{x-1}{3} + \frac{y+2}{2} = 3 \tag{1}$$

$$\frac{2x+1}{6} - \frac{3y-2}{9} = \frac{19}{18} \tag{2}$$

It is usually easier to change both equations into the straight line form, $y = mx + c$, before plotting.

Hence, multiplying equation (1) by 6 gives:

$$6\left[\frac{x-1}{3}\right] + 6\left[\frac{y+2}{2}\right] = 6(3)$$

i.e. $2x - 2 + 3y + 6 = 18$

$$3y = 18 - 2x + 2 - 6$$

$$y = -\frac{2}{3}x + \frac{14}{3}$$

Multiplying equation (2) by 18 gives:

$$18\left[\frac{2x+1}{6}\right] - 18\left[\frac{3y-2}{9}\right] = 18\left[\frac{19}{18}\right]$$

i.e. $6x + 3 - 6y + 4 = 19$

$$-6y = -6x - 3 - 4 + 19$$

$$y = x - 2$$

The straight line graphs $y = -\frac{2}{3}x + \frac{14}{3}$ and $y = x - 2$ are shown

plotted in Fig. 6.4. It is seen that the two graphs intersect at the point $x = 4$, $y = 2$ (i.e. coordinates $(4, 2)$). **The point of intersection is the only point which lies simultaneously on the two straight lines and hence (4, 2) is the solution to the simultaneous equations.**

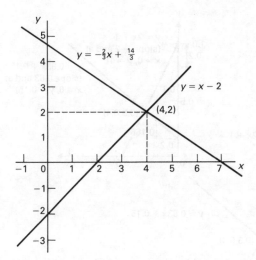

Fig. 6.4 Graphs of $y = -\dfrac{2}{3}x + \dfrac{14}{3}$ and $y = x - 2$

Checking in the original equations:

In equation (1): $\dfrac{x-1}{3} + \dfrac{y+2}{2} = 3$

When $x = 4$, $y = 2$, L.H.S. $= \dfrac{4-1}{3} + \dfrac{2+2}{2} = \dfrac{3}{3} + \dfrac{4}{2} = 1 + 2 = 3$

$= \text{R.H.S.}$

In equation (2):

$$\frac{2x+1}{6} - \frac{3y-2}{9} = \frac{19}{18}$$

When $x = 4$, $y = 2$, L.H.S. $= \dfrac{8+1}{6} - \dfrac{6-2}{9} = \dfrac{3}{2} - \dfrac{4}{9} = \dfrac{27-8}{18} = \dfrac{19}{18}$

$$= \text{R.H.S.}$$

Problem 2. Solve graphically the equations:

$$2.5x + 0.45 - 3y = 0$$

$$1.6x + 0.8y - 0.8 = 0$$

Writing these equations in the form $y = mx + c$ gives:

$$y = 0.8\dot{3}x + 0.15 \tag{1}$$

and $y = -2x + 1$ (2)

In Fig. 6.5(a) a rough graph of these two lines shows that the

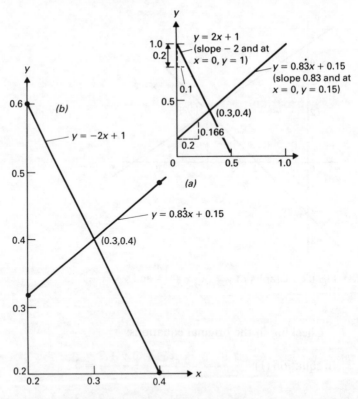

Fig. 6.5 (a) Rough graph to approximately locate the point of intersection. (b) Main graph to accurately locate point of intersection of equations $y = 0.8\dot{3}x + 0.15$ and $y = -2x + 1$

point of intersection is at about $x = 0.3$. Hence the range selected for the main graph is from $x = 0.2$ to $x = 0.4$.

For equation (1), when $x = 0.2$, $y = 0.167 + 0.15 = 0.317$

and when $x = 0.4$, $y = 0.333 + 0.15 = 0.483$

For equation (2), when $x = 0.2$, $y = -0.4 + 1 = 0.6$

and when $x = 0.4$, $y = -0.8 + 1 = 0.2$

The main graph is shown in Fig. 6.5(b).

It can be seen from the graph that the point of intersection has coordinates (0.30, 0.40). Hence the required solution is $x = 0.30$, $y = 0.40$

Checking in either of the original equations, say the first one:

L.H.S. is $2.5(0.30) + 0.45 - 3(0.40) = 0 = $ R.H.S.

Hence $x = 0.30$, $y = 0.40$ **is the required solution to the given equations.**

Further problems on solving simultaneous equations graphically may be found in Section 6.5 (Problems 1–10), page 105.

6.2 Graphical solution of quadratic equations

A graph of $y = 3x^2 + 2x - 2$ over a range $x = -2$ to $x = 1$ is shown in Fig. 6.6. The only places where y is equal to $3x^2 + 2x - 2$ and is also equal to 0 are at the points marked P and Q, where the curve cuts the straight line $y = 0$ (the x-axis). At these points, since $y = 3x^2 + 2x - 3 = 0$, then $3x^2 + 2x - 2 = 0$

Thus the x-values at P and Q are the roots of the equation $3x^2 + 2x - 2 = 0$. From Fig. 6.6, P $= -1.2$ and Q $= 0.55$

Checking: When $x = -1.2$, $3(-1.2)^2 + 2(-1.2) - 2 = -0.08$, which is very nearly equal to 0 and when $x = 0.55$, $3(0.55)^2 + 2(0.55) - 2 = 0.008$, which is also very nearly equal to 0. Hence $x = -1.2$, $x = 0.55$ are the roots of the equation $3x^2 + 2x - 2 = 0$, obtained graphically. Greater accuracy can be obtained by drawing two graphs having a much smaller range around the two points $x = -1.2$ and $x = 0.55$. Suitable ranges would be, say, a graph with x varying from -1.3 to -1.1 and another graph ranging from $x = 0.5$ to $x = 0.6$

An alternative method of finding the roots of the quadratic equation graphically is to split the equation into two parts, plot the two curves separately on the same axes, and find the points of intersection.

For example, if $2x^2 - x - 3 = 0$, then $2x^2 = x + 3$

Now, instead of plotting $y = 2x^2 - x - 3$ and finding where the curve cuts the $y = 0$ axis, separate graphs of $y_1 = 2x^2$ and $y_2 = x + 3$ are plotted as

$x =$	-2	-1.5	-1	-0.5	0	0.5	1
$3x^2 =$	12	6.75	3	0.75	0	0.75	3
$2x =$	-4	-3	-2	-1	0	1	2
$-2 =$	-2	-2	-2	-2	-2	-2	-2
Adding, $y =$	6	1.75	-1	-2.25	-2	-0.25	3

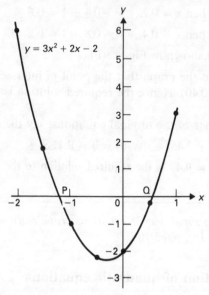

Fig. 6.6 Graph showing the curve $y = 3x^2 + 2x - 2$

shown in Fig. 6.7. The only points where $y_1 = y_2$, i.e. $2x^2 = x + 3$, are at the intersection of the two curves at $(-1, 2)$ and $(1\frac{1}{2}, 4\frac{1}{2})$.

Thus the roots of the equation $2x^2 = x + 3$ (i.e. $2x^2 - x - 3 = 0$) are $x = -1$ and $x = 1\frac{1}{2}$

Figure 6.8 shows the quadratic curve $y = 2x^2 + 5x - 3$. The curve cuts $y = 0$ (i.e. the x-axis) at M and N, i.e. at $x = -3$ and $x = \frac{1}{2}$, and these are thus the solutions of the quadratic equation $2x^2 + 5x - 3 = 0$

Figure 6.8 also shows the straight line $y = 4$ intersecting the curve $y = 2x^2 + 5x - 3$ at points P and Q, i.e. at $x = -3\frac{1}{2}$ and $x = 1$, and these are the solutions of the equation $2x^2 + 5x - 3 = 4$, i.e. $2x^2 + 5x - 7 = 0$

It is possible to use the curve $y = 2x^2 + 5x - 3$ to find the solution of another equation, say for example $2x^2 + x - 6 = 0$. This is achieved by making a comparison of the two equations:

$$y = 2x^2 + 5x - 3 \tag{1}$$
$$0 = 2x^2 + x - 6 \tag{2}$$

In order to make the right-hand side of equation (2) the same as in equation (1), $4x + 3$ is added to both sides of equation (2).

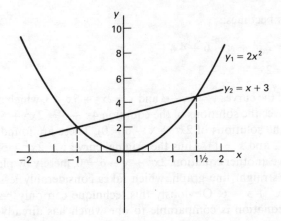

Fig. 6.7

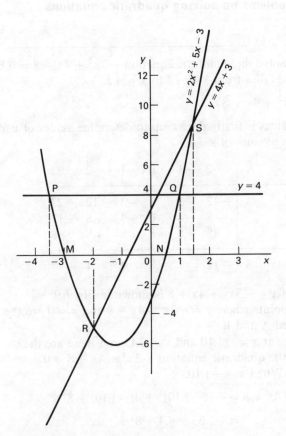

Fig. 6.8

Equation (2) thus becomes:

$$4x + 3 = 2x^2 + x - 6 + 4x + 3$$

$$\text{or } 4x + 3 = 2x^2 + 5x - 3$$

The intersection of the curves $y = 4x + 3$ and $y = 2x^2 + 5x - 3$, which occurs at points R and S, gives the solutions of the equation $4x + 3 = 2x^2 + 5x - 3$ (which, in fact, are the solutions of $2x^2 + x - 6 = 0$). These are found from Fig. 6.8 to be $x = -2$ and $x = 1\frac{1}{2}$. Thus the original curve $y = 2x^2 + 5x - 3$ has been used to solve another equation, $2x^2 + x - 6 = 0$, merely by plotting on the same axes a straight line graph, which takes considerably less time than plotting $y = 2x^2 + x - 6$. Obviously this technique can only be used when the required equation is comparable to one which has already been plotted.

Worked problems on solving quadratic equations graphically

Problem 1. Solve the quadratic equation $-3x^2 + 4x + 8 = 0$ by plotting values of x from $x = -2$ to $x = +3$

Let $y = -3x^2 + 4x + 8$

A table of values is firstly drawn up to determine values of y for corresponding values of x.

x	-2	-1	0	1	2	3
$-3x^2$	-12	-3	0	-3	-12	-27
$+4x$	-8	-4	0	$+4$	$+8$	$+12$
$+8$	$+8$	$+8$	$+8$	$+8$	$+8$	$+8$
$y = -3x^2 + 4x + 8$	-12	1	8	9	4	-7

A graph of $y = -3x^2 + 4x + 8$ is shown in Fig. 6.9

The only points where $y = -3x^2 + 4x + 8$ and $y = 0$ are the points marked A and B.

This occurs at $x = -1.10$ and $x = 2.43$ and these are the solutions of the quadratic equation $-3x^2 + 4x + 8 = 0$

Checking: When $x = -1.10$,

$$-3x^2 + 4x + 8 = -3(-1.10)^2 + 4(-1.10) + 8$$

$$= -3.63 - 4.4 + 8$$

$$= -0.03 \text{ which is nearly zero.}$$

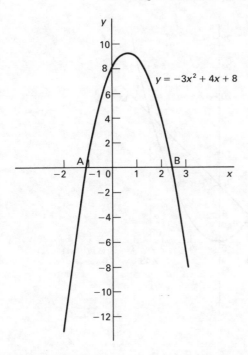

Fig. 6.9

When $x = 2.43$,

$$-3x^2 + 4x + 8 = -3(2.43)^2 + 4(2.43) + 8$$

$$= -17.715 + 9.72 + 8$$

$$= +0.005 \text{ which is nearly equal to zero.}$$

Hence $x = -1.10$ and $x = 2.43$ are the roots of the quadratic equation correct to 3 significant figures. If greater accuracy is needed two further graphs around the points $x = -1$ and $x = 2.5$ having a much smaller range should be plotted.

An alternative graphical method of solving the quadratic equation $-3x^2 + 4x + 8 = 0$ is to split the equation into two parts, i.e. $3x^2 = 4x + 8$, and plot two separate graphs. In this case if $y = 3x^2$ and $y = 4x + 8$ are plotted on the same axes then the points of intersection will give the roots of the equation $3x^2 = 4x + 8$, i.e. $-3x^2 + 4x + 8 = 0$. This is shown in Fig. 6.10 with the same results as above.

Problem 2. Solve the equation $2(x^2 + 1) = 5x$ by a graphical method given that the solutions lie between $x = 0$ and $x = 3$.

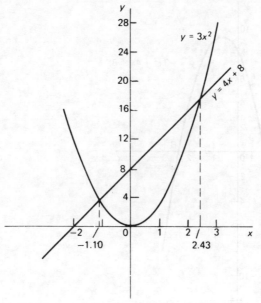

Fig. 6.10

The equation can be rewritten as $2x^2 - 5x + 2 = 0$. A graph of $y = 2x^2 - 5x + 2$ over a range $x = 0$ to $x = 3$ is shown in Fig. 6.11.

$x =$	0	0.5	1.0	1.5	2.0	2.5	3.0
$2x^2$	0	0.5	2	4.5	8	12.5	18
$-5x$	0	-2.5	-5	-7.5	-10	-12.5	-15
$+2$	+2	+2	+2	+2	+2	+2	+2
$y =$	2	0	-1	-1	0	2	5

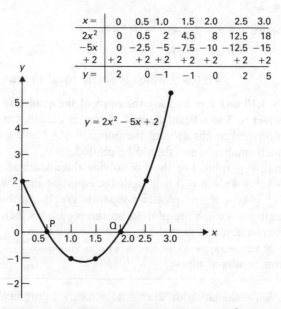

Fig. 6.11   Graph showing the curve $y = 2x^2 - 5x + 2$

The only points where $y = 2x^2 - 5x + 2$ and $y = 0$ are the points marked P and Q. Thus the roots of the equation $2x^2 - 5x + 2 = 0$ are the x-values of these points, i.e. 0.5 and 2.0

Checking: When $x = 0.5$, $2x^2 - 5x + 2 = 2(0.5)^2 - 5(0.5) + 2 = 0$, and the right-hand side is also equal to 0

When $x = 2.0$, $2x^2 - 5x + 2 = 2(2.0)^2 - 5(2.0) + 2 = 0$, and the right-hand side is also equal to 0

Hence **0.5 and 2.0 are the roots of the equation $2x^2 - 5x + 2 = 0$** (i.e. $2(x^2 + 1) = 5x$)

Problem 3. By plotting a graph of $y = 2x^2$, solve graphically the equations: (a) $2x^2 - 4 = 0$; (b) $2x^2 - 2x + \frac{1}{2} = 0$; and (c) $2x^2 + x - 6 = 0$.

(a) The graph depicting the curve of the quadratic equation $y = 2x^2$ is shown in Fig. 6.12.

When $x =$	-2.0	-1.5	-1.0	-0.5	0	0.5	1.0	1.5	2.0
$y = 2x^2 =$	8.0	4.5	2.0	0.5	0	0.5	2.0	4.5	8.0

The equation $2x^2 - 4 = 0$ can be represented by the curve $y = 2x^2$ and the straight line $y = 4$. At the points of intersection of these two equations:

$$y = 2x^2 = 4 \text{ or } 2x^2 = 4$$

i.e. $2x^2 - 4 = 0$

Thus the simultaneous solution of these two equations is at points where the curve and straight line intersect, marked P and Q in Fig. 6.12. These points have coordinates $(-1.4, 4)$ and $(1.4, 4)$

Checking: When $x = \pm 1.4$,
$$2x^2 - 4 = 2(\pm 1.4)^2 - 4 = 3.92 - 4 \approx 0$$

Hence **$x = \pm 1.4$ are the approximate solutions to the equation $2x^2 - 4 = 0$**

(b) The equation $2x^2 - 2x + \frac{1}{2} = 0$ can be written as $2x^2 = 2x - \frac{1}{2}$ and for a simultaneous solution, $y = 2x^2 = 2x - \frac{1}{2}$

Thus the point of intersection of the curve $y = 2x^2$ and the straight line $y = 2x - \frac{1}{2}$ gives the solution to these simultaneous equations. In Fig. 6.12 these curves intersect at $x = 0.5$, marked R.

Checking: When $x = 0.5$, $2x^2 - 2x + \frac{1}{2} = 2(0.5)^2 - 2(0.5) + \frac{1}{2}$
$$= \frac{1}{2} - 1 + \frac{1}{2}$$
$$= 0$$

Hence **$x = 0.5$ is the solution to the equation $2x^2 - 2x + \frac{1}{2} = 0$**

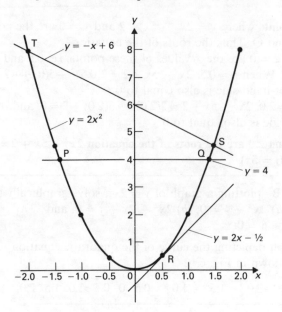

Fig. 6.12 Graph of $y = 2x^2$ and $y = 2x - \frac{1}{2}$, $y = 4$ and $y = -x + 6$

(c) The equation $2x^2 + x - 6 = 0$ can be written as $2x^2 = -x + 6$ and for a simultaneous solution, $y = 2x^2 = -x + 6$. Thus the points of intersection of the curves depicting $y = 2x^2$ and $y = -x + 6$, marked S and T in Fig. 6.12, give the solutions to the given equation. The values of x at these points are $x = 1.5$ and $x = -2$

Checking: When $x = 1.5$, $2x^2 + x - 6 = 2(1.5)^2 + 1.5 - 6$
$$= 4.5 + 1.5 - 6$$
$$= 0$$

When $x = -2$, $2x^2 + x - 6 = 2(-2)^2 + (-2) - 6$
$$= 8 - 2 - 6$$
$$= 0$$

Hence $x = 1.5$ and $x = -2$ are the two solutions to the equation $2x^2 + x - 6 = 0$

Problem 4. Plot the graph of $y = -2x^2 + 4x + 7$ for values of x from $x = -2$ to $x = 4$. Determine, from the graph, the coordinates and nature of the turning-point. Use the graph to find the roots of the following equations:

(i) $-2x^2 + 4x + 7 = 0$ (ii) $-2x^2 + 4x + 4 = 0$

(iii) $-2x^2 + 4x + 12 = 0$ (iv) $-2x^2 + x + 6 = 0$

A table of values is firstly produced to determine values of y for corresponding values of x.

x	-2	-1	0	1	2	3	4
$-2x^2$	-8	-2	0	-2	-8	-18	-32
$+4x$	-8	-4	0	$+4$	$+8$	$+12$	$+16$
$+7$	$+7$	$+7$	$+7$	$+7$	$+7$	$+7$	$+7$
$y = -2x^2 + 4x + 7$	-9	$+1$	$+7$	$+9$	$+7$	$+1$	-9

A graph of $y = -2x^2 + 4x + 7$ is shown in Fig. 6.13

The turning-point on the curve occurs at $x = 1$, $y = 9$, i.e. at coordinates $(1, 9)$, and is a maximum value since the values of y on either side of the turning-point are less than at the turning-point.

(i) $-2x^2 + 4x + 7 = 0$

The curve $y = -2x^2 + 4x + 7$ and the straight line $y = 0$ (i.e. the x-axis) intersect at $x = -1.12$ and $x = 3.12$

Hence **the roots of $-2x^2 + 4x + 7 = 0$ are -1.12 and 3.12**

(ii) $-2x^2 + 4x + 4 = 0$

This equation is compared with $y = -2x^2 + 4x + 7$

$$y = -2x^2 + 4x + 7 \qquad (1)$$
$$0 = -2x^2 + 4x + 4 \qquad (2)$$

If $+3$ is added to both sides of equation (2) then the right-hand side of each equation will be the same, i.e. $-2x^2 + 4x + 7$

Hence $3 = -2x^2 + 4x + 4 + 3$

   i.e. $3 = -2x^2 + 4x + 7$

The solution of this latter equation may be found from the points of intersection of $y = -2x^2 + 4x + 7$ and $y = 3$. From Fig. 6.13 these points are at $x = -0.73$ and 2.73

Hence **the roots of $-2x^2 + 4x + 4$ are -0.73 and 2.73**

(iii) $-2x^2 + 4x + 12 = 0$

Using the same procedure as in (ii) above:

$-2x^2 + 4x + 12 - 5 = -5$

   i.e. $-2x^2 + 4x + 7 = -5$

The solution of this equation may be found from the points of intersection of $y = -2x^2 + 4x + 7$ and $y = -5$. From Fig. 6.13 these points are at $x = -1.65$ and $x = 3.65$

Hence **the roots of $-2x^2 + 4x + 12 = 0$ are -1.65 and 3.65**

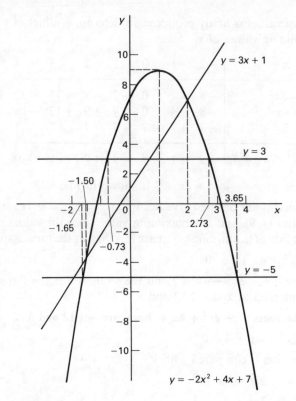

Fig. 6.13

(iv) $-2x^2 + x + 6 = 0$

Using the same procedure as in (ii) and (iii) above:

$-2x^2 + x + 6 + 3x + 1 = 3x + 1$

i.e. $-2x^2 + 4x + 7 = 3x + 1$

The solution of this equation may be found from the points of intersection of $y = -2x^2 + 4x + 7$ and $y = 3x + 1$. From Fig. 6.13 these points are at $x = -1.50$ and $x = 2.00$

Hence **the roots of $-2x^2 + x + 6 = 0$ are -1.50 and 2.00**

Further problems on solving quadratic equations graphically may be found in Section 6.5 (Problems 11–30), page 105.

6.3   Graphical solution of linear and quadratic equations simultaneously

A graph showing the curve depicting the quadratic equation, $y = 2x^2 - 5x + 2$ and the straight line $y = 2x - 3$ over a range $x = 0$ to $x = 3$ is shown in

When x =	0	0.5	1	1.5	2	2.5	3
$2x^2 =$	0	0.5	2	4.5	8	12.5	18
$-5x =$	0	-2.5	-5	-7.5	-10	-12.5	-15
$+2 =$	+2	+2	+2	+2	+2	+2	+2
Adding, y =	2	0	-1	-1	0	2	5

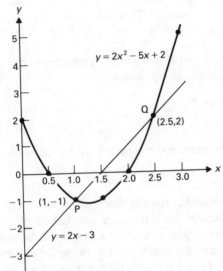

Fig. 6.14   Graph showing the curve $y = 2x^2 - 5x + 2$ and the straight line $y = 2x - 3$

Fig. 6.14. At points P and Q, the curve and straight line both have the same coordinates, i.e. these coordinates satisfy both functions. Hence the points of intersection give the solution to the simultaneous equations $y = 2x^2 - 5x + 2$ and $y = 2x - 3$. From the graph it can be seen that these points are at $x = 2.5$, $y = 2$ and $x = 1$, $y = -1$

Thus to solve linear and quadratic equations simultaneously, graphs of each equation are plotted on the same axes and their points of intersection noted.

Further problems on the graphical solution of linear and quadratic equations simultaneously may be found in Section 6.5 (Problems 31–36), page 107.

6.4   The graphical solution of cubic equations

A graph depicting the curve $y = x^3 - 4x^2 - 3x + 8$ over a range $x = -2$ to $x = 5$ is constructed from the table below and is shown in Fig. 6.15.

When $x =$	-2	-1	0	1	2	3	4	5
$x^3 =$	-8	-1	0	1	8	27	64	125
$-4x^2 =$	-16	-4	0	-4	-16	-36	-64	-100
$-3x =$	6	3	0	-3	-6	-9	-12	-15
$+8 =$	$+8$	$+8$	$+8$	$+8$	$+8$	$+8$	$+8$	$+8$
Adding, $y =$	-10	6	8	2	-6	-10	-4	18

At the points marked P, Q and R, where the curve crosses the x-axis, that is, the line $y = 0$, the simultaneous equations

$$y = x^3 - 4x^2 - 3x + 8$$

and $y = 0$

are satisfied. Thus, these points are the solutions to the equation:

$$x^3 - 4x^2 - 3x + 8 = 0$$

From Fig. 6.15 it can be seen that the solutions are $x = -1.6$, $x = 1.2$ and $x = 4.3$. Because only relatively few points are selected to plot the graph, these solutions are only very approximate and a graph can be produced for each of these values of x, having a much smaller range for accurate results. For example, to get the value of the $x = 1.2$ solution more accurately, a graph of $y = x^3 - 4x^2 - 3x + 8$ over a range, say $x = 1.1$ to $x = 1.3$, can be produced and this will show that $x = 1.16$ is a more accurate result for this root.

The number of solutions or roots to a cubic equation depends on how

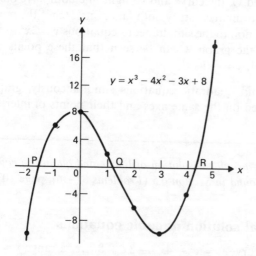

Fig. 6.15   Graph of the curve $y = x^3 - 4x^2 - 3x + 8$

many times the curve crosses the x-axis and there can be one, or two, or three solutions to any cubic equation. The sketches in Fig. 6.16 illustrate each of these three possibilities.

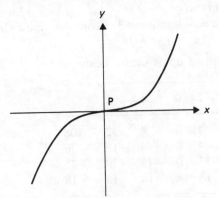

(*a*) A cubic equation having one root at P

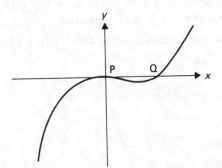

(*b*) A cubic equation having two roots, at P and Q

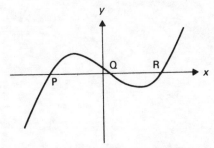

(*c*) A cubic equation having three roots, at P, Q and R

Fig. 6.16

Worked problem on the graphical solution of a cubic equation

Problem 1. Plot the graph of $y = 4x^3 - 4x^2 - 15x + 18$ for values of x between $x = -3$ and $x = 3$. Hence determine the roots of the equation $4x^3 - 4x^2 - 15x + 18 = 0$

A table of values is drawn up as shown below.

x	-3	-2	-1	0	1	2	3
$4x^3$	-108	-32	-4	0	4	32	108
$-4x^2$	-36	-16	-4	0	-4	-16	-36
$-15x$	45	30	15	0	-15	-30	-45
$+18$	18	18	18	18	18	18	18
y	-81	0	25	18	3	4	45

A graph of $y = 4x^3 - 4x^2 - 15x + 18$ is shown in Fig. 6.17. The graph crosses the x-axis at $x = -2$ and touches the x-axis at $x = 1\frac{1}{2}$

Hence the solutions of the equation $4x^3 - 4x^2 - 15x + 18 = 0$ are $x = -2$ and $x = 1\frac{1}{2}$

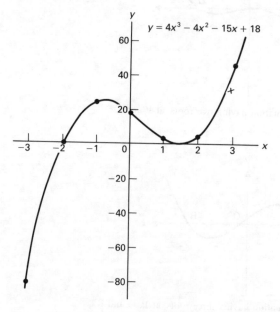

Fig. 6.17

Further problems on the graphical solution of cubic equations may be found in the following Section (6.5) (Problems 37–48), page 108.

6.5 Further problems

Graphical solution of simultaneous equations

In Problems 1–10 plot graphs to solve the simultaneous equations.

1. $\dfrac{x}{2} - 11 = -2y$

   $\dfrac{3}{5}y = 9 - 3x$ $[x = 2, y = 5]$

2. $6a - 19 = 3b$
   $13 = 5a + 6b$ $[a = 3.0, b = -0.33]$

3. $3s - 2t = 0$
   $4s + t + 11 = 0$ $[s = -2, t = -3]$

4. $6.9 + 1.7d = 0.9e$
   $4.3 + 1.3e + 2.3d = 0$ $[d = -3.0, e = 2.0]$

5. $1.4x - 3.2y = 7.06$
   $-2.1x + 6.7y = -12.87$ $[x = 2.3, y = -1.2]$

6. $1.20p + q = 1.80$
   $p - 1.20q = 3.94$ $[p = 2.50, q = -1.20]$

7. $x + 3y = 11$
   $x + 2y = 8$ $[x = 2, y = 3]$

8. $y = 1.1x - 3.0$
   $x = 2.0 + 1.05y$ $[x = 7.42, y = 5.16]$

9. $1.2x - 1.8y = -21$
   $2.5x + 0.6y = 65$ $[x = 20, y = 25]$

10. $\dfrac{1}{2a} + \dfrac{3}{5b} = 7$

    $\dfrac{4}{a} + \dfrac{1}{2b} = 13$ $[a = 0.5, b = 0.1]$

Graphical solutions of quadratic equations

Solve graphically the quadratic equations in Problems 11–20, by plotting the curves between the given limits. Give answers correct to 1 decimal place.

11. $2x^2 - 6x - 9 = 0$; $x = -2$ to $x = 5$ $[-1.1 \text{ or } 4.1]$
12. $-(2p^2 + 10p + 8) = 0$; $p = -6$ to $p = 0$ $[-4.0 \text{ or } -1.0]$
13. $4y^2 = y + 1$; $y = -1$ to $y = 1$ $[-0.4 \text{ or } 0.6]$
14. $2x^2 + 5x + 1 = 0$; $x = -3$ to $x = 0$ $[-2.3 \text{ or } -0.2]$
15. $x^2 + 17 = -10x$; $x = -9$ to $x = 0$ $[-7.8 \text{ or } -2.2]$

16. $4(t^2 + 2t) = -3$; $t = -3$ to $t = 0$ $[-1.5 \text{ or } -0.5]$

17. $\dfrac{1}{6} y^2 = \dfrac{1}{2} y + 4\dfrac{1}{2}$; $y = -6$ to $y = 8$ $[-3.9 \text{ or } 6.9]$

18. $3x^2 + 4x = 9$; $x = -3$ to $x = 2$ $[-2.5 \text{ or } 1.2]$

19. $2r^2 + 5r = 3$; $r = -4$ to $r = 2$ $[-3.0 \text{ or } 0.5]$

20. $a^2 = 2(a + 1)$; $a = -2$ to $a = 3$ $[-0.7 \text{ or } 2.7]$

21. Plot on the same axes graphs of $y = 2x^2$ and $y = 6 - x$ for values of x between $x = -3$ and $x = 3$. Find the coordinates of the points of intersection of the two graphs and hence solve the quadratic equation $2x^2 + x - 6 = 0$.
     $[(-2, 8), (1.5, 4.5); x = -2 \text{ or } 1.5]$

22. Plot the graphs $y = 4x^2$ and $y = 2 - 3x$ on the same axes and find the coordinates of the points of intersection. Hence determine the roots of the equation $4x^2 + 3x - 2 = 0$.
     $[(0.43, 0.74), (-1.18, 5.57); x = 0.43 \text{ or } -1.18]$

23. Solve the equation $3(x^2 + \frac{1}{2}) = 7x$ by a graphical method, given that the solutions lie between $x = 0$ and $x = 3$.
     $[x = 0.24 \text{ or } 2.09]$

24. Plot a graph of $3x^2 + 3.5x = 7.25 + y$ for values of x between $x = -3$ and $x = 3$.
     (a) From the graph find the roots of the equation
     $\qquad 3x^2 + 3.5x = 7.25 + y$
     (b) Use the graph to solve the equation $3x^2 + 4x - 5 = 0$. Give answers correct to 2 decimal places.
     (a) $[-2.24 \text{ or } 1.08]$ (b) $[-2.12 \text{ or } 0.79]$

25. Draw the graph of $y = x^2$ for values of x between $x = -8$ and $x = 8$. Use the graph to solve the following equations, giving answers correct to 1 decimal place.
     (a) $x^2 - 7x - 2 = 0$; (b) $x^2 + 6x = 4$; and (c) $x^2 + 2 = 5x$.
     (a) $[-0.3 \text{ or } 7.3]$ (b) $[-6.6 \text{ or } 0.6]$ (c) $[0.2 \text{ or } 4.8]$

26. Plot a graph of $y = 2x^2 + 5 - 8x$ for values of x between $x = -4$ and $x = 4$. Use the graph to solve the following equations, giving answers correct to 1 decimal place.
     (a) $2x^2 + 5 = 8x$; (b) $4 - 7x = -2x^2$; (c) $2x^2 + 8x + 5 = 0$.
     (a) $[0.8 \text{ or } 3.2]$ (b) $[0.7 \text{ or } 2.8]$ (c) $[-0.8 \text{ or } -3.2]$

27. Plot a graph of $S = 3t^2 + 3 - 7t$ for values of t between $t = -2$ and $t = 3$. Use the graph to solve the following equations, giving answers correct to 1 decimal place.
     (a) $3t^2 + 3 = 7t$; (b) $3t^2 = 6t - 2$; (c) $3t^2 - 7t - 2 = 0$.
     (a) $[0.6 \text{ or } 1.8]$ (b) $[0.4 \text{ or } 1.6]$ (c) $[-0.3 \text{ or } 2.6]$

28. Plot a graph of $y = 2x^2 - 4x - 8$ for values of x between $x = -2$ and $x = 4$. From the graph:

(a) find the value of y when $x = 1.75$;

(b) find the values of x when $y = 2.4$;

(c) determine the coordinates of the turning-point and its nature;

(d) solve the equation $2x^2 - 4x - 8 = 0$;

(e) find the roots of the equation $2x^2 - 7x + 4 = 0$.

(a) $[-6.13]$ (b) $[-1.49 \text{ and } 3.49]$ (c) $[(1, -10); \text{ minimum}]$

(d) $[-1.24 \text{ or } 3.24]$ (e) $[0.72 \text{ or } 2.78]$

29. Plot the graph of $y = -x^2 + 5x + 2$ for values of x from $x = -3$ to $x = 7$. From the graphs find the coordinates and nature of the turning-point. By making use of the graph and drawing suitable straight lines find the solutions to the following equations:

(a) $-x^2 + 5x + 2 = 0$; (b) $-x^2 + 5x = 8 = 0$;

(c) $-x^2 + 5x - 4 = 0$; (d) $-x^2 + x + 3 = 0$.

[Turning-point (2.50, 8.25); maximum]

(a) $[-0.37 \text{ or } 5.37]$ (b) $[-1.27 \text{ or } 6.27]$

(c) $[1.00 \text{ or } 4.00]$ (d) $[-1.30 \text{ or } 2.30]$

30. Plot on the same axes graphs of $y = 5x^2 + 3x - 6$ and $y = -4x^2 - 5x + 3$ for values of x between $x = -3$ and $x = 2$, and find the coordinates of their points of intersection. Hence solve the equation $9x^2 + 8x - 9 = 0$.

$[(-1.54, 1.22), (0.65, -1.94); x = -1.54 \text{ or } 0.65]$

Graphical solution of linear and quadratic equations simultaneously

31. Determine graphically the values of x which simultaneously satisfy the equations $y = x^2 + 5x - 3$ and $y = 3x - 2$. Give answers correct to 2 decimal places. $[-2.41 \text{ or } 0.41]$

32. By drawing a graph of $y = x^2$, solve the equations shown, correct to one decimal place.

(a) $x^2 - 7x - 2 = 0$ $[7.3, -0.3]$

(b) $x^2 + 6x \quad = 4$ $[-6.6, 0.6]$

and (c) $x^2 + 1 \quad = 5x$ $[4.8, 0.2]$

33. Solve graphically the equations shown, correct to one decimal place:

(a) $2x^2 + 5 \quad = 8x$ $[3.2, 0.8]$

(b) $4 - 7x \quad = -2x^2$ $[2.8, 0.7]$

and (c) $2x^2 + 8x + 5 = 0$ $[-3.2, -0.8]$

34. Solve graphically the equations shown, correct to one decimal place.

      (a) $3x^2 + 3$ $= 7x$ [1.8, 0.6]

      (b) $3x^2$ $= 6x - 2$ [1.6, 0.4]

    and (c) $3x^2 - 7x - 2 = 0$ [2.6, −0.3]

35. Solve graphically for x the simultaneous equations, correct to two significant figures:

    $y = 3x - 4$ and $y = 2x^2 - 4x - 8$ [4.00, −0.50]

36. Solve graphically the simultaneous equations for x, correct to two significant figures:

    $3x^2 + 3.5x = 7.25 + y$ and $y + 0.5x + 2.25 = 0$

    [−2.1, 0.79]

Graphical solution of cubic equations

Solve graphically the cubic equations given in Problems 37–46, correct to two significant figures.

37. $x^3 - x^2 - 8x + 12 = 0$ [2.0, −3.0]

38. $2x^3 + 5 = 7x^2 + 7x$ [−1.2, 0.5, 4.2]

39. $x^3 + 2 = x^2 + 5x$ [2.6, 0.38, 2.0]

40. $\dfrac{x^3}{2} + 1 = x^2 + x$ [0.69, 2.5]

41. $x^3 + 4x^2 = 6 - x$ [1.0, −2.0, −3.0]

42. $2(4x^3 - 5x^2 - 1) = 11x$ [−0.50, −0.25, 2.0]

43. $12x^2 + 1 = 2x(3 + 4x^2)$ [0.50]

44. $2x^3 - 12 + 5x^2 = 4x$ [1.5, −2.0]

45. $4.2 + 2x^3 - 2.6x^2 = 3.6x$ [1.6, −1.3, 1.0]

46. $2x^3 - x^2 = 1.16x - 0.54$ [0.45, 0.80, −0.75]

47. The volume (V) of a salt solution varies with concentration (C) at 277 K as follows:

    $C^3 - 2C^2 - 14C + 836 = 0.8V$

    Draw a graph of V against C and determine C when $V = 1\,120$.   [$C = 6$]

48. Show that the cubic equation $8x^3 + 12x^2 + 6x + 1 = 0$ has only one real root and determine its value.   [$x = -0.5$]

Chapter 7

Reduction of equations of the form $y = ax^n$ to linear form

7.1   Introduction

The relationship between two quantities, believed to be of the linear form $y = mx + c$, can be proved by plotting the measured values and seeing if a straight line graph results. However, frequently the relationship between x and y is not a linear one. Many laws exist where y is related to x^2 or $\frac{1}{x}$ or x to some other power. These laws can be expressed in the general form $y = ax^n$ where a and n are constants. In order to verify a law of this form, it is useful to be able to reduce a set of coordinates to a straight line form to predict whether such a law exists, and if the law does exist, to determine the values of a and n.

7.2   Reducing equations of the form $y = ax^n$ to linear form

The equation $y = ax^n$ has to be altered in some way to get it into the general straight line form $Y = mX + c$. The technique used is to take logarithms (usually to a base of 10) of each side of the equation. Then $\lg y = \lg(ax^n)$ and $\lg y = \lg a + \lg x^n$, since by the rules of logarithms, $\lg(A.B) = \lg A + \lg B$. Also by the rules of logarithms $\lg x^n = n \lg x$, hence

$$\lg y = \lg a + n \lg x$$

or $\lg y = n \lg x + \lg a$.

Comparing this with the general equation of a straight line gives

and
$$\boxed{\lg y} = n \boxed{\lg x} + \lg a$$
$$\boxed{Y} = m \boxed{X} + c$$

Thus by plotting $\lg y$ vertically (equivalent to y) and $\lg x$ horizontally (equivalent to X), the equation $y = ax^n$ is reduced to a linear form. The slope of the graph joining the coordinates so related gives the value of n (equivalent to m) and the vertical axis intercept value gives $\lg a$ (equivalent to c). The antilogarithm of this value gives the value of the constant a.

Worked problems on reducing equations of the $y = ax^n$ form to linear form

---

Problem 1. The power dissipated by a resistor was measured for various values of current flowing in the resistor and the results are as shown:

Current I amperes 1.0 1.5 2.5 4.0 5.5   7.0
Power P watts   50 112 310 800 1 510 2 450

Prove that the law relating current and power is of the form $P = RI^n$ where R and n are constants and determine the law.

To express the law $P = RI^n$ in a linear form, logarithms are taken of each side of the equation. This gives: $\lg P = \lg(RI^n)$

$$\text{or } \lg P = \lg R + \lg I^n,$$

$$\text{that is, } \lg P = n \lg I + \lg R$$

This is now in the form $Y = mX + c$ and by plotting $\lg P$ vertically (since it corresponds to Y) and $\lg I$ horizontally (since it corresponds to X), if a straight line graph is produced then the law is verified.

When I is equal to	1.0	2.5	2.5	4.0	5.5	7.0
$\lg I$ is equal to	0	0.176	0.398	0.602	0.740	0.845
When P is equal to	50	112	310	800	1 510	2 450
$\lg P$ is equal to	1.699	2.049	2.491	2.903	3.179	3.389

The graph of $\lg I$ against $\lg P$ is shown in Fig. 7.1, and since a straight line graph is produced the law $P = RI^n$ is verified.
By selecting two points on the graph, say T having coordinates (0.8, 3.3) and Q having coordinates (0.1, 1.9) we can determine the gradient and hence obtain the value of n.

Hence $n = \dfrac{3.3 - 1.9}{0.8 - 0.1} = \dfrac{1.4}{0.7} = 2$

The vertical axis intercept value when $\lg I$ is equal to zero is 1.7, this being the value of $\lg R$. By finding the antilogarithm of 1.7, i.e. 50.1, the value of R is ascertained.

Thus the required law is $P = 50.1I^2$

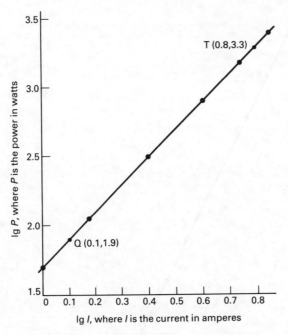

Fig. 7.1 Variation of lg I with lg P (Problem 1)

Problem 2. The pressure p and volume v of a gas at constant temperature are related by the law $pv = c$, where c is a constant. Show that the given values of p and v follow this law and determine the value of c.

Pressure, p, bar 10.6 8.0 6.4 5.3 4.6 4.0
Volume, v, m^3 1.5 2.0 2.5 3.0 3.5 4.0

Since $pv = c$, then $p = c \cdot \dfrac{1}{v}$ or $p = cv^{-1}$. That is, the law is of the form $y = ax^n$. Then, lg $p = -1$ lg $v + $ lg c and by plotting the logarithms of the coordinates, the resulting graph is a straight line and has a gradient of -1, if, and only if, the coordinates are related by the law $pv = c$.

When $p =$ 10.6   8.0   6.4   5.3   4.6   4.0
then lg $p =$ 1.025 0.903 0.806 0.724 0.663 0.602
When $v =$ 1.5   2.0   2.5   3.0   3.5   4.0
then lg $v =$ 0.176 0.301 0.398 0.477 0.544 0.602

The graph of lg p against lg v is shown in Fig. 7.2.

Selecting two points on the graph, say, R (0.1, 1.1) and S (0.5, 0.7)

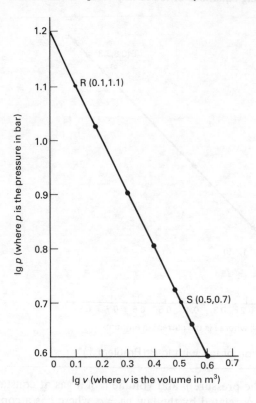

Fig. 7.2   Variation of lg **p** with lg **v** (Problem 2)

gives a gradient of $\dfrac{1.1 - 0.7}{0.1 - 0.5} = \dfrac{0.4}{-0.4} = -1$. Hence the given

values do follow the law $p = cv^{-1}$ or $pv = c$. The vertical axis intercept value when lg v is equal to zero is 1.2, hence lg $c = 1.2$ and $c = $ **15.85**

Problem 3. Two quantities Q and H are believed to be related by the equation $Q = kH^n$. The experimental values obtained for Q and H are as shown:

$Q = 0.16$ 0.20 0.27 0.34 0.40 0.47 0.55
$H = 1.14$ 1.78 3.24 5.14 7.11 9.82 13.44

Determine the law connecting Q and H, and the value of Q when H is 6.00

Since $Q = kH^n$, then lg $Q = n$ lg $H + $ lg k and to determine the law connecting Q and H, lg Q is plotted against lg H.

$Q =$	0.16	0.20	0.27	0.34	0.40	0.47	0.55
$\lg Q =$	-0.796	-0.699	-0.569	-0.469	-0.398	-0.328	-0.260
$H =$	1.14	1.78	3.24	5.14	7.11	9.82	13.44
$\lg H =$	0.057	0.250	0.511	0.711	0.852	0.922	1.128

The graph produced by plotting $\lg Q$ against $\lg H$ is shown in Fig. 7.3. Selecting two points which lie on the graph, say R $(1.1, -0.275)$ and S $(0.1, -0.775)$ gives a gradient of

$$\frac{-0.755 - (-0.275)}{0.1 - 1.1} \text{ or } \frac{-0.5}{-1} \text{ i.e. } \frac{1}{2}$$

The vertical axis intercept value, $\lg k = -0.825$ and by finding the antilogarithm, $k \approx 0.15$. Hence the law connecting Q and H is

$$Q = 0.15H^{\frac{1}{2}} \text{ or } Q = 0.15\sqrt{H}.$$

When H is 6.00, $\lg H$ is 0.778. From the graph, the corresponding value of $\lg Q$ is -0.435, shown as point T. Taking the antilogarithm gives $H = \mathbf{6.00}, Q = \mathbf{0.367}$

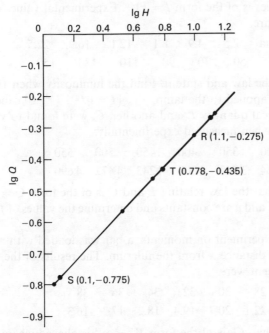

Fig. 7.3 Variation of $\lg Q$ with $\lg H$ (Problem 3)

Further problems on the reduction of equations of the $y = ax^n$ *form to linear form may be found in the following Section (7.3) (Problems 1 to 7).*

7.3   Further problems

1. **The periodic time,** T, of oscillation of a pendulum is believed to be related to the length, l, by a law of the form $T = kl^n$ where k and n are constants. Values of T were measured for various lengths of the pendulum and the results are as shown:

    Periodic time, T s   1.0 1.2 1.4 1.6 1.8 2.0 2.4

    Length, l m   2.4 3.5 4.8 6.3 7.9 9.8 14.1

    Prove the law is true. Determine the values of k and n and hence state the law. $[T = 0.64\sqrt{l}]$

2. Current I and resistance R were measured experimentally and the results relating these quantities were:

    I 3.7 5.9 7.4 9.1 11.6 13.8

    R 6.2 3.9 3.1 2.5   2.0   1.7

    Show that I and R are related by a law of the form $R = aI^n$ and determine the approximate values of a and n. $[23, -1]$

3. The luminosity, I, of a lamp varies with the applied voltage V and it is anticipated that the relationship between these two quantities is of the form $I = kV^n$. Experimental values of I and V are:

    I candela   2.5   4.9   8.1   12.1   16.9   22.5

    V volts   50   70   90   110   130   150

    Verify the law and state it. Find the luminosity when 100 volts is applied to the lamp. $[I = 0.001\ V^2;$ 10 candela$]$

4. A physical quantity L, and another, C, were found to vary as shown when measured experimentally:

L	300	350	400	450	500	550
C	3464	3742	4000	4243	4472	4690

    Show that the law relating L and C is of the form $C = fL^n$ where f and n are constants and determine the values of f and n. $[200, \frac{1}{2}]$

5. In an experiment on moments, a bar was loaded with a mass, W, at a distance x from the fulcrum. The results of the experiment were:

x cm	28	30	32	34	36	38
W kg	22.1	20.7	19.4	18.2	17.2	16.3

    Verify that a law of the form $W = ax^n$ is obeyed where a and n are constants and determine the law.

    $$\left[W = \frac{620}{x} \text{ or } W = 620x^{-1} \right]$$

6. Experimental values of I and r were measured and are shown below. Prove that the law relating I and r is of the form $I = mr^n$ where m and n are constants and determine the law. What will be the value of r when I is 0.006 0?

r	0.1	0.3	0.5	0.7	0.9	1.1	1.3
I	0.000 17	0.001 5	0.004 3	0.008 3	0.014	0.021	0.029

$[I = 0.017r^2; r = 0.594]$

7. Variation of the admittance Y of an electrical circuit with the applied voltage V is as shown:

Voltage V volts	0.37	0.51	0.74	0.98	1.13	1.34
Admittance Y siemens	2.24	1.62	1.12	0.85	0.73	0.62

Show that the law connecting V and Y is of the form $V = RY^n$ where R and n are constants and determine the approximate values of R and n. From the graph, determine the value of admittance when the voltage is 0.86 volts.
$[0.83, -1, 0.965 \text{ siemens}]$

Chapter 8

Exponential growth and decay

8.1 The exponential function

In calculus and more advanced mathematics, a mathematical constant e is frequently used. This constant is called the exponent and has a value of approximately 2.718 3. A function containing e^x is called an **exponential function** and can also be written as exp x. e^x is a function which increases at a rate proportional to its own magnitude.

All the natural laws of growth and decay are of the form $y = a\,e^x$ and therefore the exponent is of considerable importance in science and engineering.

In the same way that logarithms to the base 10 were introduced to facilitate calculations based on powers of 10, logarithms to the base e were developed by Napier to simplify calculations involving the exponental function.

8.2 Evaluating exponential functions

The value of e^x may be determined by using either 4-figure tables of exponential functions, or more likely, by using a calculator.

Most **scientific notation calculators** contain an 'e^x' function which enables all practical values of e^x and e^{-x} to be determined, correct to 8 or 9 significant figures.

For example, $e^1 = 2.718\,281\,83$,

$$e^{2.5} = 12.182\,494\,0,$$

and $e^{-1.732} = 0.176\,930\,195$, correct to 9 significant figures.

In practical situations the degree of accuracy given by a calculator is often far greater than is appropriate. The accepted convention is that the final result is stated to one significant figure greater than the least significant measured value.

Use your calculator to check the following worked problems.

Worked problems on evaluating exponential functions

Problem 1. Using a calculator, evaluate, correct to 5 significant figures:

(a) $e^{1.729}$ (b) $e^{-2.462}$ (c) $\dfrac{2}{3} e^{3.1785}$

(a) $e^{1.729} = 5.635\,016\,07\ldots = \mathbf{5.635\,0}$, correct to 5 significant figures

(b) $e^{-2.462} = 0.085\,264\,25\ldots = \mathbf{0.085\,264}$, correct to 5 significant figures

(c) $\dfrac{2}{3} e^{3.1785} = \dfrac{2}{3}(24.010\,710\,4\ldots) = 16.007\,140\,3\ldots = \mathbf{16.007}$,

correct to 5 significant figures

Problem 2. Using a calculator, evaluate, correct to 4 significant figures:

(a) $\dfrac{1}{5} e^{4.9823}$ (b) $0.015\,e^{-1.112}$ (c) $\dfrac{e^{0.25} - e^{-0.25}}{e^{0.25} + e^{-0.25}}$

(a) $\dfrac{1}{5} e^{4.9823} = \dfrac{1}{5}(145.809\,357\ldots) = 29.161\,871\,5\ldots = \mathbf{29.16}$,

correct to 4 significant figures

(b) $0.015\,e^{-1.112} = 0.015(0.328\,900\,50\ldots) = 0.004\,933\,507\ldots$
$= \mathbf{0.004\,934}$, correct to 4 significant figures

(c) $\dfrac{e^{0.25} - e^{-0.25}}{e^{0.25} + e^{-0.25}} = \dfrac{(1.284\,02\ldots) - (0.778\,800\ldots)}{(1.284\,02\ldots) + (0.778\,800\ldots)}$

$= \dfrac{0.505\,22\ldots}{2.062\,82\ldots}$

$= \mathbf{0.2449}$, correct to 4 significant figures

Further problems on exponential functions may be found in Section 8.7 (Problems 1–6), page 135.

8.3   Graphs of exponential functions

A graph of the curves $y = e^x$ and $y = e^{-x}$ over the range $x = -3$ to $x = 3$ is shown in Fig. 8.1. The values of e^x and e^{-x}, correct to 2 decimal places, are obtained by calculator, and are shown below.

x	−3.0	−2.5	−2.0	−1.5	−1.0	−0.5	0	0.5	1.0	1.5	2.0	2.5	3.0
e^x	0.05	0.08	0.14	0.22	0.37	0.61	1	1.65	2.72	4.48	7.39	12.18	20.09
e^{-x}	20.09	12.18	7.39	4.48	2.72	1.65	1	0.61	0.37	0.22	0.14	0.08	0.05

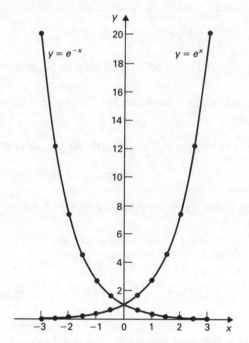

Fig. 8.1   Graphs depicting $y = e^x$ and $y = e^{-x}$

A graph of $y = 1 - e^{-x}$ over the range $x = 0$ to $x = 3.5$ is shown in Fig. 8.2
A table of values is shown below.

x	0	0.5	1.0	1.5	2.0	2.5	3.0	3.5
e^{-x}	1	0.61	0.37	0.22	0.14	0.08	0.05	0.03
$1 - e^{-x}$	0	0.39	0.63	0.78	0.86	0.92	0.95	0.97

For graphs of the form $y = e^{kx}$ where k is constant and can be positive
or negative, 'k' has the effect of altering the scale of x. For graphs of the
form $y = A\,e^{kx}$, where A is a constant, 'A' has the effect of altering the
scale of y. Hence every curve of the form $y = A\,e^{kx}$ has the same general
shape as shown in Fig. 8.1 and A and k are called **scale factors** of the graph.
Their only function is to alter the values of x and y shown on the axes. Thus
similar curves can be obtained for every function of the form $y = A\,e^{kx}$ by
selecting appropriate scale factors. For example, the curve of $y = 2\,e^{3x}$
becomes identical to the curve $y = e^x$ shown in Fig. 8.1 by making the
y-axis markings 4, 8, 12, 16,... instead of 2, 4, 6, 8,... and the x-axis
markings $\frac{1}{3}, \frac{2}{3}$..., instead of 1, 2, 3 ...

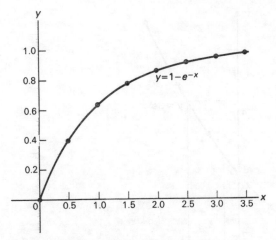

Fig. 8.2 Graph depicting $y = 1 - e^{-x}$

By similar reasoning, all curves of the form $y = a(1 - e^{-kx})$, where a and k are constants, have the same general shape as that shown in Fig. 8.2.

Curves of $y = A e^{kx}$, $y = A e^{-kx}$ and $y = a(1 - e^{-kx})$ have many applications since they define mathematically the laws of growth and decay which are discussed in Section 8.4.

Worked problems on the graphs of exponential functions

Problem 1. Draw a graph of $y = 3 e^{0.2x}$ over a range of $x = -3$ to $x = 3$ and hence determine the approximate value of y when $x = 1.7$ and the approximate value of x when $y = 3.3$

The values of y are calculated for integer values of x over the range required and are shown in the table below.

The values of the exponential functions are obtained using a calculator. The points are plotted and the curve drawn as shown in Fig. 8.3.

x	-3	-2	-1	0	1	2	3
$0.2x$	-0.6	-0.4	-0.2	0	0.2	0.4	0.6
$e^{0.2x}$	0.549	0.670	0.819	1	1.221	1.492	1.822
$3 e^{0.2x}$	1.65	2.01	2.46	3	3.66	4.48	5.47

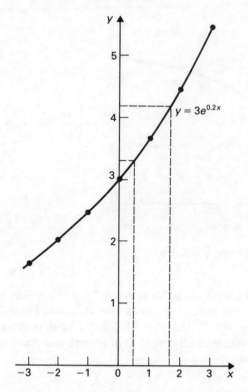

Fig. 8.3   Graph depicting $y = 3\,e^{0.2x}$ (Problem 1)

From the graph, when $x = 1.7$, the corresponding value of y is **4.2** and when y is **3.3**, the corresponding value of x is **0.48**

Problem 2. Draw a graph of $y = e^{-x^2}$ over a range $x = -2$ to $x = 2$

The values of the coordinates are calculated as shown below:

x	-2	-1.5	-1	-0.5	0	0.5	1.0	1.5	2.0
$-x^2$	-4	-2.25	-1	-0.25	0	-0.25	-1	-2.25	-4
e^{-x^2}	0.02	0.11	0.37	0.78	1.0	0.78	0.37	0.11	0.02

The graph is shown plotted in Fig. 8.4.

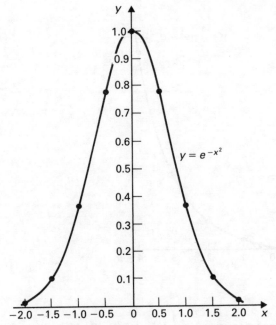

Fig. 8.4 Graph depicting $y = e^{-x^2}$ (Problem 2)

Problem 3. Draw a graph of $y = \dfrac{1}{5}(e^x - e^{-2x})$ over a range $x = -1$ to $x = 4$. Determine from the graph the value of x when $y = 6.6$

The values of the coordinates are calculated as shown below. Since the values used to determine $\dfrac{1}{5}(e^x - e^{-2x})$ range from zero to over 50, only 1 decimal place accuracy is taken.

x	-1	-0.5	0	0.5	1	2	3	4
e^x	0.4	0.6	1	1.6	2.7	7.4	20.1	54.6
$-2x$	2	1	0	-1	-2	-4	-6	-8
e^{-2x}	7.4	2.7	1	0.4	0.1	0.0	0.0	0.0
$e^x - e^{-2x}$	-7.0	-2.1	0	1.2	2.6	7.4	20.1	54.6
$\dfrac{1}{5}(e^x - e^{-2x})$	-1.4	-0.4	0	0.2	0.5	1.5	4.0	10.9

Using these values, the graph shown in Fig. 8.5 is drawn.

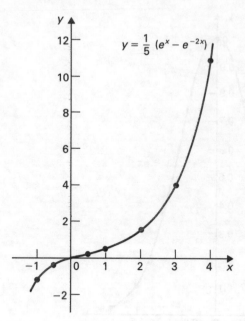

Fig. 8.5 Graph depicting $y = \dfrac{1}{5}(e^x - e^{-2x})$ (Problem 3)

From the graph, when $y = 6.6$, $x = 3.5$

Problem 4. Plot the curves $y = 2 e^{-1.5x}$ and $y = 1.2 (1 - e^{-2x})$ on the same axes from $x = 0$ to $x = 1$ and determine their point of intersection.

A table of values is drawn up as shown below, values being taken correct to 2 decimal places.

x	0	0.2	0.4	0.6	0.8	1.0
$2 e^{-1.5x}$	2.00	1.48	1.10	0.81	0.60	0.45
$1.2(1 - e^{-2x})$	0	0.40	0.66	0.84	0.96	1.04

The curves are shown in Fig. 8.6 and are seen to intersect at $(0.59, 0.83)$. (Hence the solution of the simultaneous equations $y = 2 e^{-1.5x}$ and $y = 1.2(1 - e^{-2x})$ is $x = 0.59$ and $y = 0.83$)

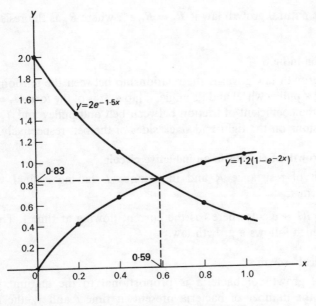

Fig. 8.6   Graph depicting $y = 2\,e^{-1.5x}$ and $y = 1.2(1 - e^{-2x})$

Further problems on graphs of exponential functions may be found in Section 8.7 (Problems 7–15), page 135.

8.4   Laws of growth and decay

The laws of exponential growth or decay occur frequently in engineering and science and are always of the form $y = A\,e^{kx}$ and $y = A(1 - e^{kx})$, where A and k are constants and can be either positive or negative. The natural law $y = A\,e^{kx}$ relates quantities in which the rate of increase of y is proportional to y itself for the growth law, or in which the rate of decrease of y is proportional to y itself for the decay law. Some of the quantities following exponential laws are given below.

(i)   Linear expansion

A rod length l at temperature $\theta°C$ and having a positive coefficient of linear expansion of α will get longer when heated. The natural growth law is:

$l = l_0\,e^{\alpha\theta}$, where l_0 is the length of the rod at $0°C$.

(ii)   Change of electrical resistance with temperature

A resistor of resistance R_θ at temperature $\theta°C$ and having a positive temperature coefficient of resistance of α increases in resistance when heated. The natural growth law is $R_\theta = R_0\, e^{\alpha\theta}$, where R_0 is the resistance at $0°C$.

(iii) Tension in belts

A natural growth law governs the relationship between the tension T_1 in a belt around a pulley wheel and its angle of lap α. It is of the form $T_1 = T_0\, e^{\mu\alpha}$, where μ is the coefficient of friction between belt and pulley and T_1 and T_0 are the tensions on the tight and slack sides of the belt respectively.

(iv) The growth of current in an inductive circuit

In a circuit of resistance R and inductance L having a final value of steady current I,

$$i = I(1 - e^{-\frac{Rt}{L}}),$$ where i is the current flowing at time t. This is an equation which follows a growth law.

(v) Biological growth

The rate of growth of bacteria is proportional to the amount present. When y is the number of bacteria present at time t and y_0 the number present at time $t = 0$ then

$$y = y_0\, e^{kt},$$ where k is the growth constant.

(vi) Newton's law of cooling

The rate at which a body cools is proportional to the excess of its temperature above that of its surroundings. The law is: $\theta = \theta_0\, e^{-kt}$, where the excess of temperature at time $t = 0$ is θ_0 and at time t is θ. The negative power of the exponent indicates a decay curve when k is positive.

(vii) Discharge of a capacitor

When a capacitor of capacitance C, having an initial charge of Q is discharged through a resistor R, then

$$q = Q\, e^{-\frac{t}{CR}},$$ where q is the charge after time t.

(viii) Atmospheric pressure

The pressure p at height h above ground level is given by

$$p = p_0\, e^{\frac{h}{c}}$$

where p_0 is the pressure at ground level and c is a constant.

(ix) The decay of current in an inductive circuit

When a circuit having a resistance R, inductance L and initial current I is allowed to decay, it follows a natural law of the form

$$i = I\, e^{-\frac{Rt}{L}},$$ where i is the current flowing after time t.

(x) Radioactive decay

The rate of disintegration of a radioactive nucleus having 'N_0' radioactive atoms present and a decay constant of λ is given by:

$$N = N_0\, e^{-\lambda t}$$

where N is the number of radioactive atoms present after time t.

These are just some of the relationships which exist which follow the laws of growth or decay.

Worked problems on the laws of growth and decay

Problem 1. A belt is in contact with a pulley for a sector of $\theta = 1.073$ radians and the coefficient of friction between these two surfaces is $\mu = 0.27$. Determine the tension on the taut side of the belt, T newtons, when the tension on the slack side is given by $T_0 = 23.8$ newtons, given that these quantities are related by the law $T = T_0\, e^{\mu\theta}$.

$$T = T_0\, e^{\mu\theta} = 23.8\, e^{(0.27 \times 1.073)}$$

$$= 23.8\, e^{0.289\,71}$$

$$= 23.8 \times 1.336\,039\,9\ldots$$

$$= \mathbf{31.80\ newtons}$$

Problem 2. The instantaneous current, i amperes at time t seconds, is given by: $i = 6.0\, e^{-\frac{t}{CR}}$, when a capacitor is being charged. The capacitance C is 8.3×10^{-6} farads and the resistance R has a value of 0.24×10^6 ohms. Determine the instantaneous current when t is 3.0 seconds. Sketch a curve of current against time from $t = 0$ to $t = 6$ s.

$$i = 6.0\, e^{-\frac{3}{8.3 \times 10^{-6} \times 0.24 \times 10^6}}$$

$$= 6.0\, e^{-\frac{3}{8.3 \times 0.24}}$$

$$= 6.0\, e^{-1.506\,0}$$

$$= 6.0 \times 0.221\,8$$

$$= 1.33\ \text{amperes}$$

That is, the current flowing when t is 3.0 seconds is 1.33 amperes.

Since $i = 6.0\, e^{-\frac{t}{CR}}$ and $C = 8.3 \times 10^{-6}$ and $R = 0.24 \times 10^6$, then $i = 6.0\, e^{-0.502t}$. At time $t = 0$, $i = 6.0\, e^0 = 6.0$ (since $e^0 = 1$), and when $t = \infty$, $i = 6.0\, e^{(-0.502)\infty} = 6.0(0) = 0$

A decay curve representing $i = 6.0\, e^{-0.502t}$ is shown in Fig. 8.7

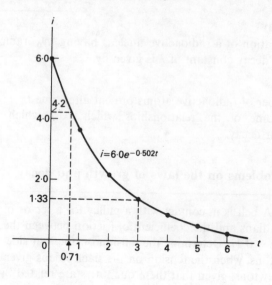

Fig. 8.7   Decay curve of current against time

Problem 3. The temperature θ_2 degrees Celsius of a winding which is being heated electrically, at time t seconds, is given by:
$\theta_2 = \theta_1(1 - e^{-\frac{t}{T}})$, where θ_1 is the temperature at time $t = 0$ seconds and T seconds is a constant. Calculate θ_1 in degrees Celsius when θ_2 is $45°C$, t is 28 s and T is 73 s

Transposing the formula to make θ_1 the subject gives:

$$\theta_1 = \frac{\theta_2}{1 - e^{-\frac{t}{T}}}$$

Substituting the values of θ_2, t and T gives:

$$\theta_1 = \frac{45}{1 - e^{-\frac{28}{73}}} = \frac{45}{1 - e^{-0.383\,56}}$$

i.e. $\theta_1 = \dfrac{45}{1 - 0.681\,43} = \dfrac{45}{0.318\,57}$

$\theta_1 = 141.26°C$

That is, the initial temperature is 141°C, correct to the nearest degree.

Problem 4. The current i flowing in a coil is given by
$i = \dfrac{E}{R}(1 - e^{-\frac{Rt}{L}})$ amperes, where R is 10 ohms, L is 2.5 henrys, E is 20 volts and t is the time in seconds. Draw a graph of current

against time from $t = 0$ to $t = 1$ second. The time constant of the circuit is given by $\tau = \dfrac{L}{R}$. Show that in a time τ s the current rises to 63% of its final value.

Since $R = 10$, $L = 2.5$ and $E = 20$, then $i = \dfrac{20}{10}(1 - e^{\frac{-10t}{2.5}})$ i.e.

$i = (1 - e^{-4t})$

A table of values is drawn up as shown below.

t	0	0.05	0.10	0.15	0.20	0.30	0.40	0.60	0.80	1.00
i	0	0.36	0.66	0.90	1.10	1.40	1.60	1.82	1.92	1.96

A graph of current i against time t is shown in Fig. 8.8

When $t = \infty$, $i = 2(1 - e^{-4\infty}) = 2(1 - 0) = 2$ A

Thus the curve is tending towards a final value of $i = 2$ A $\left(\text{i.e. } \dfrac{E}{R}\right)$

Time constant $\tau = \dfrac{L}{R} = \dfrac{2.5}{10} = 0.25$ s. From Fig. 8.8, when

$t = 0.25$ s, $i = 1.26$ A, which is the same as 63% of 2 A

Hence in time $\tau = \dfrac{L}{R}$ the current rises to 63% of its final value.

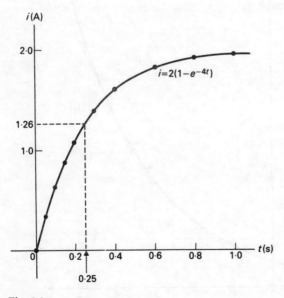

Fig. 8.8

Further problems on the laws of growth and decay may be found in Section 8.7 (Problems 16–25), page 136.

8.5   Graphs of natural laws of growth and decay

A law of natural growth of the form $y = A\,e^{kx}$ is depicted in Fig. 8.9 for values of the constants A and k of 3 and $\dfrac{1}{2}$ respectively and values of x from -1 to 5. The values of y are determined as shown below.

x	-1	0	1	2	3	4	5
$\dfrac{1}{2}x$	-0.5	0	0.5	1.0	1.5	2.0	2.5
$e^{\frac{1}{2}x}$	0.606 5	1.000 0	1.648 7	2.718 3	4.481 7	7.389 1	12.182 5
$y = 3\,e^{\frac{1}{2}x}$	1.82	3.0	4.95	8.15	13.45	22.17	36.55

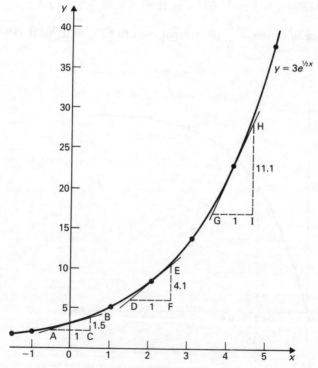

Fig. 8.9   Graph depicting the natual growth law $y = 3e^{\frac{1}{2}x}$

The slope of the curve at any point is denoted by $\dfrac{dy}{dx}$. To find the value of $\dfrac{dy}{dx}$ at any point on the curve, a tangent is drawn to the curve at that point and its slope is measured (see Chapter 21). For the growth law depicted in Fig. 8.9, i.e. $y = 3\,e^{\frac{1}{2}x}$, the relationship which exists between the slope of the graph, $\dfrac{dy}{dx}$ and y, is as follows:

When $x = 0$, $\dfrac{dy}{dx} = $ the slope of the graph at $x = 0$,

$$= \frac{BC}{AC} = 1.5, \text{ from Fig. 8.9}$$

Also, $y = 3\,e^{\frac{1}{2}x} = 3\,e^{\frac{1}{2} \times 0} = 3\,e^0 = 3$

When $x = 2$, $\dfrac{dy}{dx} = $ the slope of the graph at $x = 2$,

$$= \frac{EF}{DF} = 4.1$$

Also, $y = 3\,e^{\frac{1}{2}x} = 3\,e^{\frac{1}{2} \times 2} = 3\,e^1 = 8.2$

When $x = 4$, $\dfrac{dy}{dx} = $ the slope of the graph at $x = 4$,

$$= \frac{HI}{GI} = 11.1$$

Also, $y = 3\,e^{\frac{1}{2}x} = 3\,e^{\frac{1}{2} \times 4} = 3\,e^2 = 22.2$

Summary of results:

When	x is	0	2	4
	y is	3	8.2	22.2
and	$\dfrac{dy}{dx}$ is	1.5	4.1	11.1

These results show that $\dfrac{dy}{dx} = \dfrac{1}{2}y$. If this process is repeated for all values of x, it can be shown that in every case $\dfrac{dy}{dx}$ is proportional to y, that is, $\dfrac{dy}{dx} = (\text{a constant}) \times y$, although problems may arise due to difficulties of accurately drawing tangents to curves. It can also be shown that the constant of proportionality is the constant k in the equation $y = A\ e^{kx}$.

Thus in the case of $y = e^x$, $k = 1$, $A = 1$ and $\dfrac{dy}{dx} = e^x$

The same procedure can be repeated for all natural growth and decay laws and in all cases it can be shown that for natural laws having an equation of the form $y = A\ e^{kx}$, where k is a positive constant for growth laws and a negative constant for decay laws, $\dfrac{dy}{dx} = Ak\ e^{kx}$

$$\frac{dy}{dx} = Ak\ e^{kx} = k(A\ e^{kx}) = ky$$

This confirms the statement in Section 8.4 that for any natural law of growth or decay, the rate of change of the variable is proportional to the variable itself.

Worked problems on graphs of natural laws of growth and decay

Problem 1. A natural law of growth is of the form $y = 5\ e^{0.2x}$. Draw a graph depicting this law for values of x from $x = -3$ to $x = 3$. From the graph determine the value of y when x is 2.4 and the value of x when y is 3.19. Find the rate of change of y with respect to x at $x = -2$ and $x = 1$.

Using a calculator to determine the values of the exponential functions, the values of y are determined for various values of x as shown below and the graph shown in Fig. 8.10 drawn.

x	-3	-2	-1	0	1	2	3
$0.2x$	-0.6	-0.4	-0.2	0	0.2	0.4	0.6
$e^{0.2x}$	0.548 8	0.670 3	0.818 7	1	1.221 4	1.491 8	1.822 1
$y = 5\ e^{0.2x}$	2.74	3.35	4.09	5.00	6.11	7.46	9.11

From the graph, when x = 2.4, **y = 8.1,**

and when y = 3.19, **x = −2.25**

The slope of the graph at $x = -2$ is given by the ratio $\dfrac{BC}{AC}$

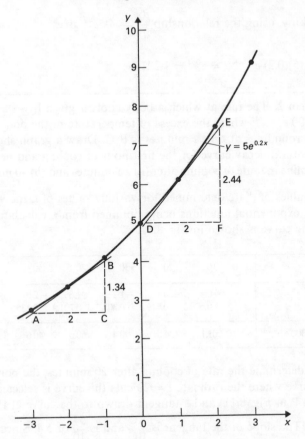

Fig. 8.10 Graph depicting the natural growth law $y = 5\,e^{0.2x}$
(Problem 1)

(Fig. 8.10), i.e. $\dfrac{1.34}{2}$ or 0.67. Hence the rate of change of y with respect

to x, i.e. $\dfrac{dy}{dx} = 0.67$ at $x = -2$. This can be checked using the

relationship $\dfrac{dx}{dy} = Ak\,e^{kx}$. In this case,

$\dfrac{dy}{dx} = (5)(0.2)\,e^{0.2(-2)} = e^{-0.4} = 0.670\,3$ from the table on page 130.

The slope of the graph when $x = 1$ is given by the ratio $\dfrac{\text{EF}}{\text{DF}}$ (Fig. 8.10),

i.e. $\dfrac{2.44}{2} = 1.22$. Hence the value of $\dfrac{dy}{dx}$ at $x = 1$ is 1.22

Checking, using the relationship $\dfrac{dy}{dx} = Ak \, e^{kx}$ gives:

$$\frac{dy}{dx} = (5)(0.2) \, e^{(0.2)(1)} = e^{0.2} = 1.221 \, 4$$

Problem 2. The rate at which a body cools is given by $\theta = 300 \, e^{-0.04t}$ where the excess of temperature of the body above its surroundings at time t minutes is $\theta°C$. Draw a graph showing this natural decay curve for the first hour of cooling and hence determine the rate of cooling after (a) 20 minutes and (b) 40 minutes.

The values of θ are determined for various values of t, the values of the exponential functions being obtained from a calculator. The cooling curve is shown in Fig. 8.11.

t minutes	0	10	20	30	40	50	60
$-0.04t$	0	-0.4	-0.8	-1.2	-1.6	-2.0	-2.4
$e^{-0.04t}$	1	0.670 3	0.449 3	0.301 2	0.201 9	0.135 3	0.090 7
$\theta°C = 300 \, e^{-0.04t}$	300.0	201.1	134.8	90.4	60.6	40.6	27.2

To determine the rate of cooling after 20 minutes, the point on the curve where the ordinate $t = 20$ cuts the curve is selected (point P in Fig. 8.11) and a tangent drawn to the curve at this point. The slope of the tangent is $\dfrac{AC}{BC}$ and is $\dfrac{54}{10}$ or 5.4. Since θ is decreasing as t increases, the slope is negative. Hence **the rate of cooling at $t = 20$ minutes is $-5.4°C \, min^{-1}$**. Applying a similar procedure at point Q gives the **rate of cooling after 40 minutes as** $\dfrac{DF}{EF}$, i.e. $\dfrac{24}{10}$ or $-2.4°C \, min^{-1}$.

These values can be checked using the general relationship that

when $y = A \, e^{kx}$, $\dfrac{dy}{dx} = Ak \, e^{kx}$. In this problem the rate of cooling

at $t = 20$ min,

$$\frac{d\theta}{dt} = 300(-0.04) \, e^{-0.04(20)}$$

$$= -5.391 \, 9°C \, min^{-1}$$

or $-5.4°C \, min^{-1}$ correct to 2 significant figures.

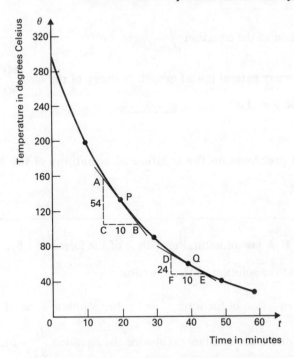

Fig. 8.11   A cooling curve following the natural law of decay
$\theta = 300\, e^{-0.04t}$ (Problem 2)

The rate of cooling when $t = 40$ min, $\dfrac{d\theta}{dt} = 300(-0.04)\, e^{-0.04(40)}$

$$= -2.4228°\text{C min}^{-1}$$

or $-2.4°\text{C min}^{-1}$ correct to 2 significant figures.

Further problems on graphs of natural laws of growth and decay may be found in Section 8.7 (Problems 26–31), page 137.

8.6   Solution of equations of the form $\dfrac{dy}{dx} = ky$

For any natural law of growth or decay, the rate of change of a variable is proportional to the variable itself, i.e. $\dfrac{dy}{dx} \propto y$. It is shown in Section 8.5 that when

$y = A\, e^{kx}$, $\dfrac{dy}{dx} = Ak\, e^{kx}$, i.e. $y = k(A\, e^{kx})$, $= ky$. Since the process of integration

reverses the process of differentiation, then when $\dfrac{dy}{dx} = ky$, $y = A\, e^{kx}$.

That is, the solution of the equation $\dfrac{dy}{dx} = ky$ is $A\,e^{kx}$.

It follows that **for any natural law of growth or decay of the form** $\dfrac{dy}{dx} = ky$, **the solution will be** $y = A\,e^{kx}$.

Worked problems on the solution of equations of the form $\dfrac{dy}{dx} = ky$

Problem 1. A law of natural growth is of the form $\dfrac{dy}{dx} = 6y$. Determine the solution of this equation.

For any equation of the form $\dfrac{dy}{dx} = ky$, the solution is $y = A\,e^{kx}$, where A and k are constants. Thus for the equation $\dfrac{dy}{dx} = 6y$, the constant k is 6 and the solution is $y = A\,e^{6x}$.

Problem 2. Find the solutions of the equations: (a) $\dfrac{dm}{di} - 4m = 0$;

(b) $\dfrac{1}{15}\dfrac{dl}{dv} + \dfrac{l}{4} = 0$;   (c) $0.74v - 3.81\dfrac{dv}{dt} = 0$.

(a)   Expressing the equation in the form $\dfrac{dy}{dx} = ky$ gives:

$\dfrac{dm}{di} = 4m$. The solution is $m = A\,e^{4i}$

(b)   Multiplying throughout by 15 and expressing in the form $\dfrac{dy}{dx} = ky$ gives:

$\dfrac{dl}{dv} = -\dfrac{15}{4}l$

Hence $l = A\,e^{-\frac{15}{4}v}$ or $A\,e^{-3\frac{3}{4}v}$

(c)   Dividing throughout by 3.81 and expressing in the form $\dfrac{dy}{dx} = ky$ gives:

$$\frac{dv}{dt} = \frac{0.74}{3.81} v = 0.194\,2v$$

Hence $v = A\,e^{0.194\,2t}$

Further problems on the solution of equations of the form $\dfrac{dy}{dx} = ky$ *may be found in the following Section (8.7) (Problems 32–43), page 138.*

8.7   Further problems

Exponental functions

In Problems 1–5, evaluate, using a calculator, correct to 4 significant figures.

1. (a) $y = e^2$; (b) $y = e^{0.4}$; and (c) $y = e^{0.1}$
    (a) [7.389]   (b) [1.492]   (c) [1.105]
2. (a) $y = e^{-0.3}$; (b) $y = e^{-0.1}$; and (c) $y = e^{-2}$
    (a) [0.740 8]   (b) [0.904 8]   (c) [0.135 3]
3. (a) $y = 3\,e^4$; (b) $y = 0.6\,e^{-0.04}$; and (c) $y = -5\,e^{-0.75}$
    (a) [163.8]   (b) [0.576 5]   (c) [−2.362]
4. (a) $e^{5.2}$; (b) $e^{-0.37}$; and (c) $e^{0.86}$
    (a) [181.3]   (b) [0.690 7]   (c) [2.363]
5. (a) $e^{-0.58}$; (b) $e^{-0.17}$; and (c) e^{12}
    (a) [0.559 9]   (b) [0.843 7]   (c) [162 800]
6. Evaluate, using a calculator, correct to 5 significant figures:
    (a) $\dfrac{e^{6.925}}{25}$; (b) $\dfrac{e^{0.892} - e^{-0.892}}{2}$; and (c) $73.29\,e^{-2.591}$
    (a) [40.696]   (b) [1.015 1]   (c) [5.492 7]

Graphs of exponential functions

In Problems 7 to 14, draw the graphs of the exponential functions given and use the graphs to determine the approximate values of x and y required.

7. $y = 5\,e^{0.4x}$ over a range $x = -3$ to $x = 3$ and determine the value of y when $x = 2.7$ and the value of x when $y = 10$
    [14.7, 1.7]
8. $y = 0.35\,e^{2.5x}$ over a range $x = -2$ to $x = 2$ and determine the value of y when $x = 1.8$ and the value of x when $y = 40$
    [31.5, 1.9]

9. $y = \frac{1}{3} e^{-2x}$ over a range $x = -2$ to $x = 2$ and determine the

    value of y when $x = -1.75$ and the value of x when $y = 5$
    $[11, -1.35]$

10. $y = 0.46 \, e^{-0.27x}$ over a range $x = -10$ to $x = 10$ and
    determine the value of y when $x = -8.5$ and the value of x
    when $y = 6.1$ $[4.6, -9.6]$

11. $y = 4 \, e^{-2x^2}$ over a range $x = -1.5$ to $x = 1.5$ and determine
    the value of y when $x = -1.2$ and the value of x when
    $y = 2.9$ $[0.22, \pm 0.4]$

12. $y = 100 \, e^{\frac{x^2}{3}}$ over a range $x = -3$ to $x = 3$ and determine the

    value of y when $x = \frac{7}{3}$ and the value of x when $y = 40$

    $[614, \pm 1.66]$

13. $y = \frac{1}{2}(e^x - e^{-x})$ over a range $x = -3$ to $x = 3$ and

    determine the value of y when $x = -2.3$ and the value of x
    when $y = 5$ $[-4.94, 2.31]$

14. $y = 3(e^{2x} - 4 \, e^{-x})$ over a range $x = 2$ to $x = -2$ and
    determine the value of y when $x = -1.6$ and the value of x
    when $y = 35$ $[-59, 1.28]$

15. Plot the curves $y = 3.2 \, e^{-1.4x}$ and $y = 1.7(1 - e^{-2x})$ on the
    same axes from $x = 0$ to $x = 1$ and determine their point of
    intersection $[0.67, 1.25]$

Laws of growth and decay

16. The instantaneous voltage v in a capacitive circuit is related
    to time t by the equation $v = V \, e^{-\frac{t}{CR}}$, where V, C and R are
    constants. Determine v when $t = 27 \times 10^{-3}$, $C = 8.0 \times 10^{-6}$,
    $R = 57 \times 10^3$ and $V = 100$. Sketch a curve of v against t
    $[94.3]$

17. The length of a bar, l, at temperature θ is given by $l = l_0 \, e^{\alpha\theta}$,
    where l_0 and α are constants. Determine l_0 when $l = 2.738$,
    $\theta = 315.7$ and $\alpha = 1.771 \times 10^{-4}$ $[2.589]$

18. Two quantities x and y are found to be related by the
    equation $y = a \, e^{-kx}$, where a and k are constants. Determine
    y when $a = 1.671 \times 10^4$, $k = -4.60$ and $x = 1.537$
    $[1.966 \times 10^7]$

19. Quantities p and q are related by the equation
    $p = 7.413(1 - e^{\frac{kq}{t}})$, where k and t are constants. Determine p
    when $k = 3.7 \times 10^{-2}$, $q = 712.8$ and $t = 5.747$ $[-722]$

20. When quantities I and C are related by the equation $I = BT^2 e^{-\frac{C}{T}}$, and B and T are constants, determine I when $B = 14.3$, $T = 1.27$ and $C = 8.15$ [0.037 66]

21. The current i amperes flowing in a capacitor at time t seconds is given by $i = 7.51(1 - e^{-\frac{t}{CR}})$, where the circuit resistance R is 27.4 kilohms and the capacitance C is 14.71 microfarads. Determine the current flow after 0.458 seconds [5.099 A]

22. The voltage drop, V volts, across an inductor of L henrys at time t seconds is given by $V = 125 e^{-\frac{Rt}{L}}$. Determine the voltage when t is 14.7 microseconds, given that the circuit resistance R is 128 ohms and its inductance is 10.3 millihenrys [104.1 V]

23. The amount A after n years of a sum invested P is given by the compound interest law $A = P e^{\frac{rn}{100}}$ when the interest rate r is added continuously. Determine the amount after 10 years for a sum of £1 000 invested if the interest rate is 4% per annum [£1 491.82]

24. The amount of product (x in mol cm^{-3}) formed in a chemical reaction starting with 3 mol cm^{-3} of reactant is given by $x = 3(1 - e^{-4t})$ where $t =$ time to form x in minutes. Plot a graph at 30-second intervals up to 3 minutes for this equation and determine x after one minute [2.945 mol cm^{-3}]

25. Find the equilibrium constant (K_p) for the reaction at 300 K $C_2H_6 = C_2H_4 + H_2$ where $K_p = e^{-\Delta H/RT}$, given that $\Delta H = 132$ kJ mol^{-1} and $R = 8.314 \times 10^{-3}$ kJ K^{-1} mol^{-1} [$1.037\,4 \times 10^{-23}$]

Graphs of natural laws of growth and decay

26. In a given chemical reaction the amount of starting material (C) in cm^3 left after t minutes is given by:

$$C = 50 e^{-0.005t}$$

Draw a graph of C against t and determine: (a) the concentration after 1 hour; (b) the time taken for the concentration to decrease by half.
(a) [37]   (b) [139 minutes]

27. The tension in two sides of a belt, T and T_0 newtons, passing round a pulley wheel and in contact with the pulley for an angle of θ radians, is given by:

$$T = 47.4 e^{0.31\theta}$$

Draw a graph depicting this relationship over a range $\theta = 0$ to $\theta = 2.0$ radians and hence find the value of $\dfrac{dT}{d\theta}$ when $\theta = 1.3$ radians   [22.0 N rad^{-1}]

28. The voltage drop across an inductor, v volts, is related to time t milliseconds by the equation:

$$v = 24.1 \times 10^3\, e^{-\frac{t}{7.25}}$$

Draw a graph showing how the values of v vary with time, for $t = 0$ to $t = 10$ milliseconds. Use this graph to find the rate of change of voltage when $t = 4.3$ milliseconds
[$-1\,840$ volts per millisecond]

29. The decay of voltage across a capacitor, v volts, at time t seconds is given by:

$$v = 200\, e^{-\frac{t}{3}}$$

Draw a graph depicting the natural decay curve over the first five seconds. Determine the rate of change of voltage at: (a) two seconds; and (b) four seconds
(a) $[-34.2\text{ V s}^{-1}]$ (b) $[-17.6\text{ V s}^{-1}]$

30. The resistance of a material, R_t ohms, at a temperature of $t°C$ is given by:

$$R_t = 250\, e^{-0.002t}$$

By drawing a graph for values of t over a range 50°C to 500°C, determine the change in resistance with respect to temperature at: (a) 150°C; and (b) 400°C
(a) $[-0.37 \text{ ohms }°C^{-1}]$ (b) $[-0.22 \text{ ohms }°C^{-1}]$

31. The temperature of a body $\theta°C$ is related to time t minutes by a natural law of decay of the form:

$$\frac{\theta}{100} = e^{-0.02t}$$

Draw a graph of the cooling curve depicting the first 90 minutes of cooling and hence determine the rate of cooling after 20 minutes and after 40 minutes
$[-1.34°C\text{ min}^{-1},\ -0.90°C\text{ min}^{-1}]$

Solution of equations of the form $\dfrac{dy}{dx} = ky$

In Problems 31 to 36 determine the solution of the equations given.

32. $\dfrac{dT}{d\theta} = 0.25T$ $[T = A\, e^{0.25\theta}]$

33. $\dfrac{dv}{dx} + 9.81v = 0$ $\qquad [v = A\,e^{-9.81x}]$

34. $-\dfrac{d\theta}{dt} = k\theta$ when $k = 0.047$ $\qquad [\theta = A\,e^{-0.047t}]$

35. $\dfrac{1}{6}\dfrac{dm}{d\theta} - \dfrac{1}{4}m = 0$ $\qquad [m = A\,e^{\frac{3}{2}\theta}]$

36. $0.37l + 2.3\dfrac{dl}{dt} = 0$ $\qquad [l = A\,e^{-0.160\,9t}]$

37. $10^{-5}\dfrac{dN}{dt} = 0.7 \times 10^4 N$ $\qquad [N = A\,e^{0.7 \times 10^9 t}]$

38. The rate of change of charge on a capacitor, $\dfrac{dQ}{dt}$, is proportional to the charge Q and is given by the equation:

$$R\dfrac{dQ}{dt} + \dfrac{Q}{C} = 0,\ \text{where } R \text{ and } C \text{ are constants.}$$

Solve this equation for Q $\qquad [Q = A\,e^{-\frac{t}{CR}}]$

39. The rate of decay of radioactive atoms, $\dfrac{dN}{dt}$, is proportional to the number of radioactive atoms N and is given by:

$$\dfrac{dN}{dt} + 0.7 \times 10^4 N = 0$$

Solve this equation for N $\qquad [N = A\,e^{-0.7 \times 10^4 t}]$

40. The change of length of a bar with respect to temperature is directly proportional to its length and may be represented by the equation:

$$\dfrac{dl}{dt} - \alpha l = 0$$

where α is a constant, and equal to 2.4×10^{-6}.
Determine an equation for l $\qquad [l = A\,e^{2.4 \times 10^{-6} t}]$

41. The rate of change of current in an inductive circuit is proportional to the current flowing, such that:

$$0.01\dfrac{di}{dt} + 2\,000i = 0$$

Solve the differential equation for i $\qquad [i = A\,e^{-2 \times 10^5 t}]$

42. The rate of change of charge on a capacitor, $\dfrac{dQ}{dt}$, is directly proportional to the charge Q, and is given by:

$$3Q - \frac{4}{7}\frac{dQ}{dt} = 0$$

Determine the solution of this equation $\qquad [Q = A\,e^{5.25t}]$

43. The rate of change of voltage across an electrical circuit, $\dfrac{dv}{dt}$, is directly proportional to the applied voltage, v, such that:

$$8.74v - 6.3\frac{dv}{dt} = 0$$

Solve this differential equation for v $\qquad [v = A\,e^{1.387t}]$

Chapter 9

Curve sketching and the determination of laws

9.1   Introduction

When a mathematical equation is known, co-ordinates may be calculated for a limited range of values, and the equation may be represented pictorially as a graph, within this range of values calculated. Sometimes, it it useful to show all the characteristic features of an equation, and in this case a sketch depicting the equation can be drawn, in which all the important features are shown, but the accurate plotting of points is less important. This technique is called 'curve sketching' and involves the use of differential calculus, hence is better left until a knowledge of differentiation and its applications has been covered. However, at this stage, certain basic curves such as the circle, parabola, ellipse and hyperbola can be readily drawn and are dealt with in this chapter. The quadratic graph is introduced in Section 9.2, and in Section 9.3 simple equations of the circle, parabola, ellipse and hyperbola, together with the characteristic curves associated with these equations are considered.

9.2   The quadratic graph

A general quadratic equation is given by $y = ax^2 + bx + c$, where a, b and c are constants, and $a \neq 0$.

(i) $y = ax^2$

The simplest quadratic equation is $y = x^2$. In order to plot a graph of $y = x^2$ a table of values is drawn up.

x	-3	-2	-1	0	1	2	3
$y = x^2$	9	4	1	0	1	4	9

Figure 9.1(a) shows a graph of $y = x^2$ which is symmetrical about the y-axis. (Note that the co-ordinates on the graph are joined by a smooth curve and not from point to point.) A curve of this shape is called a **parabola**. At the origin, i.e. at $(0, 0)$, the lowest point on the curve is reached. This is called a **turning-point**. The values of y on either side of this point are greater than

141

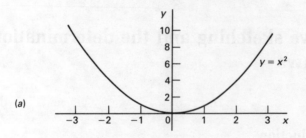

(a)

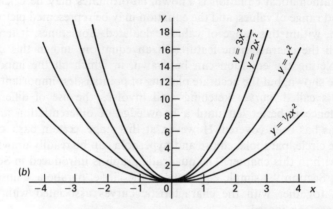

(b)

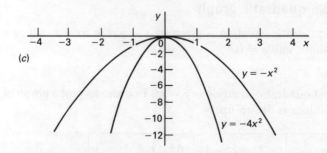

(c)

Fig. 9.1

at the turning-point. Such a point is called a **minimum value**, and can be thought of as being the 'bottom of a valley'.

The curve depicting $y = ax^2$ could have been sketched by considering the following, easily ascertained values:

(a) when x is 0, y is 0,

(b) when x is large and positive, y is large and positive,

and (c) when x is large and negative, y is large and positive.

It can be seen that the curve shown in Fig. 9.1(a) meets these three criteria.

Figure 9.1(b) shows the effect of the constant a where $a = 1, 2, 3$ and $\frac{1}{2}$, i.e. it shows graphs of $y = x^2$, $y = 2x^2$, $y = 3x^2$ and $y = \frac{1}{2}x^2$. As the value of a increases, the curves become more steep; as a decreases the curves become less steep. All the curves of the form $y = ax^2$ are symmetrical about the y-axis, and the y-axis is called the axis of symmetry.

In addition to points (a) to (c) given above, the inclusion of one other pair of points will give the steepness of the curve, e.g. if $y = x^2$ and $x = \pm 2$, then $y = 4$.

(ii) $y = -ax^2$

Figure 9.1(c) shows graphs of $y = -x^2$ and $y = -4x^2$. This shows the effect of a being negative. The numerical values of y will be the same as previously obtained but will all be negative, giving the inverted parabolas shown. At the origin $(0, 0)$ a turning-point again exists. This time the values of y on either side of this point are less than at the turning-point. Such a point is called a **maximum value** and can be thought of as being the 'crest of a wave'.

(iii) $y = ax^2 + c$

A table of values for $y = x^2$ is shown on p. 141. If a table of values is drawn up for say, $y = x^2 + 3$, then all the values of y in the original table will be increased by 3. Similarly, if $y = x^2 - 2$ then all the values of y will be decreased by 2.

Thus when the constant c is a positive value the parabola is raised by c units; when c is a negative value then the parabola is lowered by c units.

Figure 9.2(a) shows graphs of $y = x^2$, $y = 2x^2 + 3$ and $y = 3x^2 - 6$ and Fig. 9.2(b) shows graphs of $y = -x^2 + 5$ and $y = -3x^2 - 5$.

(iv) $y = ax^2 + bx + c$

A graph of $y = p(x + q)^2 + r$ where p, q and r are constants, is the same general shape as $y = ax^2$, i.e., a parabola.

When the constant r is a positive value the parabola is raised by r units from the zero position; when r is a negative value the parabola is lowered by r units (as in case (iii) above). When the constant q is a positive value the parabola is moved q units to the left from its zero position; when q is a negative value it is moved q units to the right.

For example, the graph of $y = 2(x + 3)^2 + 4$ is shown in Fig. 9.3. This is a parabola whose turning-point (a minimum value) is situated at $(-3, 4)$. Similarly, the graph of $y = -3(x - 2)^2 - 4$ is shown in Fig. 9.3. This is also a parabola, whose turning-point (a maximum value) is situated at $(2, -4)$. Of course, a quadratic expression is not normally expressed in the form $y = p(x + q)^2 + r$. If, however, the general expression $y = ax^2 + bx + c$ can be changed into the form $y = p(x + q)^2 + r$ then the curve can be readily sketched.

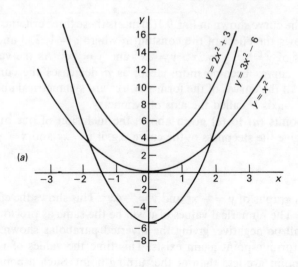

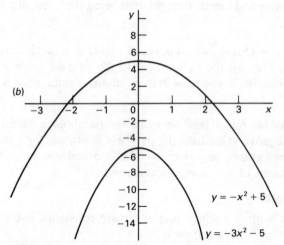

Fig. 9.2

For example, let $y = x^2 + 8x + 7$

To obtain the q term in $(x + q)^2$, half the coefficient of the x term is taken, since when expanding $(x + q)^2$, the coefficient of x is $2q$.

In this case $q = \dfrac{8}{2}$ or 4

However, $(x + 4)^2 = x^2 + 8x + 16$, which is 16 more than is needed. The 16 is therefore subtracted.

Hence $y = x^2 + 8x + 7$

$\qquad = [(x + 4)^2 - 16] + 7$

$\qquad y = (x + 4)^2 - 9$, which is a parabola with a minimum value at $(-4, -9)$ as shown in Fig. 9.4(a).

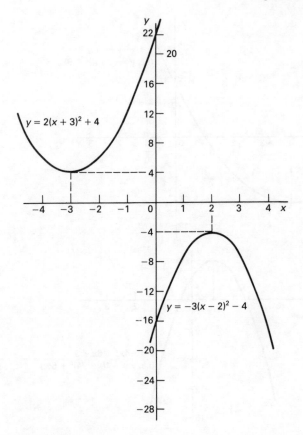

Fig. 9.3

Similarly, if $y = -4x^2 - 6x + 3$

then $y = -4\left(x^2 + \dfrac{6}{4}x\right) + 3$

$= -4\left\{\left[x + \dfrac{3}{4}\right]^2 - \dfrac{9}{16}\right\} + 3$

$= -4\left[x + \dfrac{3}{4}\right]^2 + 2\dfrac{1}{4} + 3$

i.e. $y = -4\left[x + \dfrac{3}{4}\right]^2 + 5\dfrac{1}{4}$

Comparing this with $y = p(x + q)^2 + r$ shows the parabola to be inverted (since p is negative), moved horizontally $\frac{3}{4}$ units to the left from the zero position (since q is $+\frac{3}{4}$) and raised by $5\frac{1}{4}$ units from the zero position (since r is $+5\frac{1}{4}$). The turning-point on the curve (a maximum value) thus occurs at $(-\frac{3}{4}, 5\frac{1}{4})$ as shown in Fig. 9.4(b).

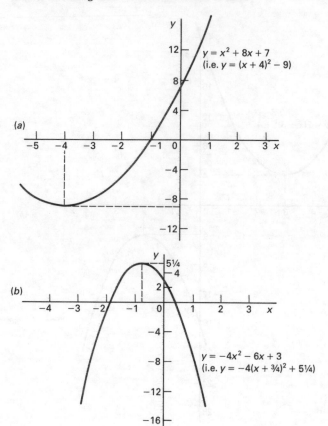

Fig. 9.4

Worked problems on sketching and plotting quadratic graphs

Problem 1. Sketch on the same axes the following curves:

(a) $y = 2x^2$; (b) $\dfrac{y}{4} = x^2$; (c) $2\sqrt{y} = x$; (d) $-y = 3x^2$

(a) $y = 2x^2$ is a parabola with its turning-point (a minimum value) at $(0, 0)$. When $x = \pm 2$, $y = 2(\pm 2)^2 = 8$

(b) $\dfrac{y}{4} = x^2$ or $y = 4x^2$ is a parabola with its turning-point

(a minimum value) at $(0, 0)$. This curve is steeper than $y = 2x^2$ by a factor of 2. When $x = \pm 2$, $y = 4(\pm 2)^2 = 16$

(c) $2\sqrt{y} = x$ or $y = \dfrac{x^2}{4}$ is a parbola with its turning-point (a minimum value) at $(0, 0)$. This curve is less steep than $y = 2x^2$ by a factor of $\frac{1}{8}$.

When $x = \pm 2$, $y = \dfrac{(\pm 2)^2}{4} = 1$

(d) $-y = 3x^2$ or $y = -3x^2$ is an inverted parabola with its turning-point (a maximum value) at $(0, 0)$

When $x = \pm 2$, $y = -3(\pm 2)^2 = -12$

Figure 9.5 shows the four curves.

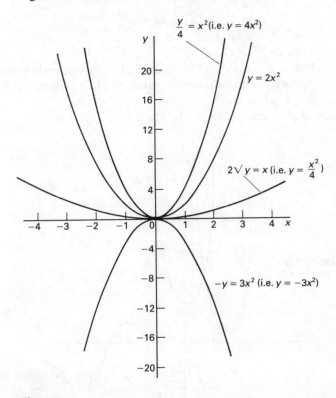

Fig. 9.5

Problem 2. Sketch the following curves and determine the co-ordinates of their turning-points: (a) $y = 4x^2 + 5$; (b) $y = 4x^2 + 24x + 37$; (c) $y = -3x^2 - 2$; (d) $y = -2x^2 + 16x - 29$

(a) $y = 4x^2 + 5$ is a parabola with its turning-point (a minimum value) at **(0, 5)**. When $x = \pm 1$, $y = 4(\pm 1)^2 + 5 = 9$

(b) $y = 4x^2 + 24x + 37$
$\quad = 4(x^2 + 6x) + 37$
$\quad = 4[(x + 3)^2 - 9] + 37$

$$= 4(x + 3)^2 - 36 + 37$$
i.e., $y = 4(x + 3)^2 + 1$

Hence $y = 4x^2 + 24x + 37$ is a parabola with its turning-point (a minimum value) at $(-3, 1)$. When $x = -2$, $y = 5$. When $x = -4$, $y = 5$

(c) $y = -3x^2 - 2$ is an inverted parabola with its turning-point (a maximum value) at $(0, -2)$. When $x = \pm 2$, $y = -14$

(d) $y = -2x^2 + 16x - 29$
$$= -2(x^2 - 8x) - 29$$
$$= -2[(x - 4)^2 - 16] - 29$$
$$= -2(x - 4)^2 + 32 - 29$$
i.e., $y = -2(x - 4)^2 + 3$

Hence $y = -2x^2 + 16x - 29$ is an inverted parabola with its turning-point (a maximum value) at $(4, 3)$. When $x = 3$, $y = 1$, when $x = 5$, $y = 1$

Each of the curves is shown in Fig. 9.6

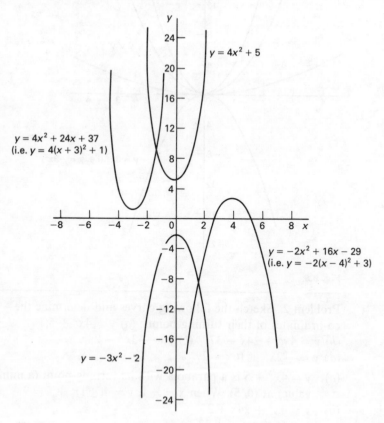

Fig. 9.6

Problem 3. Plot the quadratic curve $y = 3x^2 - 2x + 4$ from $x = -3$ to $x = +4$ and find the co-ordinates of its turning-point.

To plot the quadratic curve a table is drawn up.

x	-3	-2	-1	0	1	2	3	4
$3x^2$	27	12	3	0	3	12	27	48
$-2x$	6	4	2	0	-2	-4	-6	-8
$+4$	4	4	4	4	4	4	4	4
$y = 3x^2 - 2x + 4$	37	20	9	4	5	12	25	44

The curve is plotted in Fig. 9.7. The turning-point (a minimum value) occurs at (0.3, 3.6)

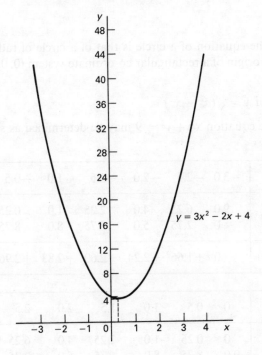

Fig. 9.7

The co-ordinates of the turning-point may be checked as follows:

Since $y = 3x^2 - 2x + 4$

then $y = 3\left(x^2 - \dfrac{2}{3}x\right) + 4$

$$= 3\left[\left(x - \frac{1}{3}\right)^2 - \frac{1}{9}\right] + 4$$

$$= 3\left(x - \frac{1}{3}\right)^2 - \frac{1}{3} + 4$$

i.e., $y = 3\left(x - \frac{1}{3}\right)^2 + 3\frac{2}{3}$

Hence the turning-point is at $\left(\dfrac{1}{3}, 3\dfrac{2}{3}\right)$

Further problems on the quadratic graph may be found in Section 9.5 (Problems 1–23), page 169.

9.3 Simple curves of the circle, parabola, ellipse and hyperbola

(i) The circle

The simplest form of the equation of a circle is that of a circle of radius a, having its centre at the origin of a rectangular co-ordinate system $(0, 0)$. The equation is:

$$x^2 + y^2 = a^2 \text{ or } y = \sqrt{(a^2 - x^2)}$$

The co-ordinates of the equation $x^2 + y^2 = 9$ may be determined as shown below.

x	-3.0	-2.5	-2.0	-1.5	-1.0	-0.5
x^2	9.0	6.25	4.0	2.25	1.0	0.25
$9 - x^2$	0	2.75	5.0	6.75	8.0	8.75
$y = \sqrt{(9 - x^2)}$	0	± 1.66	± 2.24	± 2.60	± 2.83	± 2.96

x	0	0.5	1.0	1.5	2.0	2.5	3.0
x^2	0	0.25	1.0	2.25	4.0	6.25	9.0
$9 - x^2$	9.0	8.75	8.0	6.75	5.0	2.75	0
$y = \sqrt{(9 - x^2)}$	± 3	± 2.96	± 2.83	± 2.60	± 2.24	± 1.66	0

A graph of these values is as shown in Fig. 9.8. The graph produced confirms that $x^2 + y^2 = 9$ is a circle, centre at the origin and of radius 3 units.

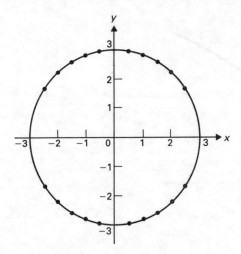

Fig. 9.8   Circle, $x^2 + y^2 = 9$

(ii)   The parabola

Parabolas having as equation of the form $y = ax^2 + bx + c$ are introduced in Section 9.2, these parabolas having their axes of symmetry parallel to or coinciding with the y-axis. The simplest form of the equation of a parabola is:

$$y = a\sqrt{x} \text{ or } y = ax^{\frac{1}{2}}$$

Parabolas having equations of this form have their axes of symmetry coinciding with the x-axis, and their vertices at the origin $(0, 0)$. The co-ordinates of the equation $y = 2\sqrt{x}$, for example, may be determined as shown below for positive values of x (negative values of x give a complex result).

x	0	0.5	1.0	1.5	2.0	2.5	3.0
$\sqrt{x}$	0	± 0.71	± 1.0	± 1.22	± 1.41	± 1.58	± 1.73
$y = 2\sqrt{x}$	0	± 1.42	± 2.00	± 2.44	± 2.82	± 3.16	± 3.46

A graph of these values is shown in Fig. 9.9, the resulting curve being a parabola with its axis of symmetry coinciding with the x-axis and its vertex at the origin $(0, 0)$.

(iii)   The ellipse

An ellipse has two axes at right angles to one another, corresponding to the maximum and minimum lengths which can be obtained within the ellipse.

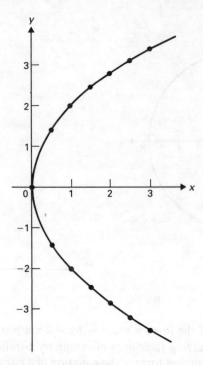

Fig. 9.9   Parabola, $y = 2x^{\frac{1}{2}}$

These axes are called the major axis for the greatest length and the minor axis for the least length.

The equation of an ellipse having its centre at the origin of a rectangular co-ordinate system and its axes, coinciding with the axes of the co-ordinate system is:

$$\frac{x^2}{a^2} + \frac{y^2}{b^2} = 1 \text{ or } y = b\sqrt{\left(1 - \frac{x^2}{a^2}\right)}$$

When $a = b$, this equation becomes $x^2 + y^2 = a^2$, i.e., a circle, centre at the origin of the co-ordinate system and of radius a (see (i) above).

The co-ordinates of the equation $\dfrac{x^2}{4} + \dfrac{y^2}{9} = 1$, i.e., $y = 3\sqrt{\left(1 - \dfrac{x^2}{4}\right)}$ may be determined as shown in the table on page 153.

A graph of these values is shown in Fig. 9.10, the resulting curve being an ellipse with its centre at the origin. The width of the ellipse along the x-axis is 4 units and the total height along the y-axis is 6 units, i.e., the major axis is from $+3$ to -3 along the y-axis. The width AB in Fig. 9.10 is called the **minor axis** and is given by $2a$ in the general equation $\dfrac{x^2}{a^2} + \dfrac{y^2}{b^2} = 1$. The height

x	-2.0	-1.5	-1.0	-0.5	0	0.5	1.0	1.5	2.0
$\dfrac{x^2}{4}$	1.0	0.563	0.25	0.063	0	0.063	0.25	0.563	1.0
$1-\dfrac{x^2}{4}$	0	0.437	0.75	0.937	1.0	0.937	0.75	0.437	0
$\sqrt{\left(1-\dfrac{x^2}{4}\right)}$	0	±0.661	±0.866	±0.968	±1.0	±0.968	±0.866	±0.661	0
$y=3\sqrt{\left(1-\dfrac{x^2}{4}\right)}$	0	±1.98	±2.60	±2.90	±3	±2.90	±2.60	±1.98	0

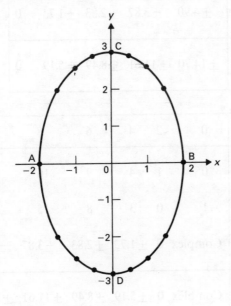

Fig. 9.10   Ellipse, $\dfrac{x^2}{4}+\dfrac{y^2}{9}=1$

CD in Fig. 9.10 is called the **major** (or longer) **axis** and is given by $2b$ in the
equation $\dfrac{x^2}{a^2}+\dfrac{y^2}{b^2}=1$

(iv) The Hyperbola

The equation of a hyperbola is of the form

$$\frac{x^2}{a^2}-\frac{y^2}{b^2}=1 \text{ or } y=b\sqrt{\left(\frac{x^2}{a^2}-1\right)}$$

The co-ordinates of the equation $\dfrac{x^2}{4} - \dfrac{y^2}{9} = 1$, i.e., $y = 3\sqrt{\left(\dfrac{x^2}{4} - 1\right)}$ may be determined as shown below.

x	-10	-8	-6	-4	-2
$\dfrac{x^2}{4}$	25	16	9	4	1
$\dfrac{x^2}{4} - 1$	24	15	8	3	0
$\sqrt{\left(\dfrac{x^2}{4} - 1\right)}$	± 4.90	± 3.87	± 2.83	± 1.73	0
$y = 3\sqrt{\left(\dfrac{x^2}{4} - 1\right)}$	± 14.70	± 11.61	± 8.49	± 5.19	0

x	0	2	4	6	8	10
$\dfrac{x^2}{4}$	0	1	4	9	16	25
$\dfrac{x^2}{4} - 1$	-1	0	3	8	15	24
$\sqrt{\left(\dfrac{x^2}{4} - 1\right)}$	Complex	0	± 1.73	± 2.83	± 3.87	± 4.90
$y = 3\sqrt{\left(\dfrac{x^2}{4} - 1\right)}$	Complex	0	± 5.19	± 8.49	± 11.61	± 14.70

[For complex numbers and their representation, see *Technician mathematics 3*.] The graph of these values is shown in Fig. 9.11, the resulting curve being a hyperbola which is symmetrical about both the x- and the y-axis. The distance AB in Fig. 9.11 is given by 2a in the general equation

$$\frac{x^2}{a^2} - \frac{y^2}{b^2} = 1$$

The distance OB is the value of x when y is equal to zero, i.e., when $\dfrac{x^2}{a^2} = 1$ or when $x = a$. Thus, due to symmetry, distance AB is 2a as stated.

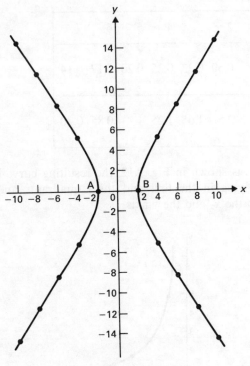

Fig. 9.11   Hyperbola, $\dfrac{x^2}{4} - \dfrac{y^2}{9} = 1$

The simplest equation of a hyperbola is of the form $y = \dfrac{a}{x}$, this being the equation of a rectangular hyperbola which is symmetrical about both the x- and y-axis and lies entirely in the first and third quadrants. The co-ordinates of the equation $y = \dfrac{5}{x}$ may be determined as shown below:

x	-7	-6	-5	-4	-3	-2	-1
$\dfrac{1}{x}$	-0.14	-0.17	-0.20	-0.25	-0.33	-0.50	-1
$y = \dfrac{5}{x}$	-0.70	-0.85	-1	-1.25	-1.65	-2.5	-5

x	0	1	2	3	4	5	6	7
$\dfrac{1}{x}$	∞	1	0.50	0.33	0.25	0.20	0.17	0.14
$y = \dfrac{5}{x}$	∞	5	2.5	1.65	1.25	1		0.85	0.70

A graph of these values is shown in Fig. 9.12, the resulting curve being a rectangular hyperbola lying in the first and third quadrants and being symmetrical about both the x- and the y-axis.

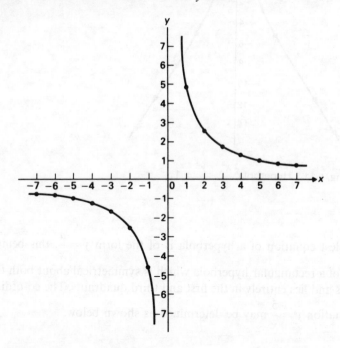

Fig. 9.12   Rectangular hyperbola, $y = \dfrac{5}{x}$

Worked problems on simple curves of the circle, parabola, ellipse and hyperbola

Problem 1. Sketch curves depicting the following equations, showing the significant values where possible.

(a) $x = \sqrt{(5 - y^2)}$; (b) $y^2 = 9x$; (c) $5x^2 = 35 - 7y^2$;
(d) $3y^2 = 5x^2 - 15$; (e) $xy = 7$

(a) Squaring both sides of the equation and transposing gives $x^2 + y^2 = 5$. Comparing this with the standard equation of a circle, centre origin, radius a, i.e., $x^2 + y^2 = a^2$, shows that $x^2 + y^2 = 5$ represents a circle, centre origin, radius $\sqrt{5}$. A sketch of this circle is shown in Fig. 9.13(a).

(b) One form of the equation of a parabola is $y = a\sqrt{x}$. Squaring both sides of this equation gives $y^2 = a^2x$. The equation $y^2 = 9x$ is of this form and thus represents a

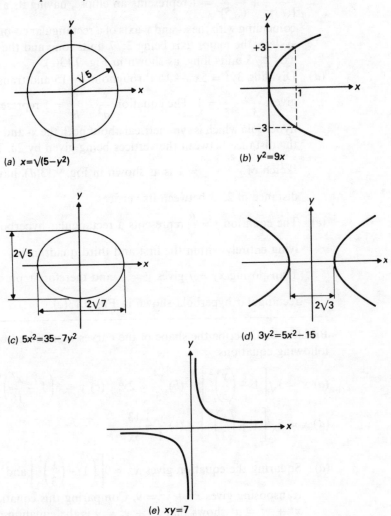

(a) $x=\sqrt{(5-y^2)}$

(b) $y^2=9x$

(c) $5x^2=35-7y^2$

(d) $3y^2=5x^2-15$

(e) $xy=7$

Fig. 9.13

parabola which is symmetrical about the x-axis andhaving its vertex at the origin, $(0, 0)$. Also, when $x = 1$, $y = \pm 3$. A sketch of this parabola is shown in Fig. 9.13(b).

(c)   By dividing throughout by 35 and transposing, the equation $5x^2 = 35 - 7y^2$ can be written as $\dfrac{x^2}{7} + \dfrac{y^2}{5} = 1$. The equation

of an ellipse is of the form $\dfrac{x^2}{a^2} + \dfrac{y^2}{b^2} = 1$, where $2a$ and $2b$

represent the length of the axes of the ellipse. Thus $\dfrac{x^2}{(\sqrt{7})^2} + \dfrac{y^2}{(\sqrt{5})^2} = 1$ represents an ellipse, having its axes

coinciding with the x- and y-axis of a rectangular co-ordinate system, the major axis being $2\sqrt{7}$ units long and the minor axis $2\sqrt{5}$ units long, as shown in Fig. 9.13(c).

(d)   Dividing $3y^2 = 5x^2 - 15$ throughout by 15 and transposing gives: $\dfrac{x^2}{3} - \dfrac{y^2}{5} = 1$. The equation $\dfrac{x^2}{a^2} - \dfrac{y^2}{b^2} = 1$ represents a

hyperbola which is symmetrical about both the x- and y-axis, the distance between the vertices being given by $2a$. Thus a sketch of $\dfrac{x^2}{3} - \dfrac{y^2}{5} = 1$ is as shown in Fig. 9.13(d), having a

distance of $2\sqrt{3}$ between its vertices.

(e)   The equation $y = \dfrac{a}{x}$ represents a rectangular hyperbola

lying entirely within the first and third quadrants.

Transposing $xy = 7$ gives $y = \dfrac{7}{x}$, and therefore represents the

rectangular hyperbola shown in Fig. 9.13(e).

Problem 2. Describe the shape of the curves represented by the following equations:

(a) $x = 3\sqrt{\left[1 - \left(\dfrac{y}{3}\right)^2\right]}$;   (b) $\dfrac{y^2}{5} = 2x$;   (c) $y = 6\left(1 - \dfrac{x^2}{16}\right)^{1/2}$

(d) $x = 5\sqrt{\left[1 + \left(\dfrac{y}{2}\right)^2\right]}$;   (e) $\dfrac{y}{4} = \dfrac{13}{3x}$

(a)   Squaring the equation gives $x^2 = 9\left[\left(1 - \left(\dfrac{y}{3}\right)^2\right)\right]$ and

transposing gives $x^2 + y^2 = 9$. Comparing this equation with $x^2 + y^2 = a^2$ shows that $x^2 + y^2 = 9$ is the equation of a circle having centre at the origin $(0, 0)$, and of radius 3 units.

(b) Tranposing $\dfrac{y^2}{5} = 2x$ gives $y = \sqrt{(10)}\sqrt{x}$. Thus $\dfrac{y^2}{5} = 2x$ is the

equation of a parabola having its axis of symmetry coinciding with the x-axis and its vertex at the origin of a rectangular co-ordinate system.

(c) $y = 6\left(1 - \dfrac{x^2}{16}\right)^{1/2}$ can be transposed to $\dfrac{y}{6} = \left(1 - \dfrac{x}{16}\right)^{1/2}$ and

squaring both sides gives $\dfrac{y^2}{36} = 1 - \dfrac{x^2}{16}$, i.e., $\dfrac{x^2}{16} + \dfrac{y^2}{36} = 1$

This is the equation of an ellipse, centre at the origin of a rectangular co-ordinate system, the major axis coinciding with the y-axis and being $2\sqrt{36}$, i.e., 12 units long. The minor axis coincides with the x-axis and is $2\sqrt{16}$, i.e., 8 units long.

(d) Since $x = 5\sqrt{\left[1 + \left(\dfrac{y}{2}\right)^2\right]}$

$$x^2 = 25\left[1 + \left(\dfrac{y}{2}\right)^2\right]$$

i.e. $\dfrac{x^2}{25} - \dfrac{y^2}{4} = 1$

This is a hyperbola which is symmetrical about both the x- and y-axis, the vertices being $2\sqrt{25}$, i.e., 10 units apart. (With reference to Section 9.3(iv) page 153, a is equal to ± 5.)

(e) The equation $\dfrac{y}{4} = \dfrac{13}{3x}$ is of the form $y = \dfrac{a}{x}$ where

$a = \dfrac{52}{3} = 17.3$. This represents a rectangular hyperbola,

symmetrical about both the x- and y-axis and lying entirely in the first and third quadrants, similar in shape to the curves shown in Fig. 9.13(e).

Further problems on simple curves of the circle, parabola, ellipse and hyperbola may be found in Section 9.5 (Problems 24–39), page 170.

9.4 Reduction of non-linear laws to a linear form (i.e. determination of laws)

In experimental work, when two variables are believed to be connected, a set of corresponding measurements can be made. Such results can then be used to discover if there is a mathematical law relating the two variables. If such a law is found it can be used to predict further values. Usually the results obtained from experiments are plotted as a graph and an attempt is made to deduce the law from the graph. Now the relationship between two variables believed to be of the linear form $y = mx + c$ can be proved by plotting the measured values x and y and seeing if a straight line graph results. If a straight line does fit the plotted points then the slope m and y-axis intercept c can be found, which establishes the law relating x and y for all values **in the given range**. However, frequently the relationship between variables x and y is not a linear one, i.e., when x is plotted against y a curve results. In such cases the non-linear equation is modified to the linear form $y = mx + c$ so that the constants can be found and the law relating the variables determined. This process is called the **determination of laws**. Some common forms of non-linear equations are given below. These are rearranged into a linear form by making a direct comparison with the straight line form $y = mx + c$, that is, arranging in the form:

a variable = (a constant × a variable) + a constant.

In the conversion of non-linear laws to a linear form it is useful to isolate the constant term first.

(i) $y = ax^2 + b$

$$\boxed{y} = a \boxed{x^2} + b$$

compares with $\boxed{y} = m \boxed{x} + c$

Hence plot y against x^2 to produce a straight line graph of slope a and y-axis intercept b (see worked Problem 1).

(ii) $y = \dfrac{a}{x} + b$

$$\boxed{y} = a \boxed{\dfrac{1}{x}} + b$$

compares with $\boxed{y} = m \boxed{x} + c$

Hence plot y against $\dfrac{1}{x}$ to produce a straight line graph of slope a and y-axis intercept b (see worked Problem 2).

(iii) $y = \dfrac{a}{x^2} + b$

$$\boxed{y} = a \boxed{\dfrac{1}{x^2}} + b$$

compares with $\boxed{y} = m \boxed{x} + c$

Hence plot y against $\dfrac{1}{x^2}$ to produce a straight line graph of slope a and y-axis intercept b.

(iv) $y = ax^2 + bx$

In this case there is no constant term as it stands, but by dividing throughout by x a constant term b is produced. Dividing both sides of the equation by x gives:

$$\boxed{\frac{y}{x}} = a \boxed{x} + b$$

compares with $\boxed{y} = m \boxed{x} + c$

Hence plot $\dfrac{y}{x}$ against x to produce a straight line graph of slope a and $\dfrac{y}{x}$ axis intercept b (see worked Problem 3).

(v) $xy = ax + by$

Dividing both sides of the equation by y gives:

$$\boxed{x} = a \boxed{\frac{x}{y}} + b$$

compares with $\boxed{y} = m \boxed{x} + c$

Hence plot x against $\dfrac{x}{y}$ to produce a straight line graph of slope a and x-axis intercept b.

(vi) $y = ax^n$

Taking logarithms (usually to a base 10) of each side of the equation gives:

$$\lg y = \lg (ax^n)$$
$$\lg y = \lg a + \lg x^n$$
$$\lg y = \lg a + n \lg x$$

compares with $\boxed{\lg y} = n \boxed{\lg x} + \lg a$
$\boxed{y} = m \boxed{x} + c$

Hence plot $\lg y$ against $\lg x$ to produce a straight line graph of slope n and $\lg y$-axis intercept $\lg a$ (a is obtained by taking antilogarithms of $\lg a$). This type is discussed separately in Chapter 7.

(vii) $y = ab^x$

Taking logarithms of each side of the equation gives:

$$\lg y = \lg ab^x$$
$$\lg y = \lg a + \lg b^x$$
$$\lg y = \lg a + x \lg b$$

i.e. $\boxed{\lg y} = (\lg b)\;\boxed{x} + \lg a$

compares with $\boxed{y} = m\;\boxed{x} + c$

Hence plot $\lg y$ against x to produce a straight line graph of slope $\lg b$ and $\lg y$-axis intercept $\lg a$ (a and b are obtained by taking antilogarithms of $\lg a$ and $\lg b$ respectively) (see worked Problem 4).

Worked problems on determination of laws

Problem 1. The following experimental values of x and y are believed to be related by the law $y = ax^2 + b$. By plotting a suitable graph test if this is so and find the approximate values of a and b.

x	0	1	2	3	4	5	6
y	3.0	4.8	12.0	23.3	38.1	59.5	83.5

If y is plotted against x the non-linear curve shown in Fig. 9.14 is produced.

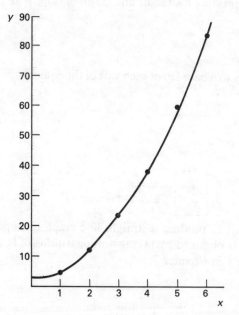

Fig. 9.14 Graph of y/x

It is not possible to determine the values of a and b from such a curve.

Comparing $\boxed{y} = a \boxed{x^2} + b$ with the straight line form
$\boxed{y} = m \boxed{x} + c$

shows that y is to be plotted vertically against x^2 horizontally to produce a straight line of slope a and y-axis intercept b. Thus an extension to the above table of values needs to be produced. This is shown below.

y	3.0	4.8	12.0	23.3	38.1	59.5	83.5
x	0	1.0	2.0	3.0	4.0	5.0	6.0
x^2	0	1.0	4.0	9.0	16.0	25.0	36.0

The best straight line that fits the points is shown in the graph of Fig. 9.15

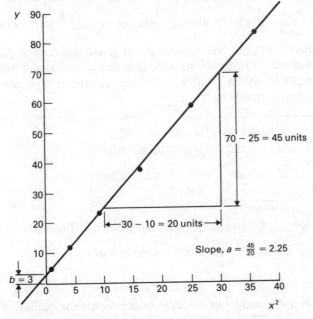

Fig. 9.15 Graph of y/x^2

Note that not every point lies on the straight line drawn; rarely in experimental results will all points lie exactly on a straight line. In such cases there is likely to be a difference of opinion as to the exact position of the line. Hence any results are only approximate. From the graph the slope a is found to be **2.25** and the y-axis intercept b is found to be **3**

Hence the law relating the variables x and y is $y = 2.25x^2 + 3$

Problem 2. In an experiment the following values of resistance R and voltage V were taken:

R ohms	45.3	49.8	52.4	57.6	62.3
V millivolts	113	102	96	86	79

It is thought that R and V are connected by a law of the form $R = \dfrac{d}{V} + e$ where d and e are constants. Verify the law and find approximate values of d and e.

$$R = \frac{d}{V} + e \qquad \text{i.e.} \quad \boxed{R} = d \; \boxed{\frac{1}{V}} + e$$
$$\text{Comparing with} \quad \boxed{y} = m \; \boxed{x} + c$$

shows that R is to be plotted vertically against $\dfrac{1}{V}$ horizontally to produce a straight line of slope d and R-axis intercept e. Another table of values is drawn up with V in this case changed from millivolts to volts so that when taking reciprocals of V, more manageable numbers result.

R	45.3	49.8	52.4	57.6	62.3
V	0.113	0.102	0.096	0.086	0.079
$\dfrac{1}{V}$	8.85	9.80	10.42	11.63	12.66

A graph of R against $\dfrac{1}{V}$ is shown plotted in Fig. 9.16.

A straight line fits the points which verifies that the law $R = \dfrac{d}{V} + e$ is obeyed.

It is not practical in this case to commence the scaling of each axis at zero. Thus it is not possible to find the R-axis intercept $\left(\text{i.e. at } \dfrac{1}{V} = 0\right)$ from the graph. A simultaneous equation approach is therefore necessary. Any two points such as A and B may be used.

At A, $R = 59.3$ and $\dfrac{1}{V} = 12.0$

At B, $R = 50.5$ and $\dfrac{1}{V} = 10.0$

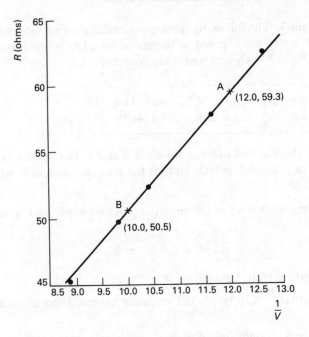

Fig. 9.16 Graph of $R \left/ \dfrac{1}{V} \right.$

Hence, since $R = \dfrac{1}{V} d + e$:

$$59.3 = 12.0\,d + e \qquad\qquad (1)$$
$$\text{and } 50.5 = 10.0\,d + e \qquad\qquad (2)$$

Subtracting equation (2) from equation (1) gives:

$$8.8 = 2.0\,d$$
$$d = 4.4$$

Substituting $d = 4.4$ in equation (2) gives:

$$50.5 = (10.0)(4.4) + e$$
$$e = 6.5$$

Checking in equation (1): R.H.S. $= (12.0)(4.4) + 6.5$
$$= 52.8 + 6.5$$
$$= 59.3 = \text{L.H.S.}$$

Hence the law connected R and V is $R = \dfrac{4.4}{V} + 6.5$

Problem 3. The following table gives corresponding values of p and V which are believed to be related by a law of the form $p = aV^2 + bV$ where a and b are constants.

V	0.5	2.6	5.3	7.7	9.2	11.4	12.7
p	4.5	38.5	121.4	231.8	318.3	469.7	565.2

Verify the law and find the values of a and b. Hence find: (a) the value of p when V is 10.6, and (b) the positive value of V when $p = 150$.

Dividing both sides of the equation $p = aV^2 + bV$ by V gives:

$$\boxed{\frac{p}{V}} = a\boxed{V} + b$$

Comparing with $\boxed{y} = m\boxed{x} + c$

shows that $\dfrac{p}{V}$ is to be plotted vertically against V horizontally to

produce a straight line of slope a and $\dfrac{p}{V}$-axis intercept b.

Another table of values is drawn up.

V	0.5	2.6	5.3	7.7	9.2	11.4	12.7
p	4.5	38.5	121.4	231.8	318.3	469.7	565.2
$\dfrac{p}{V}$	9.0	14.8	22.9	30.1	34.6	41.2	44.5

A graph of $\dfrac{p}{V}$ against V is shown in Fig. 9.17 where a straight line fits the points. Thus the law is verified.

The $\dfrac{p}{V}$-axis intercept, $b = 7$

The slope, or gradient, $a = 3$
Hence the law relating p and V is: $p = 3V^2 + 7V$.

(a)   When $V = 10.6$, $p = 3(10.6)^2 + 7(10.6)$
$$= 337.1 + 74.2 = 411.3$$
(b)   When $p = 150$, $150 = 3V^2 + 7V$
i.e., $3V^2 + 7V - 150 = 0$

Thus: $V = \dfrac{-7 \pm \sqrt{[(7)^2 - 4(3)(-150)]}}{2(3)}$

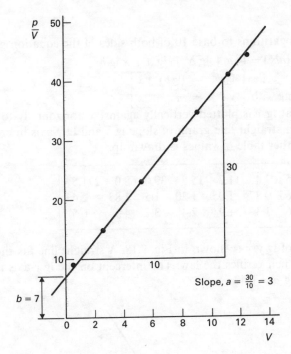

Fig. 9.17 Graph of $\frac{p}{V}/V$

$$= \frac{-7 \pm \sqrt{[49 + 1\,800]}}{6} = \frac{-7 \pm 43}{6}$$

$$= 6 \text{ or } -8\frac{1}{3}$$

Thus for the law $p = 3V^2 + 7V$, $V = 6$ when $p = 150$

Problem 4. Values of x and y are believed to be related by a law of the form $y = ab^x$ where a and b are constants. The values of y and corresponding values of x are shown:

y	4.5	7.4	11.2	15.8	39.0	68.0	271.5
x	0.6	1.3	1.9	2.4	3.7	4.5	6.5

Verify that the law relating y and x is as stated and determine the approximate values of a and b.

Hence determine: (a) the value of y when x is 3.2, and (b) the value of x when y is 126.7

$y = ab^x$

Taking logarithms to base 10 of both sides of the equation gives:

$\lg y = \lg(ab^x) = \lg a + \lg b^x = \lg a + x \lg b$

i.e. $\boxed{\lg y} = (\lg b)\ \boxed{x}$

Comparing with $\boxed{y} = m\ \boxed{x}$

shows that $\lg y$ is plotted vertically against x horizontally to produce a straight line graph of slope $\lg b$ and $\lg y$-axis intercept $\lg a$. Another table of values is drawn up.

y	4.5	7.4	11.2	15.8	39.0	68.0	271.5
$\lg y$	0.65	0.87	1.05	1.20	1.5	1.83	2.43
x	0.6	1.3	1.9	2.4	3.7	4.5	6.5

A graph of $\lg y/x$ is shown in Fig. 9.18. A straight line fits the points, which verifies the law. The intercept on the $\lg y$-axis is 0.47, i.e. $\lg a = 0.47$

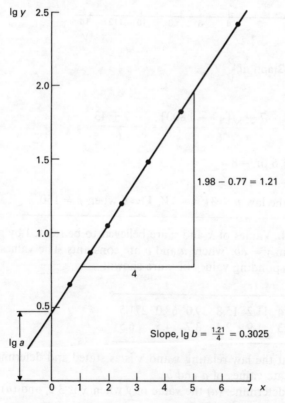

Fig. 9.18

Taking antilogarithms, $a = 2.951 = 3.0$, correct to 2 significant figures.

Slope, or gradient, $\lg b = \dfrac{1.21}{4} = 0.302\,5$

Taking antilogarithms, $b = 2.006 = 2.0$ correct to 2 significant figures.

Hence the law relating x and y is $y = (3.0)(2.0)^x$

From the graph, when $x = 3.2$, $\lg y \approx 1.44$, giving $y \approx 27.5$
and when $y = 126.7$, $\lg y = 2.10$ giving $x \approx 5.4$

Alternatively: (a) When $x = 3.2$, $y = (3.0)(2.0)^{3.2} = \mathbf{27.6}$
(b) When $y = 126.7$, $126.7 = (3.0)(2.0)^x$

$$\text{i.e. } \frac{126.7}{3.0} = (2.0)^x$$

$$42.23 = (2.0)^x$$

Taking logarithms: $\lg 42.23 = x \lg 2.0$

$$x = \frac{\lg 42.23}{\lg 2.0} = \frac{1.625\,6}{0.301\,0} = \mathbf{5.40}$$

Further problems on reduction of non-linear laws to a linear form (i.e. determination of laws) may be found in the following Section (9.5) (Problems 40–66), page 172.

9.5   Further problems

The quadratic graph

In Problems 1 to 10, sketch the quadratic graphs and determine the coordinates of the turning-points.

1. $y = 5x^2$ $[(0, 0)]$

2. $\dfrac{y}{3} = x^2$ $[(0, 0)]$

3. $y = 2x^2 + 4$ $[(0, 4)]$

4. $y = 4x^2 - 1$ $[(0, -1)]$

5. $y + 2x^2 = 6$ $[(0, 6)]$

6. $y = -(5x^2 + 7)$ $[(0, -7)]$

7. $y = 2x^2 + 5x + 2$ $[(-1\frac{1}{4}, -1\frac{1}{8})]$

8. $y - 19 = 3x^2 - 12x$ $[(2, 7)]$

9. $y + 5 = 2x(2 - x)$ $[(1, -3)]$

10. $y + 3x^2 = 1 - 6x$ $[(-1, 4)]$

In Problems 11 to 19, plot the graphs and determine the coordinates of the turning-points, stating whether they are maximum or minimum values.

11. $y = 3x^2 + 4x + 8$ $\left[\left(-\dfrac{2}{3}, 6\dfrac{2}{3}\right); \text{Minimum}\right]$

12. $y = x^2 - 3x + 6$ $\left[\left(1\dfrac{1}{2}, 3\dfrac{3}{4}\right); \text{Minimum}\right]$

13. $y + \dfrac{97}{24} = 7x - 6x^2$ $\left[\left(\dfrac{7}{12}, -2\right); \text{Maximum}\right]$

14. $3x^2 + 4x + y = 5$ $\left[\left(-\dfrac{2}{3}, 6\dfrac{1}{3}\right); \text{Maximum}\right]$

15. $y + \dfrac{2}{3} = x(3x + 16)$ $\left[\left(-2\dfrac{2}{3}, -22\right); \text{Minimum}\right]$

16. $\dfrac{3}{16} + y = x(3 - 4x)$ $\left[\left(\dfrac{3}{8}, \dfrac{3}{8}\right); \text{Maximum}\right]$

17. $y + 7x = x^2 + 2\dfrac{1}{4}$ $\left[\left(3\dfrac{1}{2}, -10\right); \text{Minimum}\right]$

18. $15x + y = 18x^2 + 7\dfrac{1}{8}$ $\left[\left(\dfrac{5}{12}, 4\right); \text{Minimum}\right]$

19. $20x^2 + 1 = 4x - y$ $\left[\left(\dfrac{1}{10}, -\dfrac{4}{5}\right); \text{Maximum}\right]$

20. Find the point of intersection of the two curves $y = 3x^2$ and $y + 15x = 3x^2 + 5$ by plotting the two curves. $\left[\left(\dfrac{1}{3}, \dfrac{1}{3}\right)\right]$

21. Determine the coordinates of the points of intersection of the two curves $y + 2 = 5x^2$ and $y = 2x^2 - 6x + 7$ by plotting the two curves. $[(1, 3); (-3, 43)]$

22. Plot the graph of $y = 5x^2 - 3x + 4$. From the graph find: (a) the value of y when $x = 2.4$; and (b) the values of x when $y = 11.6$ (a) $[25.6]$ (b) $[1.57 \text{ and } -0.97]$

23. Draw up a table of values and plot the graph $y = -3x^2 + 4x - 6$ and from the graph find: (a) the value of y when $x = 1.7$; and (b) the values of x when $y = -8.8$ (a) $[-7.87]$ (b) $[1.84 \text{ and } -0.51]$

Simple curves of the circle, parabola, ellipse and hyperbola

In Problems 24 to 31, sketch the curves depicting the equations given, showing significant values where possible.

24. $x = 7\sqrt{\left[1 - \left(\dfrac{y}{7}\right)^2\right]}$ [Circle, centre $(0, 0)$, radius 7 units]

25. $\dfrac{y^2}{3x} = 4$ [Parabola, symmetrical about x-axis, vertex at $(0, 0)$]

26. $\sqrt{x} = \dfrac{y}{5}$ [Parabola, symmetrical about x-axis, vertex at $(0, 0)$]

27. $y^2 = \dfrac{x^2 - 12}{4}$ [Hyperbola, symmetrical about x- and y-axis, distance between vertices $4\sqrt{3}$ units along x-axis]

28. $\dfrac{y^2}{5} = 5 - x^2$ [Ellipse, centre $(0, 0)$, major axis 10 units along y-axis, minor axis $2\sqrt{5}$ units along x-axis]

29. $x = \dfrac{1}{3}\sqrt{[99 - 11y^2]}$ [Ellipse, centre $(0, 0)$, major axis $2\sqrt{11}$ units along x-axis, minor axis 6 units along y-axis]

30. $x = 2\sqrt{(1 + y^2)}$ [Hyperbola, symmetrical about x- and y-axis, distance between vertices 4 units along x-axis]

31. $x^2 y^2 = 5$ [Rectangular hyperbola, lying in first and third quadrants only]

In Problems 32 to 39, describe the shape of the curves represented by the equations given.

32. $y = \sqrt{[2(x^2 - 1)]}$ [Ellipse, centre $(0, 0)$, major axis $2\sqrt{2}$ units along y-axis, minor axis 2 units along x-axis]

33. $y = \sqrt{[2(1 + x^2)]}$ [Hyperbola, symmetrical about x- and y-axis, vertices 2 units apart along x-axis]

34. $y = \sqrt{(2 - x^2)}$ [Circle, centre $(0, 0)$, radius $\sqrt{2}$ units]

35. $y = 2x^{-1}$ [Rectangular hyperbola, lying in first and third quadrants, symmetrical about x- and y-axis]

36. $y = (2x)^{\frac{1}{2}}$ [Parabola, vertex at $(0, 0)$, symmetrical about the x-axis]

37. $2y^2 - 6 = -3x^2$ [Ellipse, centre $(0, 0)$, major axis $2\sqrt{3}$ units along the y-axis, minor axis $2\sqrt{2}$ units along the x-axis]

38. $y^2 = 13 - x^2$ [Circle, centre $(0, 0)$, radius $\sqrt{13}$]

39. $4x^2 - 3y^2 = 12$ [Hyperbola, symmetrical about x- and y-axis, vertices on x-axis distance $2\sqrt{3}$ apart]

Determination of laws

In Problems 40 to 51, x and y are variables and all other letters denote constants. For the stated laws to be verified it is necessary to modify them into a straight line form. In order to plot a straight line graph state: (a) what should be plotted on the vertical axis; (b) what should be plotted on the horizontal axis; (c) the slope or gradient; and (d) the vertical-axis intercept.

40. $y = ax + b$ (a) $[y]$ (b) $[x]$ (c) $[a]$ (d) $[b]$

41. $y = cx^2 + d$ (a) $[y]$ (b) $[x^2]$ (c) $[c]$ (d) $[d]$

42. $y - f = e\sqrt{x}$ (a) $[y]$ (b) $[\sqrt{x}]$ (c) $[e]$ (d) $[f]$

43. $y = \dfrac{g}{x} + h$ (a) $[y]$ (b) $\left[\dfrac{1}{x}\right]$ (c) $[g]$ (d) $[h]$

44. $y = \sqrt{\left(\dfrac{j}{x}\right)} + k$ (a) $[y]$ (b) $\left[\dfrac{1}{\sqrt{x}}\right]$ (c) $[\sqrt{j}]$ (d) $[k]$

45. $y = lx^2 + mx$ (a) $\left[\dfrac{y}{x}\right]$ (b) $[x]$ (c) $[l]$ (d) $[m]$

46. $y - p = \dfrac{n}{x^2}$ (a) $[y]$ (b) $\left[\dfrac{1}{x^2}\right]$ (c) $[n]$ (d) $[p]$

47. $y = \dfrac{q}{x} + rx$ (a) $\left[\dfrac{y}{x}\right]$ (b) $\left[\dfrac{1}{x^2}\right]$ (c) $[q]$ (d) $[r]$

48. $y = sx^t$ (a) $[\lg y]$ (b) $[\lg x]$ (c) $[t]$ (d) $[\lg s]$

49. $y = uv^x$ (a) $[\lg y]$ (b) $[x]$ (c) $[\lg v]$ (d) $[\lg u]$

50. $y = \dfrac{a}{x - b}$ (a) $\left[\dfrac{1}{y}\right]$ (b) $[x]$ (c) $\left[\dfrac{1}{a}\right]$ (d) $\left[-\dfrac{b}{a}\right]$

51. $y = \dfrac{1}{wx + z}$ (a) $\left[\dfrac{1}{y}\right]$ (b) $[x]$ (c) $[w]$ (d) $[z]$

52. The following experimental values of x and y are believed to be related by the law $y = mx^2 + c$ where m and c are constants. By plotting a suitable graph verify this law and find the approximate values of m and c.

x	2.3	4.1	6.0	8.4	9.9	11.3
y	13.9	31.2	60.2	111.8	153.0	197.5

$[m = 1.5, c = 6.0]$

53. Show that the following values of p and q obey a law of the form $p = m\sqrt{q} + n$, where m and n are constants.

p	5.6	8.0	10.0	12.8	14.9	16.7	18.8
q	0.64	3.61	9.0	18.5	30.3	39.7	54.8

Determine approximate values of m and n.

$[m = 2.0, n = 4.1]$

54. Experimental values of load W newtons and distance l metres are shown in the following table:

W	33.2	30.4	27.4	23.2	18.6	14.0	9.4	5.5
l	0.741	0.364	0.233	0.163	0.116	0.093	0.076	0.067

Verify that W and l are related by a law of the form

$W = \dfrac{a}{l} + b$ and find the approximate values of a and b.

$[a = -2.05, b = 36]$

55. The solubility S of potassium chlorate is shown by the following table:

$t\,(°C)$	10	20	30	40	50	60	80	100
S	4.5	7	10	14	19	25	39	57

The relation between S and t is thought to be of the form $S = 3 + at + bt^2$. Plot a graph in order to test the supposition, and, if correct, use your graph to find probable values of a and b, explaining your methods.
$[a = 0.12, b = 0.004]$

56. The pressure p and volume V of a gas at constant temperature are related by the law $pV = c$, where c is a constant. Show that the given values of p and V follow this law and determine the value of c.

Pressure, p (bar)	10.6	8.0	6.4	5.3	4.6	4.0
Volume, V (m³)	1.5	2.0	2.5	3.0	3.5	4.0

$[c = 15.85]$

57. Measurements of the resistance (R ohms) of varying diameters (d mm) of wire were made in an experiment with the following results:

R	1.44	1.13	0.88	0.73	0.66
d	1.13	1.38	1.78	2.26	2.76

It is suspected that R is related to d by the law $R = \dfrac{c}{d^2} + a$.

Verify this and find approximate values for c and a. Estimate also the cross-sectional area of wire needed for a resistance reading of 0.52 ohms. $[c = 1.2, a = 0.5, 47.1 \text{ mm}^2]$

58. Show that the following values of s and t follow a law of the type $s = at^3 + b$ (where a and b are constants) and find approximate values for a and b.

s	49.7	47.4	41.4	29.5	10.0
t	1.0	2.0	3.0	4.0	5.0

$[a = -0.32, b = 50]$

59. The pH of a one-twentieth molar solution of potassium hydrogen phthalate varies with temperature (t) as shown:

pH 4.011   4.001   4.001   4.011   4.031   4.061
$t°C$ 0   10   20   30   40   50

Show that these results obey an equation of the form pH $= a + (t - 15)^2/b$ and determine the constants a and b. [$a = 4.0, b = 20\,000$]

60. The following table gives corresponding values of two quantities x and y which are believed to be related by a law of the form $y = ax^2 + bx$ where a and b are constants.

y 28.6 49.0 63.5 81.4 115.7 136.3
x 2.8 4.6 5.8 7.2 9.7   11.1

Verify the law and estimate the values of a and b. Hence find: (i) the value of y when x is 6.6; and (ii) the value of x when y is 102.5   [$a = 0.25, b = 9.5$; (i) 73.3, (ii) 8.8]

61. Quantities p and q are believed to be related by a law of the form $q = ab^p$. The value of p and corresponding values of q are as shown:

p 0   0.5   1.0   1.5   2.0   2.5   3.0
q 1.0 3.2   10.0   31.6   100.0 316.2   1\,000.0

Verify the law and find approximate values of a and b. [$a = 1, b = 10$]

62. Two variables x and y are believed to be related by the law $xy = ax + by$ where a and b are constants. Results obtained by experiment are as follows:

x 1.5 3.0 4.5 6.0 7.5 9.0
y 20.0 6.7 5.5 5.0 4.8 4.6

Verify the law and find approximate values of a and b. Hence find (i) the value of y when x is 5.2; and (ii) the value of x when y is 7.0   [$a = 4, b = 1.2$, (i) 5.2, (ii) 2.8]

63. The values of m and corresponding values of n are as follows:

m 0.2   0.7   1.3   1.9   2.4   3.6
n 6.0   14.2 39.4   109.7   257.2   1\,990

The law relating these quantities is of the form $n = ab^m$. Determine the approximate values of a and b. [$a = 4.3, b = 5.5$]

64. The power dissipated by a resistor was measured for various values of current flowing in the resistor and the results are as shown below:

Current, I amperes   1.3   2.4   3.7   4.9   5.8
Power, P watts   37   127   301   528   740

Prove that the law relating current and power is of the form $P = RI^n$, where R and n are constants, and determine the law. $[P = 22I^2]$

65. The vapour pressure (p) of triethylamine at various temperatures (T) is given below.

p (Pa)   0.439   10.120   93.330
$T(K)$ 100   120   140

Show that these results are related by an equation of the form $\lg p = A - \dfrac{B}{T - 3.12}$. Find A and B.

$[A = 7.6, B = 770]$

66. The values of resistance R and voltage V were measured in an experiment and the results are shown below.

R ohms   59.5   74.2   84.1   94.8   152.2
V millivolts   135   107   94   83   51

Resistance and voltage are thought to be connected by a law of the form $R = \dfrac{a}{V} + b$, where a and b are constants. Verify the law and find the approximate values of a and b. Determine the voltage when the resistance is 90.0 ohms.

$[a = 7.6, b = 3.2; 87.6$ millivolts$]$

Chapter 10

The notation of second order matrices and determinants

10.1 Introduction

Matrices are used in engineering and science for solving linear simultaneous equations.

Terms used in connection with matrices

Consider the linear simultaneous equations:

$$2x + 3y = 4 \tag{1}$$
$$\text{and } 5x - 6y = 7 \tag{2}$$

In **matrix** notation, the coefficients of x and y are written as $\begin{pmatrix} 2 & 3 \\ 5 & -6 \end{pmatrix}$, that is, occupying the same relative positions as in equations (1) and (2) above. The grouping of the coefficients of x and y in this way is called an **array** and the coefficients forming the array are called the **elements** of the matrix.

If there are m rows across an array and n columns down an array, then the matrix is said to be of order $m \times n$, called 'm by n'. Thus for the equations

$$2x + 3y - 4z = 5 \tag{3}$$
$$6x - 7y + 8z = 9 \tag{4}$$

the matrix of the coefficients of x, y and z is $\begin{pmatrix} 2 & 3 & -4 \\ 6 & -7 & 8 \end{pmatrix}$ and is a '2 by 3' matrix. A matrix having a single row is called a **row** matrix and one having a single column is called a **column** matrix. For example, in equation (3) above, the coefficients of x, y and z form a row matrix of $(2 \quad 3 \quad -4)$ and the coefficients of x in equations (3) and (4) form a column matrix of $\begin{pmatrix} 2 \\ 6 \end{pmatrix}$.

A matrix having the same number of rows as columns is called a **square** matrix. Thus the matrix for the coefficients of x and y in equations (1) and (2) above, i.e. $\begin{pmatrix} 2 & 3 \\ 5 & -6 \end{pmatrix}$, is a square matrix, and is called a second order matrix. Matrices are generally denoted by capital letters and if the matrix representing the coefficients of x and y in equations (1) and (2) above is A, then $A = \begin{pmatrix} 2 & 3 \\ 5 & -6 \end{pmatrix}$

10.2 Addition, subtraction and multiplication of second order matrices

In arithmetic, once the basic procedures associated with addition, subtraction, multiplication and division have been mastered, simple problems may be solved. With matrices, the various rules governing them have to be understood before they can be used to solve practical problems. In this section the basic laws of addition, subtraction and multiplication are introduced.

A matrix does not have a single numerical value and cannot be simplified to a particular answer. The main advantage of using matrices is that by applying the laws of matrices, given in this section, they can be simplified, and by comparing one matrix with another similar matrix, values of unknown elements can be determined. It will be seen in Chapter 11 that matrices can be used in this way for solving simultaneous equations. Matrices can be added, subtracted and multiplied and suitable definitions for these operations are formulated, so that they obey most of the laws which govern the algebra of numbers.

Addition

Only matrices of the same order may be added. Thus a 2 by 2 matrix can be added to a 2 by 2 matrix by adding corresponding elements, but a 3 by 2 matrix cannot be added to a 2 by 2 matrix, since some elements in one matrix do not have corresponding elements in the other. The sum of two matrices is the matrix obtained by adding the elements occupying corresponding positions in the matrix, and results in two matrices being simplified to a single matrix.

For example, the matrices

$$\begin{pmatrix} 1 & 3 \\ 2 & -4 \end{pmatrix} \text{ and } \begin{pmatrix} 2 & 5 \\ 6 & -7 \end{pmatrix}$$

are added as follows:

$$\begin{pmatrix} 1 & 3 \\ 2 & -4 \end{pmatrix} + \begin{pmatrix} 2 & 5 \\ 6 & -7 \end{pmatrix} = \begin{pmatrix} 1+2 & 3+5 \\ 2+6 & (-4)+(-7) \end{pmatrix}$$

$$= \begin{pmatrix} 3 & 8 \\ 8 & -11 \end{pmatrix}$$

Subtraction

Only matrices of the same order can be subtracted and the difference between two matrices, say $A - B$, is the matrix obtained by subtracting the elements of matrix B from those occupying the corresponding positions in matrix A.

For example:

$$\begin{pmatrix} 1 & 3 \\ 2 & -4 \end{pmatrix} - \begin{pmatrix} 2 & 5 \\ 6 & -7 \end{pmatrix} = \begin{pmatrix} 1-2 & 3-5 \\ 2-6 & (-4)-(-7) \end{pmatrix}$$

$$= \begin{pmatrix} -1 & -2 \\ -4 & 3 \end{pmatrix}$$

By adding the single matrix obtained by adding A and B to, say, matrix C, the single matrix representing $A + B + C$ is obtained. By taking, say, matrix D from this single matrix, $A + B + C - D$ is obtained. Thus the laws of addition and subtraction can be applied to more than two matrices, providing that they are all of the same order.

Multiplication

(a)   Scalar multiplication

When a matrix is multiplied by a number, the resultant matrix is one of the same order having each element multiplied by the number.

Thus, if matrix $A = \begin{pmatrix} 1 & 3 \\ 2 & -4 \end{pmatrix}$, then $2A = 2\begin{pmatrix} 1 & 3 \\ 2 & -4 \end{pmatrix}$

$$= \begin{pmatrix} 2 \times 1 & 2 \times 3 \\ 2 \times 2 & 2 \times (-4) \end{pmatrix}$$

$$= \begin{pmatrix} 2 & 6 \\ 4 & -8 \end{pmatrix}$$

(b)   Multiplication of matrices

Two matrices can only be multiplied together when the number of columns in the first one is equal to the number of rows in the second one. This is because the process of matrix multiplication depends on finding the sum of the products of the rows in one matrix with the columns in the other. Thus it is possible to multiply a 2 by 2 matrix by a column matrix having two elements or by another 2 by 2 matrix, but it is not possible to multiply it by a row matrix. Thus if:

$$A = \begin{pmatrix} 2 & 3 \\ 5 & 6 \end{pmatrix}, B = \begin{pmatrix} 1 \\ 8 \end{pmatrix} \text{ and } C = (4 \quad 9)$$

it is possible to find $A \times B$, since the number of columns in A is equal to the number of rows in B, but it is not possible to find $A \times C$ since there are two columns in A but only one row in C. If a 2 by 2 matrix A is multiplied by a column matrix, B, having two elements, the resulting matrix is a two-element column matrix. The top element is the sum of the products obtained by taking the elements of the top row of A with B. The bottom element is the sum of the products obtained by taking the bottom elements of A with B.

If $A = \begin{pmatrix} a & b \\ c & d \end{pmatrix}$ and $B = \begin{pmatrix} p \\ q \end{pmatrix}$

then $A \times B = \begin{pmatrix} a & b \\ c & d \end{pmatrix} \times \begin{pmatrix} p \\ q \end{pmatrix} = \begin{pmatrix} ap + bq \\ cp + dq \end{pmatrix}$

For example, to multiply the matrices, say $\begin{pmatrix} 2 & -5 \\ 4 & 3 \end{pmatrix}$ and $\begin{pmatrix} 1 \\ 6 \end{pmatrix}$, gives

$$\begin{pmatrix} 2 & -5 \\ 4 & 3 \end{pmatrix} \times \begin{pmatrix} 1 \\ 6 \end{pmatrix} = \begin{pmatrix} 2 \times 1 + (-5) \times 6 \\ 4 \times 1 + \ 3 \times 6 \end{pmatrix} = \begin{pmatrix} 2 - 30 \\ 4 + 18 \end{pmatrix}$$

$$= \begin{pmatrix} -28 \\ 22 \end{pmatrix}$$

If a 2 by 2 matrix, say A, is multiplied by a 2 by 2 matrix, say B, the resulting matrix is a 2 by 2 matrix, say C. The top elements of C are the sum of the products obtained by taking the top row of A with the columns of B. The bottom elements of C are the sum of the products obtained by taking the bottom row of A with the columns of B.

For example:

$$\begin{pmatrix} 1 & 3 \\ -2 & 4 \end{pmatrix} \times \begin{pmatrix} 5 & 0 \\ 7 & -6 \end{pmatrix} = \begin{pmatrix} 1 \times 5 + 3 \times 7 & 1 \times 0 + 3 \times (-6) \\ -2 \times 5 + 4 \times 7 & -2 \times 0 + 4 \times (-6) \end{pmatrix}$$

$$= \begin{pmatrix} 26 & -18 \\ 18 & -24 \end{pmatrix}$$

In general, when a matrix of dimension (m by n) is multiplied by a matrix of dimension (n by q), the resulting matrix is one of dimension (m by q).

Although the laws of matrices are so formulated that they follow most of the laws which govern the algebra of numbers, frequently in the multiplication of matrices

$A \times B \neq B \times A$

It is shown above that

$$A \times B = \begin{pmatrix} 1 & 3 \\ -2 & 4 \end{pmatrix} \times \begin{pmatrix} 5 & 0 \\ 7 & -6 \end{pmatrix}$$

$$= \begin{pmatrix} 26 & -18 \\ 18 & -24 \end{pmatrix}$$

However, $B \times A = \begin{pmatrix} 5 & 0 \\ 7 & -6 \end{pmatrix} \times \begin{pmatrix} 1 & 3 \\ -2 & 4 \end{pmatrix}$

$$= \begin{pmatrix} 5 \times 1 + & 0 \times (-2) & 5 \times 3 + & 0 \times 4 \\ 7 \times 1 + (-6) \times (-2) & 7 \times 3 + (-6) \times 4 \end{pmatrix}$$

i.e. $B \times A = \begin{pmatrix} 5 & 15 \\ 19 & -3 \end{pmatrix}$

That is, $A \times B \neq B \times A$ in this case. The results are said to be *non-commutative* (i.e. they are not in agreement).

Worked problems on the addition, subtraction and multiplication of second order matrices

Problem 1. If $A = \begin{pmatrix} 1 & 4 \\ -3 & 2 \end{pmatrix}$, $B = \begin{pmatrix} 5 & -1 \\ 0 & 1 \end{pmatrix}$ and $C = \begin{pmatrix} 3 & -4 \\ 7 & 2 \end{pmatrix}$

determine the single matrix for (a) $A + C$, (b) $A - C$ and (c) $A + B - C$

(a) $A + C = \begin{pmatrix} 1 & 4 \\ -3 & 2 \end{pmatrix} + \begin{pmatrix} 3 & -4 \\ 7 & 2 \end{pmatrix} = \begin{pmatrix} 1+3 & 4+(-4) \\ (-3)+7 & 2+2 \end{pmatrix}$

$$= \begin{pmatrix} 4 & 0 \\ 4 & 4 \end{pmatrix}$$

(b) $A - C = \begin{pmatrix} 1 & 4 \\ -3 & 2 \end{pmatrix} - \begin{pmatrix} 3 & -4 \\ 7 & 2 \end{pmatrix} = \begin{pmatrix} 1-3 & 4-(-4) \\ (-3)-7 & 2-2 \end{pmatrix}$

$$= \begin{pmatrix} -2 & 8 \\ -10 & 0 \end{pmatrix}$$

(c) From part (b), $A - C = \begin{pmatrix} -2 & 8 \\ -10 & 0 \end{pmatrix}$

Hence $A + B - C = \begin{pmatrix} -2 & 8 \\ -10 & 0 \end{pmatrix} + \begin{pmatrix} 5 & -1 \\ 0 & 1 \end{pmatrix}$

$$= \begin{pmatrix} -2+5 & 8+(-1) \\ -10+0 & 0+1 \end{pmatrix}$$

$$= \begin{pmatrix} 3 & 7 \\ -10 & 1 \end{pmatrix}$$

Problem 2. Determine the single matrix for (*a*) $A \cdot C$ and (*b*) $A \cdot B$, where $A = \begin{pmatrix} 2 & 4 \\ 1 & -3 \end{pmatrix}$, $B = \begin{pmatrix} 3 & -7 \\ 4 & -5 \end{pmatrix}$ and $C = \begin{pmatrix} 2 \\ -5 \end{pmatrix}$

(*a*) $A \cdot C = \begin{pmatrix} 2 & 4 \\ 1 & -3 \end{pmatrix} \times \begin{pmatrix} 2 \\ -5 \end{pmatrix} = \begin{pmatrix} 2 \times 2 + & 4 \times (-5) \\ 1 \times 2 + (-3) \times (-5) \end{pmatrix}$

$$= \begin{pmatrix} 4 - 20 \\ 2 + 15 \end{pmatrix} = \begin{pmatrix} -16 \\ 17 \end{pmatrix}$$

(2) $A \cdot B = \begin{pmatrix} 2 & 4 \\ 1 & -3 \end{pmatrix} \times \begin{pmatrix} 3 & -7 \\ 4 & -5 \end{pmatrix}$

$$= \begin{pmatrix} [2 \times 3 + 4 \times 4] & [2 \times (-7) + 4 \times (-5)] \\ [1 \times 3 + (-3) \times 4] & [1 \times (-7) + (-3) \times (-5)] \end{pmatrix}$$

$$= \begin{pmatrix} 6 + 16 & -14 + (-20) \\ 3 + (-12) & -7 + 15 \end{pmatrix}$$

$$= \begin{pmatrix} 22 & -34 \\ -9 & 8 \end{pmatrix}$$

Further problems on the addition, subtraction and multiplication of matrices may be found in Section 10.5 (Problems 1–10), page 183.

10.3 The unit matrix

A unit matrix is one in which the values of the elements in the leading diagonal, $(\backslash)$, are 1, the remaining elements being 0. Thus, a 2 by 2 unit matrix is $\begin{pmatrix} 1 & 0 \\ 0 & 1 \end{pmatrix}$, and is usually denoted by the symbol I. If A is a square matrix and I the unit matrix, then $A \times I = I \times A$, that is, this is one case in matrices where the law $A \times B = B \times A$ of the algebra of numbers is true. The unit matrix is analogous to the number 1 in ordinary algebra.

10.4 Second order determinants

The solution of the linear simultaneous equations:

$$a_1 x + b_1 y + c_1 = 0 \tag{1}$$

$$a_2 x + b_2 y + c_2 = 0 \tag{2}$$

may be found by the elimination method of solving simultaneous equations.

To eliminate y:

$$\text{Equation (1)} \times b_2: \quad a_1b_2x + b_1b_2y + c_1b_2 = 0$$

$$\text{Equation (2)} \times b_1: \quad a_2b_1x + b_1b_2y + c_2b_1 = 0$$

$$\text{Subtracting:} \quad (a_1b_2 - a_2b_1)x + (c_1b_2 - c_2b_1) = 0$$

Thus, $\quad x = \dfrac{-(c_1b_2 - c_2b_1)}{a_1b_2 - a_2b_1}$

i.e. $x = \dfrac{(b_1c_2 - b_2c_1)}{a_1b_2 - a_2b_1}$ \hfill (3)

Similarly, to eliminate x:

$$\text{Equation (1)} \times a_2: \quad a_1a_2x + a_2b_1y + a_2c_1 = 0$$

$$\text{Equation (2)} \times a_1: \quad a_1a_2x + a_1b_2y + a_1c_2 = 0$$

$$\text{Subtracting:} \quad (a_2b_1 - a_1b_2)y + (a_2c_1 - a_1c_2) = 0$$

Thus, $\quad y = \dfrac{-(a_2c_1 - a_1c_2)}{(a_2b_1 - a_1b_2)} = \dfrac{(a_1c_2 - a_2c_1)}{(a_2b_1 - a_1b_2)} = \dfrac{(a_1c_2 - a_2c_1)}{-(a_1b_2 - a_2b_1)}$

i.e. $-y = \dfrac{(a_1c_2 - a_2c_1)}{(a_1b_2 - a_2b_1)}$ \hfill (4)

Equations (3) and (4) can be written in the form

$$\frac{x}{b_1c_2 - b_2c_1} = \frac{-y}{a_1c_2 - a_2c_1} = \frac{1}{a_1b_2 - a_2b_1} \tag{5}$$

The denominators of equation (5) are all of the general form

$$pq - rs$$

Although as stated in Section 10.2 a matrix does not have a single numerical value and cannot be simplified to a particular answer, coefficients written in this form may be expressed as a special matrix, denoted by an array within vertical lines rather than brackets. In this case:

$$\begin{vmatrix} a & b \\ c & d \end{vmatrix} = ad - bc, \text{ and is called a second order } \mathbf{determinant}$$

It is shown in Chapter 11 that determinants can be used to solve linear simultaneous equations such as those given in equations (1) and (2) above.

Worked problem on second order determinants

Problem 1. Evaluate the determinants:

(a) $\begin{vmatrix} 3 & -1 \\ 4 & 2 \end{vmatrix}$ and (b) $\begin{vmatrix} a & -2b \\ 2a & -3b \end{vmatrix}$

By the definition of a determinant, $\begin{vmatrix} a & b \\ c & d \end{vmatrix} = ad - bc$, hence

(a) $\begin{vmatrix} 3 & -1 \\ 4 & 2 \end{vmatrix} = (3 \times 2) - ((-1) \times 4) = 6 + 4 = \mathbf{10}$

(b) $\begin{vmatrix} a & -2b \\ 2a & -3b \end{vmatrix} = (a \times (-3b)) - ((-2b) \times 2a) = -3ab + 4ab$

$$= ab$$

Further problems on second order determinants may be found in Section 10.5 following (Problems 11–15), page 185.

10.5 Further problems

Addition, subtraction and multiplication of second order matrices

In Problems 1 to 5, matrices A, B, C and D are given by

$$A = \begin{pmatrix} 1 & 4 \\ -3 & 2 \end{pmatrix}, B = \begin{pmatrix} 2 & 7 \\ -1 & 0 \end{pmatrix},$$

$$C = \begin{pmatrix} 5 & -1 \\ 0 & 1 \end{pmatrix} \text{ and } D = \begin{pmatrix} 3 & -4 \\ 7 & 2 \end{pmatrix}$$

Determine the single matrix for the expressions given.

1. (a) $A + B$ $\quad \left[\begin{pmatrix} 3 & 11 \\ -4 & 2 \end{pmatrix} \right]$

   (b) $A + C$ $\quad \left[\begin{pmatrix} 6 & 3 \\ -3 & 3 \end{pmatrix} \right]$

2. (a) $C + D$ $\quad \left[\begin{pmatrix} 8 & -5 \\ 7 & 3 \end{pmatrix} \right]$

   (b) $B + D$ $\quad \left[\begin{pmatrix} 5 & 3 \\ 6 & 2 \end{pmatrix} \right]$

3.   (a) $B - A$ $\left[\begin{pmatrix} 1 & 3 \\ 2 & -2 \end{pmatrix}\right]$

      (b) $D - B$ $\left[\begin{pmatrix} 1 & -11 \\ 8 & 2 \end{pmatrix}\right]$

4.   (a) $C - A$ $\left[\begin{pmatrix} 4 & -5 \\ 3 & -1 \end{pmatrix}\right]$

      (b) $B - C$ $\left[\begin{pmatrix} -3 & 8 \\ -1 & -1 \end{pmatrix}\right]$

5.   (a) $A + B + C$ $\left[\begin{pmatrix} 8 & 10 \\ -4 & 3 \end{pmatrix}\right]$

      (b) $A - B + C - D$ $\left[\begin{pmatrix} 1 & 0 \\ -9 & 1 \end{pmatrix}\right]$

In Problems 6 to 10 matrices A, B, C, D and E are given by

$$A = \begin{pmatrix} 3 & 1 \\ -2 & 4 \end{pmatrix}, B = \begin{pmatrix} 2 & -5 \\ 0 & 1 \end{pmatrix}, C = \begin{pmatrix} -1 & 6 \\ 3 & 0 \end{pmatrix},$$

$$D = \begin{pmatrix} 2 \\ 3 \end{pmatrix} \text{ and } E = \begin{pmatrix} -1 \\ 4 \end{pmatrix}$$

Determine the single matrix for the expressions given.

6.   (a) $A \cdot D$ $\left[\begin{pmatrix} 9 \\ 8 \end{pmatrix}\right]$

      (b) $B \cdot E$ $\left[\begin{pmatrix} -22 \\ 4 \end{pmatrix}\right]$

7.   (a) $C \cdot D$ $\left[\begin{pmatrix} 16 \\ 6 \end{pmatrix}\right]$

      (b) $A \cdot E$ $\left[\begin{pmatrix} 1 \\ 18 \end{pmatrix}\right]$

8.   (a) $A \cdot C$ $\left[\begin{pmatrix} 0 & 18 \\ 14 & -12 \end{pmatrix}\right]$

      (b) $C \cdot B$ $\left[\begin{pmatrix} -2 & 12 \\ 6 & -15 \end{pmatrix}\right]$

9. (a) $A \cdot B$ $\begin{bmatrix} \begin{pmatrix} 6 & -14 \\ -4 & 14 \end{pmatrix} \\ \begin{pmatrix} 16 & -18 \\ -2 & 4 \end{pmatrix} \end{bmatrix}$ (Note that $A \cdot B \neq B \cdot A$)

(b) $B \cdot A$

10. (a) $B \cdot C$ $\begin{bmatrix} \begin{pmatrix} -17 & 12 \\ 3 & 0 \end{pmatrix} \\ \begin{pmatrix} -15 & 23 \\ 9 & 3 \end{pmatrix} \end{bmatrix}$

(b) $C \cdot A$

Second order determinants

In Problems 11 to 15, evaluate the determinants given.

11. (a) $\begin{vmatrix} 2 & 3 \\ 4 & 5 \end{vmatrix}$ (b) $\begin{vmatrix} -1 & -1 \\ 7 & 2 \end{vmatrix}$ (a) $[-2]$ (b) $[5]$

12. (a) $\begin{vmatrix} 3 & -1 \\ 4 & 7 \end{vmatrix}$ (b) $\begin{vmatrix} 5 & -2 \\ 3 & 1 \end{vmatrix}$ (a) $[25]$ (b) $[11]$

13. (a) $\begin{vmatrix} -2 & 4 \\ 3 & 1 \end{vmatrix}$ (b) $\begin{vmatrix} 1 & -4 \\ 5 & 1 \end{vmatrix}$ (a) $[-14]$ (b) $[21]$

14. (a) $\begin{vmatrix} x & 2x \\ -3x & 5 \end{vmatrix}$ (b) $\begin{vmatrix} y^2 & 3y \\ 4y^2 & -2y \end{vmatrix}$ (a) $[5x + 6x^2]$ (b) $[-14y^3]$

15. (a) $\begin{vmatrix} c & -2b \\ 4c & -3b \end{vmatrix}$ (b) $\begin{vmatrix} a & c \\ 2a & 4c \end{vmatrix}$ (a) $[5bc]$ (b) $[2ac]$

Chapter 11

The solution of simultaneous equations having two unknowns using determinants and matrices

11.1 The solution of simultaneous equations having two unknowns using determinants

When introducing determinants in Section 10.4, the simultaneous equations

$$a_1 x + b_1 y + c_1 = 0 \tag{1}$$
$$a_2 x + b_2 y + c_2 = 0 \tag{2}$$

were solved using the elimination method of solving simultaneous equations, and it was shown that

$$\frac{x}{b_1 c_2 - b_2 c_1} = \frac{-y}{a_1 c_2 - a_2 c_1} = \frac{1}{a_1 b_2 - a_2 b_1} \tag{3}$$

It is also stated that the denominators of equation (3) are all of the general form

$$pq - rs$$

This algebraic expression is denoted by a special matrix having its array within vertical lines and is called a **determinant**. Thus

$$\begin{vmatrix} a & b \\ c & d \end{vmatrix} = ad - bc$$

and is called a second order determinant.

The denominators of equation (3) can be written in determinant form, giving

$$\frac{x}{\begin{vmatrix} b_1 & c_1 \\ b_2 & c_2 \end{vmatrix}} = \frac{-y}{\begin{vmatrix} a_1 & c_1 \\ a_2 & c_2 \end{vmatrix}} = \frac{1}{\begin{vmatrix} a_1 & b_1 \\ a_2 & b_2 \end{vmatrix}} \tag{4}$$

This expression is used to solve simultaneous equations by determinants and can be remembered by the 'cover-up' rule. In this rule:

(i)   the equations are written in the form $a_1 x + b_1 y + c_1 = 0$.
(ii) If equation (4) is written in the form

$$\frac{x}{|D_1|} = \frac{-y}{|D_2|} = \frac{1}{|D|}, \text{ then}$$

(iii) $|D_1|$ is obtained by covering up the x-column and writing down the remaining coefficients in determinant form in positions corresponding to the positions they occupy in the equations.

(iv) $|D_2|$ is obtained by covering up the y-column and treating the coefficients as in (iii) above.

(v) $|D|$ is obtained by covering up the constants-column and treating the coefficients as in (iii) above.

For example, to solve the equations

$$2x + 3y = 11 \tag{5}$$
$$4x + 2y = 10 \tag{6}$$

by using determinants:

(i)   the equations are written as

$$2x + 3y - 11 = 0$$
$$4x + 2y - 10 = 0$$

(ii) $\dfrac{x}{|D_1|} = \dfrac{-y}{|D_2|} = \dfrac{1}{|D|}$

(iii) $|D_1| = \begin{vmatrix} 3 & -11 \\ 2 & -10 \end{vmatrix}$, obtained by covering up the x-column in (i) above.

(iv) $|D_2| = \begin{vmatrix} 2 & -11 \\ 4 & -10 \end{vmatrix}$, obtained by covering up the y-column in (i) above.

(v) $|D| = \begin{vmatrix} 2 & 3 \\ 4 & 2 \end{vmatrix}$, obtained by covering up the constants-column in (i) above.

Thus,

$$\frac{x}{\begin{vmatrix} 3 & -11 \\ 2 & -10 \end{vmatrix}} = \frac{-y}{\begin{vmatrix} 2 & -11 \\ 4 & -10 \end{vmatrix}} = \frac{1}{\begin{vmatrix} 2 & 3 \\ 4 & 2 \end{vmatrix}}$$

i.e. $\dfrac{x}{3 \times (-10) - 2 \times (-11)} = \dfrac{-y}{2 \times (-10) - 4 \times (-11)} = \dfrac{1}{2 \times 2 - 4 \times 3}$

$$\frac{x}{-8} = \frac{-y}{24} = \frac{1}{-8}$$

giving $x = \dfrac{-8}{-8} = 1$ and $-y = \dfrac{24}{-8}$, i.e. $y = 3$

Checking in the original equations:

L.H.S. of equation (5) $= 2 \times 1 + 3 \times 3 = 11 =$ R.H.S.

L.H.S. of equation (6) $= 4 \times 1 + 2 \times 3 = 10 =$ R.H.S.

Hence, $x = 1$, $y = 3$ is the correct solution.

If, in a determinant of the form $D = \begin{vmatrix} a_1 & b_1 \\ a_2 & b_2 \end{vmatrix}$, $a_1 b_2 = a_2 b_1$, then $a_1 b_2 - a_2 b_1 = 0$, i.e. $D = 0$. This means that in the simultaneous equations on which the determinant is based (say of the form $ax + by = c$), a_1 and b_1 are each multiplied by the same constant to give a_2 and b_2. So essentially there is only one equation with two unknown quantities which is not capable of solution. An example is the determinant $\begin{vmatrix} 2 & 1 \\ 10 & 5 \end{vmatrix}$, which arises when trying to solve the equations, say

$$2x + y = 3$$
$$\text{and } 10x + 5y = 15$$

and in this case $D = 2 \times 5 - 10 \times 1 = 0$, i.e. the equations cannot be solved. The matrix of the coefficients of x and y, i.e. $\begin{pmatrix} 2 & 1 \\ 10 & 5 \end{pmatrix}$, is called a **singular matrix** when the determinant of the matrix is equal to 0.

Worked problems on the solution of simultaneous equations having two unknowns using determinants

Problem 1. Solve the simultaneous equations:

$$\tfrac{3}{2}p - 2q = \tfrac{1}{2} \tag{1}$$
$$p + \tfrac{3}{2}q = 6 \tag{2}$$

by using determinants.

(i)   Writing the equations in the form $ax + by + c = 0$ gives

$$\tfrac{3}{2}p - 2q - \tfrac{1}{2} = 0$$
$$p + \tfrac{3}{2}q - 6 = 0$$

(ii) $\dfrac{p}{|D_1|} = \dfrac{-q}{|D_2|} = \dfrac{1}{|D|}$ (note the signs are $+$, $-$, $+$)

(iii)   Covering up the p-column gives $|D_1| = \begin{vmatrix} -2 & -\tfrac{1}{2} \\ \tfrac{3}{2} & -6 \end{vmatrix}$

(iv)   Covering up the q-column gives $|D_2| = \begin{vmatrix} \tfrac{3}{2} & -\tfrac{1}{2} \\ 1 & -6 \end{vmatrix}$

(v)   Covering up the constants-column gives $|D| = \begin{vmatrix} \tfrac{3}{2} & -2 \\ 1 & \tfrac{3}{2} \end{vmatrix}$

Thus,

$$\frac{p}{\begin{vmatrix} -2 & -\frac{1}{2} \\ \frac{3}{2} & -6 \end{vmatrix}} = \frac{-q}{\begin{vmatrix} \frac{3}{2} & -\frac{1}{2} \\ 1 & -6 \end{vmatrix}} = \frac{1}{\begin{vmatrix} \frac{3}{2} & -2 \\ 1 & \frac{3}{2} \end{vmatrix}}$$

i.e.
$$\frac{p}{(-2)\times(-6) - (\frac{3}{2})\times(-\frac{1}{2})} = \frac{-q}{(\frac{3}{2})\times(-6) - (1)\times(-\frac{1}{2})}$$

$$= \frac{1}{(\frac{3}{2})\times(\frac{3}{2}) - (1)\times(-2)}$$

$$\frac{p}{12\frac{3}{4}} = \frac{-q}{-8\frac{1}{2}}$$

$$= \frac{1}{-4\frac{1}{4}}$$

Hence, $p = \dfrac{12\frac{3}{4}}{4\frac{1}{4}} = 3$ and $q = \dfrac{8\frac{1}{2}}{4\frac{1}{4}} = 2$

Checking in the original equations:

L.H.S. of equation (1) $= \frac{3}{2} \times 3 - 2 \times 2 = \frac{1}{2} =$ R.H.S.

L.H.S. of equation (2) $= 3 + \frac{3}{2} \times 2 = 6 =$ R.H.S.

Hence $p = 3$, $q = 2$ is the correct solution.

Problem 2. Use determinants to solve the simultaneous equations:

$$-0.5f + 0.4g = 0.7 \tag{1}$$
$$1.2f - 0.3g = 3.6 \tag{2}$$

Writing the equations in the form $ax + by + c = 0$ gives

$-0.5f + 0.4g - 0.7 = 0$ $\dfrac{f}{|D_1|} = \dfrac{-g}{|D_2|} = \dfrac{1}{|D|}$

$1.2f - 0.3g - 3.6 = 0$

Covering up the f-column gives

$$|D_1| = \begin{vmatrix} 0.4 & -0.7 \\ -0.3 & -3.6 \end{vmatrix}$$

$$= (0.4)\times(-3.6) - (-0.3)\times(-0.7) = -1.44 - 0.21 = -1.65$$

Covering up the g-column gives

$$|D_2| = \begin{vmatrix} -0.5 & -0.7 \\ 1.2 & -3.6 \end{vmatrix}$$

$$= (-0.5)\times(-3.6) - (1.2)\times(-0.7) = 1.8 - (-0.84) = 2.64$$

Covering up the constants-column gives

$$|D| = \begin{vmatrix} -0.5 & 0.4 \\ 1.2 & -0.3 \end{vmatrix}$$

$$= (-0.5)\times(-0.3) - (1.2)\times(0.4) = 0.15 - 0.48 = -0.33$$

Hence, $\dfrac{f}{-1.65} = \dfrac{-g}{2.64} = \dfrac{1}{-0.33}$

i.e. $f = \dfrac{-1.65}{-0.33} = 5$ and $g = \dfrac{-2.64}{-0.33} = 8$

Checking:

L.H.S. of equation (1) $= -0.5 \times 5 + 0.4 \times 8 = 0.7 =$ R.H.S.

L.H.S. of equation (2) $= \quad 1.2 \times 5 - 0.3 \times 8 = 3.6 =$ R.H.S.

Thus, $f = 5$, $g = 8$ is the correct solution.

Problem 3. Use determinants to solve the simultaneous equations:

$$\dfrac{10}{a} - \dfrac{4}{b} = 3 \tag{1}$$

$$\dfrac{6}{a} + \dfrac{8}{b} = 7 \tag{2}$$

Let $x = \dfrac{1}{a}$ and $y = \dfrac{1}{b}$, then

$$10x - 4y = 3$$

$$6x + 8y = 7$$

Writing in the $ax + by + c = 0$ form gives

$$10x - 4y - 3 = 0$$

$$6x + 8y - 7 = 0$$

Applying the cover-up rule gives

$$\frac{x}{\begin{vmatrix} -4 & -3 \\ 8 & -7 \end{vmatrix}} = \frac{-y}{\begin{vmatrix} 10 & -3 \\ 6 & -7 \end{vmatrix}} = \frac{1}{\begin{vmatrix} 10 & -4 \\ 6 & 8 \end{vmatrix}}$$

i.e. $\dfrac{x}{28 + 24} = \dfrac{-y}{-70 + 18} = \dfrac{1}{80 + 24}$ and $x = \frac{52}{104} = \frac{1}{2}$, $y = \frac{52}{104} = \frac{1}{2}$

Since $x = \dfrac{1}{a}$, $a = 2$, and since $y = \dfrac{1}{b}$, $b = 2$

Checking:

L.H.S. of equation (1) $= \frac{10}{2} - \frac{4}{2} = 3 =$ R.H.S.

L.H.S. of equation (2) $= \frac{6}{2} + \frac{8}{2} = 7 =$ R.H.S.

Hence $a = 2$, $b = 2$ is the correct solution.

Problem 4. The forces acting on a bolt are resolved horizontally and vertically, giving the simultaneous equations shown below. Use determinants to find the values of F_1 and F_2, correct to three significant figures:

$$3.4F_1 - 0.83F_2 = 3.9 \tag{1}$$

$$0.7F_1 + 1.47F_2 = -2.05 \tag{2}$$

Writing the equations in the $ax + by + c = 0$ form gives:

$$3.4F_1 - 0.83F_2 - 3.9 = 0$$

$$0.7F_1 + 1.47F_2 + 2.05 = 0$$

Applying the cover-up rule gives:

$$\frac{F_1}{\begin{vmatrix} -0.83 & -3.9 \\ 1.47 & 2.05 \end{vmatrix}} = \frac{-F_2}{\begin{vmatrix} 3.4 & -3.9 \\ 0.7 & 2.05 \end{vmatrix}} = \frac{1}{\begin{vmatrix} 3.4 & -0.83 \\ 0.7 & 1.47 \end{vmatrix}}$$

Hence,

$$\frac{F_1}{(-0.83) \times 2.05 - 1.47 \times (-3.9)} = \frac{-F_2}{3.4 \times 2.05 - 0.7 \times (-3.9)}$$

$$= \frac{1}{3.5 \times 1.47 - 0.7 \times (-0.83)}$$

that is: $\dfrac{F_1}{-1.701\ 5 + 5.733} = \dfrac{-F_2}{6.970 + 2.730} = \dfrac{1}{4.998 + 0.581}$

i.e. $\dfrac{F_1}{4.032} = \dfrac{-F_2}{9.700} = \dfrac{1}{5.579}$

Thus, $F_1 = \dfrac{4.032}{5.579} = 0.723$, correct to three significant figures,

and $F_2 = \dfrac{-9.700}{5.579} = -1.74$, correct to three significant figures.

Checking:

L.H.S. of equation (1) $= 3.4 \times 0.723 - 0.83 \times (-1.74)$

$= 3.90$, correct to three significant figures, $=$ R.H.S.

L.H.S. of equation (2) $= 0.7 \times 0.723 + 1.47 \times (-1.74)$

$= -2.05$, correct to three significant figures, $=$ R.H.S.

Hence $F_1 = 0.723$, $F_2 = -1.74$ is the correct solution.

Further problems on solving simultaneous equations having two unknowns using determinants may be found in Section 11.4 (Problems 1–15), page 198.

11.2   The inverse of a matrix

The inverse or reciprocal of matrix A is the matrix A^{-1}, such that $A \cdot A^{-1} = I = A^{-1} \cdot A$, where I is the unit matrix, introduced in Section 10.3

The process of inverting a matrix makes division possible. If three matrices, A, B and X, are such that

$$A \cdot X = B$$

$$\text{then } X = \frac{B}{A} = A^{-1} \cdot B$$

Let the inverse of matrix A be $A^{-1} = \begin{pmatrix} a & b \\ c & d \end{pmatrix}$, and let matrix A be, say, $\begin{pmatrix} 2 & 3 \\ -1 & 1 \end{pmatrix}$. By the definition of the inverse of a matrix, $\begin{pmatrix} 2 & 3 \\ -1 & 1 \end{pmatrix} \times \begin{pmatrix} a & b \\ c & d \end{pmatrix} = \begin{pmatrix} 1 & 0 \\ 0 & 1 \end{pmatrix}$, the unit matrix. Multiplying the matrices on the left-hand side gives

$$\begin{pmatrix} 2a + 3c & 2b + 3d \\ -a + c & -b + d \end{pmatrix} = \begin{pmatrix} 1 & 0 \\ 0 & 1 \end{pmatrix} \tag{1}$$

Since these two matrices are equal to one another, the corresponding elements are equal to one another, hence

$$-a + c = 0, \text{ that is, } a = c$$
$$2b + 3d = 0, \text{ that is, } b = -\tfrac{3}{2}d$$

Substituting for a and b in equation (1) above gives:

$$\begin{pmatrix} 5c & 0 \\ 0 & \dfrac{5d}{2} \end{pmatrix} = \begin{pmatrix} 1 & 0 \\ 0 & 1 \end{pmatrix}$$

Thus, $5c = 1$, that is, $c = \tfrac{1}{5}$

$$\frac{5d}{2} = 1, \text{ that is, } d = \tfrac{2}{5}$$

Since $a = c$, $a = \tfrac{1}{5}$, and since $b = -\tfrac{3}{2}d$, $b = -\tfrac{3}{5}$. Thus the inverse of

matrix $\begin{pmatrix} 2 & 3 \\ -1 & 1 \end{pmatrix}$ is $\begin{pmatrix} \tfrac{1}{5} & -\tfrac{3}{5} \\ \tfrac{1}{5} & \tfrac{2}{5} \end{pmatrix}$

There is an alternative method of finding the inverse of a matrix. If the inverses of many matrices are determined and the inverses of the matrices are compared with the matrices, a relationship is seen to exist between matrices and their inverses. This relationship for a matrix of the form $\begin{pmatrix} a & b \\ c & d \end{pmatrix}$

is that in the inverse:

(i)   the position of the a and d elements are interchanged,

(ii)   the sign of both the b and c elements is changed, and

(iii) the matrix is multiplied by $\dfrac{1}{ad-bc}$, i.e. the reciprocal of the determinant of the matrix.

Thus, the inverse of matrix $\begin{pmatrix} a & b \\ c & d \end{pmatrix}$ is $\dfrac{1}{ad-bc}\begin{pmatrix} d & -b \\ -c & a \end{pmatrix}$.

For the matrix $\begin{pmatrix} 2 & 3 \\ -1 & 1 \end{pmatrix}$ considered previously the inverse is

$$\frac{1}{2\times 1 - 3\times(-1)}\begin{pmatrix} 1 & -3 \\ 1 & 2 \end{pmatrix} = \tfrac{1}{5}\begin{pmatrix} 1 & -3 \\ 1 & 2 \end{pmatrix} = \begin{pmatrix} \tfrac{1}{5} & -\tfrac{3}{5} \\ \tfrac{1}{5} & \tfrac{2}{5} \end{pmatrix}$$

as shown previously.

Worked problem on the inverse of a matrix

Problem 1. Determine the inverse of the matrix $A = \begin{pmatrix} 5 & -3 \\ -2 & 1 \end{pmatrix}$

Let the inverse matrix be $A^{-1} = \begin{pmatrix} a & b \\ c & d \end{pmatrix}$

Since $A \cdot A^{-1} = I$, the unit matrix, then

$$\begin{pmatrix} 5 & -3 \\ -2 & 1 \end{pmatrix} \times \begin{pmatrix} a & b \\ c & d \end{pmatrix} = \begin{pmatrix} 1 & 0 \\ 0 & 1 \end{pmatrix}$$

Applying the multiplication law to the left-hand side gives:

$$\begin{pmatrix} 5a - 3c & 5b - 3d \\ -2a + c & -2b + d \end{pmatrix} = \begin{pmatrix} 1 & 0 \\ 0 & 1 \end{pmatrix} \qquad (1)$$

Equating corresponding elements gives:

$-2a + c = 0$, i.e. $a = \dfrac{c}{2}$ and $5b - 3d = 0$, i.e. $b = \tfrac{3}{5}d$

Substituting in equation (1): $\begin{pmatrix} -\dfrac{c}{2} & 0 \\ 0 & -\dfrac{d}{5} \end{pmatrix} = \begin{pmatrix} 1 & 0 \\ 0 & 1 \end{pmatrix}$

i.e. $-\dfrac{c}{2} = 1$, $c = -2$ and $-\dfrac{d}{5} = 1$, $d = -5$

Since $a = \dfrac{c}{2}$, $a = -1$ and since $b = \tfrac{3}{5}d$, $b = -3$

Thus the inverse matrix of $\begin{pmatrix} 5 & -3 \\ -2 & 1 \end{pmatrix}$ is $\begin{pmatrix} -1 & -3 \\ -2 & -5 \end{pmatrix}$

The solution may be checked, using $A \cdot A^{-1} = I$. Thus

$$\begin{pmatrix} 5 & -3 \\ -2 & 1 \end{pmatrix} \times \begin{pmatrix} -1 & -3 \\ -2 & -5 \end{pmatrix} = \begin{pmatrix} -5+6 & -15+15 \\ 2-2 & 6-5 \end{pmatrix}$$

$$= \begin{pmatrix} 1 & 0 \\ 0 & 1 \end{pmatrix}, \text{ the inverse matrix.}$$

Hence the solution $\begin{pmatrix} -1 & -3 \\ -2 & -5 \end{pmatrix}$ is correct.

Alternatively, the relationship that the inverse of matrix $\begin{pmatrix} a & b \\ c & d \end{pmatrix}$ is $\dfrac{1}{ad-bc}\begin{pmatrix} d & -b \\ -c & a \end{pmatrix}$ could have been applied.

The inverse of $\begin{pmatrix} 5 & -3 \\ -2 & 1 \end{pmatrix} = \dfrac{1}{5 \times 1 - (-2) \times (-3)}\begin{pmatrix} 1 & 3 \\ 2 & 5 \end{pmatrix}$

$$= \dfrac{1}{-1}\begin{pmatrix} 1 & 3 \\ 2 & 5 \end{pmatrix}$$

Applying the law for scalar multiplication gives $\begin{pmatrix} -1 & -3 \\ -2 & -5 \end{pmatrix}$,

as obtained previously. The alternative method of applying a formula is the easiest method of determining the inverse of a matrix.

Further problems on the inverse of a matrix may be found in Section 11.4 (*Problems 16–20*), page 200.

11.3 The solution of simultaneous equations having two unknowns using matrices

Matrices may be used to solve linear simultaneous equations. For equations having two unknown quantities there is no advantage in using a matrix method. However, for equations having three or more unknown quantities, solution by a matrix method can usually be performed more quickly and accurately.

Two linear simultaneous equations, such as:

$$2x + 3y = 4 \tag{1}$$

$$x - 5y = 6 \tag{2}$$

may be written in matrix form, as:

$$\begin{pmatrix} 2 & 3 \\ 1 & -5 \end{pmatrix}\begin{pmatrix} x \\ y \end{pmatrix} = \begin{pmatrix} 4 \\ 6 \end{pmatrix} \tag{3}$$

The inverse of the matrix $\begin{pmatrix} 2 & 3 \\ 1 & -5 \end{pmatrix}$ is obtained as shown in Section 11.2
and is $\begin{pmatrix} \frac{5}{13} & \frac{3}{13} \\ \frac{1}{13} & -\frac{2}{13} \end{pmatrix}$. Multiplying both sides of equation (3) by this inversed
matrix gives:

$$\begin{pmatrix} 1 & 0 \\ 0 & 1 \end{pmatrix}\begin{pmatrix} x \\ y \end{pmatrix} = \begin{pmatrix} \frac{5}{13} & \frac{3}{13} \\ \frac{1}{13} & -\frac{2}{13} \end{pmatrix}\begin{pmatrix} 4 \\ 6 \end{pmatrix} = \begin{pmatrix} \frac{20}{13} + \frac{18}{13} \\ \frac{4}{13} - \frac{12}{13} \end{pmatrix}$$

$$= \begin{pmatrix} \frac{38}{13} \\ -\frac{8}{13} \end{pmatrix}$$

i.e. $\begin{pmatrix} x \\ y \end{pmatrix} = \begin{pmatrix} \frac{38}{13} \\ -\frac{8}{13} \end{pmatrix}$

Equating corresponding elements gives:

$$x = \tfrac{38}{13} \text{ and } y = -\tfrac{8}{13}$$

Check: L.H.S. of equation (1) is $2 \times \tfrac{38}{13} + 3 \times (-\tfrac{8}{13}) = 4 = \text{R.H.S.}$
L.H.S. of equation (2) is $\tfrac{38}{13} - 5(-\tfrac{8}{13}) = 6 = \text{R.H.S.}$

Hence $x = \tfrac{38}{13}$, $y = -\tfrac{8}{13}$ is the correct solution.

Summary

To solve linear simultaneous equations with two unknown quantities by using matrices:

(i)   write the equations in the standard form

$$ax + by = c$$
$$dx + ey = f$$

(ii)   write this in matrix form, i.e.

$$\begin{pmatrix} a & b \\ d & e \end{pmatrix}\begin{pmatrix} x \\ y \end{pmatrix} = \begin{pmatrix} c \\ f \end{pmatrix}$$

(iii)   determine the inverse of matrix $\begin{pmatrix} a & b \\ d & e \end{pmatrix}$

(iv)   multiply each side of (ii) by the inversed matrix, and express in the form

$$\begin{pmatrix} x \\ y \end{pmatrix} = \begin{pmatrix} g \\ h \end{pmatrix}$$

(v)   solve for x and y by equating corresponding elements, and
(vi)   check the solution in the original equations.

Worked problems on solving simultaneous equations having two unknowns using matrices

---

Problem 1. Use matrices to solve the simultaneous equations:

$$4a - 3b = 18 \tag{1}$$

$$a + 2b = -1 \tag{2}$$

With reference to the above summary:

(i) The equations are in the standard form.

(ii) The matrices are $\begin{pmatrix} 4 & -3 \\ 1 & 2 \end{pmatrix}\begin{pmatrix} a \\ b \end{pmatrix} = \begin{pmatrix} 18 \\ -1 \end{pmatrix}$

(iii) The inverse matrix of

$$\begin{pmatrix} 4 & -3 \\ 1 & 2 \end{pmatrix} \text{ is } \frac{1}{4 \times 2 - (-3) \times 1}\begin{pmatrix} 2 & 3 \\ -1 & 4 \end{pmatrix}$$

i.e. $\frac{1}{11}\begin{pmatrix} 2 & 3 \\ -1 & 4 \end{pmatrix}$, that is $\begin{pmatrix} \frac{2}{11} & \frac{3}{11} \\ -\frac{1}{11} & \frac{4}{11} \end{pmatrix}$

(iv) Multiplying each side of (ii) by this inversed matrix gives:

$$\begin{pmatrix} a \\ b \end{pmatrix} = \begin{pmatrix} \frac{2}{11} & \frac{3}{11} \\ -\frac{1}{11} & \frac{4}{11} \end{pmatrix}\begin{pmatrix} 18 \\ -1 \end{pmatrix}$$

$$\begin{pmatrix} a \\ b \end{pmatrix} = \begin{pmatrix} \frac{36}{11} + (-\frac{3}{11}) \\ -\frac{18}{11} + (-\frac{4}{11}) \end{pmatrix} = \begin{pmatrix} 3 \\ -2 \end{pmatrix}$$

(v) Thus $a = 3, b = -2$

(iv) Checking:

L.H.S. of equation (1) is $4 \times 3 - 3(-2) = 18 = $ R.H.S.

L.H.S. of equation (2) is $3 + 2(-2) = -1 = $ R.H.S.

Hence **$a = 3, b = -2$** is the correct solution.

Problem 2. Solve the simultaneous equations:

$$\frac{3}{x} - \frac{2}{y} = 0 \tag{1}$$

$$\frac{1}{x} + \frac{4}{y} = 14 \tag{2}$$

by using matrices.

With reference to the summary:

(i)   The equations may be expressed in standard form by letting $\frac{1}{x}$ be p and $\frac{1}{y}$ be q. Thus equations (1) and (2) become

$$3p - 2q = 0$$
$$p + 4q = 14$$

(ii)   The matrices are $\begin{pmatrix} 3 & -2 \\ 1 & 4 \end{pmatrix}\begin{pmatrix} p \\ q \end{pmatrix} = \begin{pmatrix} 0 \\ 14 \end{pmatrix}$

(iii)   The inverse of $\begin{pmatrix} 3 & -2 \\ 1 & 4 \end{pmatrix}$ is $\frac{1}{14}\begin{pmatrix} 4 & 2 \\ -1 & 3 \end{pmatrix}$

(iv)   Multiplying each side of (ii) by (iii) gives

$$\begin{pmatrix} p \\ q \end{pmatrix} = \frac{1}{14}\begin{pmatrix} 4 & 2 \\ -1 & 3 \end{pmatrix}\begin{pmatrix} 0 \\ 14 \end{pmatrix}$$

i.e. $\begin{pmatrix} p \\ q \end{pmatrix} = \frac{1}{14}\begin{pmatrix} 4 \times 0 + 2 \times 14 \\ -1 \times 0 + 3 \times 14 \end{pmatrix}$

$$= \frac{1}{14}\begin{pmatrix} 28 \\ 42 \end{pmatrix} = \begin{pmatrix} \frac{1}{14} \times 28 \\ \frac{1}{14} \times 42 \end{pmatrix} = \begin{pmatrix} 2 \\ 3 \end{pmatrix}$$

(v)   Thus $p = 2$ and $q = 3$, i.e. $x = \frac{1}{2}$, $y = \frac{1}{3}$

(vi)   Checking:

L.H.S. of equation (1) is $\frac{3}{1/2} - \frac{2}{1/3} = 6 - 6 = 0 =$ R.H.S.

L.H.S. of equation (2) is $\frac{1}{1/2} + \frac{4}{1/3} = 2 + 12 = 14 =$ R.H.S.

Hence $x = \frac{1}{2}$, $y = \frac{1}{3}$ is the correct solution.

Problem 3. A force system is analysed, and by resolving the forces horizontally and vertically the following equations are obtained:

$$6F_1 - F_2 = 5 \tag{1}$$
$$5F_1 + 2F_2 = 7 \tag{2}$$

Use matrices to solve for F_1 and F_2.

The matrices are $\begin{pmatrix} 6 & -1 \\ 5 & 2 \end{pmatrix}\begin{pmatrix} F_1 \\ F_2 \end{pmatrix} = \begin{pmatrix} 5 \\ 7 \end{pmatrix}$

The inverse of $\begin{pmatrix} 6 & -1 \\ 5 & 2 \end{pmatrix}$ is $\frac{1}{17}\begin{pmatrix} 2 & 1 \\ -5 & 6 \end{pmatrix}$

Hence $\begin{pmatrix} F_1 \\ F_2 \end{pmatrix} = \frac{1}{17}\begin{pmatrix} 2 & 1 \\ -5 & 6 \end{pmatrix}\begin{pmatrix} 5 \\ 7 \end{pmatrix} = \frac{1}{17}\begin{pmatrix} 10 + 7 \\ -25 + 42 \end{pmatrix} = \begin{pmatrix} 1 \\ 1 \end{pmatrix}$

Thus $F_1 = 1$, $F_2 = 1$

Checking: L.H.S. of equation (1) is $6 - 1 = 5 = $ R.H.S.

L.H.S. of equation (2) is $5 + 2 = 7 = $ R.H.S.

Thus, $F_1 = 1$ and $F_2 = 1$ is the correct solution.

Any technical problems, such as the equations formed by the resolution of vector quantities or the equations relating load and effort in machines, which were previously solved by using simultaneous equations, may be solved either by using determinants or by using matrices.

Further problems on the solution of simultaneous equations having two unknowns using matrices may be found in Section 11.4 following (Problems 21–30), page 200.

11.4   Further problems

The solution of simultaneous equations having two unknowns using determinants

In Problems 1 to 11, use determinants to solve the simultaneous equations given.

1. $4v_1 - 3v_2 = 18$
   $v_1 + 2v_2 = -1$ $[v_1 = 3, v_2 = -2]$

2. $3m - 2n = -4.5$
   $4m + 3n = 2.5$ $[m = -\frac{1}{2}, n = 1\frac{1}{2}]$

3. $\dfrac{a}{3} + \dfrac{b}{4} = 8$

   $\dfrac{a}{6} - \dfrac{b}{8} = 1$ $[a = 15, b = 12]$

4. $s + t = 17$

   $\dfrac{s}{5} - \dfrac{t}{7} = 1$ $[s = 10, t = 7]$

5. $\dfrac{c}{5} + \dfrac{d}{3} = \dfrac{43}{30}$

   $\dfrac{c}{9} - \dfrac{d}{6} = -\dfrac{1}{12}$ $[c = 3, d = 2\frac{1}{2}]$

6. $0.5i_1 - 1.2i_2 = -13$
   $0.8i_1 + 0.3i_2 = 12.5$ $[i_1 = 10, i_2 = 15]$

7. $1.25L_1 - 0.75L_2 = 1$
   $0.25L_1 + 1.25L_2 = 17$ $[L_1 = 8.0, L_2 = 12.0]$

8. $$\frac{1}{2a} + \frac{3}{5b} = 7$$

    $$\frac{4}{a} + \frac{1}{2b} = 13 \qquad [a = \tfrac{1}{2}, b = \tfrac{1}{10}]$$

9. $$\frac{3}{v_1} - \frac{2}{v_2} = \frac{1}{2}$$

    $$\frac{5}{v_1} + \frac{3}{v_2} = \frac{29}{12} \qquad [v_1 = 3, v_2 = 4]$$

10. $$\frac{4}{p_1 - p_2} = \frac{16}{21}$$

    $$\frac{3}{p_1 + p_2} = \frac{4}{9} \qquad [p_1 = 6, p_2 = \tfrac{3}{4}]$$

11. $$\frac{2x + 1}{5} - \frac{1 - 4y}{2} = \frac{5}{2}$$

    $$\frac{1 - 3x}{7} + \frac{2y - 3}{5} + \frac{32}{35} = 0 \qquad [x = 2, y = 1]$$

12. A vector system to determine the shortest distance between two moving bodies is analysed, producing the following equations:

    $$11S_1 - 10S_2 = 30$$
    $$21S_2 - 20S_1 = -40$$

    Use determinants to find the values of S_1 and S_2
    $[S_1 = 7.42, S_2 = 5.16]$

13. The power in a mechanical device is given by $p = aN + \dfrac{b}{N}$, where a and b are constants. Use determinants to find the value of a and b if $p = 13$ when $N = 3$ and $p = 12$ when $N = 2$
    $[a = 3, b = 12]$

14. The law connecting friction F and load L for an experiment to find the friction force between two surfaces is of the type $F = aL + b$, where a and b are constants.
    When $F = 6.0$, $L = 7.5$ and when $F = 2.7$, $L = 2.0$
    Find the values of a and b by using determinants.
    $[a = 0.60, b = 1.5]$

15. The length L metres of an alloy at temperature $t\,°C$ is given by $L = L_0(1 + \alpha t)$, where L_0 and α are constants.
    Use determinants to find the values of L_0 and α if $L = 20$ m when t is $52°C$ and $L = 21$ m when $t = 100°C$.
    $[L_0 = 18.92$ m, $\alpha = 0.001\ 1]$

The inverse of a matrix

In Problems 16 to 20, find the inverse of the matrices given.

16. $\begin{pmatrix} 2 & -1 \\ -5 & -1 \end{pmatrix}$ $\left[\begin{pmatrix} \frac{1}{7} & -\frac{1}{7} \\ -\frac{5}{7} & -\frac{2}{7} \end{pmatrix} \right]$

17. $\begin{pmatrix} 1 & -3 \\ 1 & 7 \end{pmatrix}$ $\left[\begin{pmatrix} \frac{7}{10} & \frac{3}{10} \\ -\frac{1}{10} & \frac{1}{10} \end{pmatrix} \right]$

18. $\begin{pmatrix} 3 & 5 \\ -2 & 1 \end{pmatrix}$ $\left[\begin{pmatrix} \frac{1}{13} & -\frac{5}{13} \\ \frac{2}{13} & \frac{3}{13} \end{pmatrix} \right]$

19. $\begin{pmatrix} -2 & -1 \\ 4 & 3 \end{pmatrix}$ $\left[\begin{pmatrix} -\frac{3}{2} & -\frac{1}{2} \\ 2 & 1 \end{pmatrix} \right]$

20. $\begin{pmatrix} -4 & -3 \\ 5 & 3 \end{pmatrix}$ $\left[\begin{pmatrix} 1 & 1 \\ -\frac{5}{3} & -\frac{4}{3} \end{pmatrix} \right]$

The solution of simultaneous equations having two unknowns using matrices

In Problems 21 to 26, solve the simultaneous equations given by using matrices.

21. $p + 3q = 11$
$p + 2q = 8$ $[p = 2, q = 3]$

22. $3a + 4b - 5 = 0$
$12 = 5b - 2a$ $[a = -1, b = 2]$

23. $\dfrac{m}{3} + \dfrac{n}{4} = 6$

$\dfrac{m}{6} - \dfrac{n}{8} = 0$ $[m = 9, n = 12]$

24. $4a - 6b + 2.5 = 0$
$7a - 5b + 0.25 = 0$ $[a = \frac{1}{2}, b = \frac{3}{4}]$

25. $\dfrac{x}{8} + \dfrac{5}{2} = y$

$13 - \dfrac{y}{3} - 3x = 0$ $[x = 4, y = 3]$

26. $\dfrac{a-1}{3} + \dfrac{b+2}{2} = 3$

$\dfrac{1-a}{6} + \dfrac{4-b}{2} = \dfrac{1}{2}$ $[a = 4, b = 2]$

27. When determining the relative velocity of a system, the following equations are produced:

$$3.0 = 0.10v_1 + (v_1 - v_2)$$
$$-2.0 = 0.05v_2 - (v_1 - v_2)$$

Use matrices to find the values of v_1 and v_2
$[v_1 = 7.42, v_2 = 5.16]$

28. Applying Newton's laws of motion to a mechanical system gives the following equations:

$$14 = 0.2u + 2u + 8(u - v)$$
$$0 = -8(u - v) + 2v + 10v$$

Use matrices to find the values of u and v
$[u = 2.0, v = 0.8]$

29. Equations connecting the lens system in a position transducer are:

$$\frac{4}{u_1} + \frac{6}{v_1} + \frac{9}{v_2} = 6$$

$$\frac{15}{u_1} + \frac{11}{v_1} + \frac{2}{v_2} = 8\frac{1}{12}$$

If $v_1 = v_2$, use matrices to find the values of u_1, v_1 and v_2
$[u_1 = 4, v_1 = v_2 = 3]$

30. When an effort E is applied to the gearbox on a diesel motor it is found that a resistance R can be overcome and that E and R are connected by a formula $E = a + bR$, where a and b are constants. An effort of 3.5 newtons overcomes a resistance of 5 newtons and an effort of 5.3 newtons overcomes a resistance of 8 newtons. Use matrices to find the values of a and b $[a = 0.50, b = 0.60]$

Chapter 12

The circle

12.1 Radian measure

An angle may be measured in degrees or **radians**.

A radian is defined as the angle subtended at the centre of a circle by an arc equal in length to the radius.

Thus, in Fig. 12.1, when $r = s$ then angle θ is 1 radian. This angle is the unit of measurement in so-called circular measure. It is of constant size whatever the length of the radius.

The length of arc, s, for 1 radian is equal to r
The length of arc, s, for 2 radians is equal to $2r$
The length of arc, s, for θ radians is equal to $r\theta$

Therefore generally $\boxed{s = r\theta^c \quad \text{or} \quad \theta^c = \dfrac{s}{r}}$

'θ radians' may be written as 'θ^c', the 'c' standing for circular measure.

The number of radians subtended by the whole circumference of a circle is given by the number of times the radius is contained in the circumference.

Now the circumference $= 2\pi r$. Hence the number of radians in one complete revolution is given by:

$$\frac{2\pi r}{r} = 2\pi = 6.283 \text{ (correct to 3 decimal places)}$$

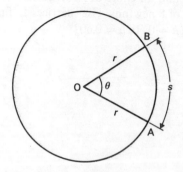

Fig. 12.1

Hence 2π **radians** $= 360°$

$$1 \text{ radian} = \left(\frac{360}{2\pi}\right)^\circ = 57.30° \text{ (correct to 2 decimal places)}$$

There are therefore just over 6 radians contained within a complete revolution.

Whenever conversion from radians to degrees or vice versa is necessary, either of the above relationships may be used. Some calculators have the facility to convert from radians to degrees and vice versa at the press of a button.

The following relationships are often used and can be derived from $2\pi \text{ rad} = 360°$:

$$\pi \text{ rad} = 180°; \quad \frac{\pi}{2} \text{ rad} = 90°; \quad \frac{\pi}{3} \text{ rad} = 60°;$$

$$\frac{\pi}{4} \text{ rad} = 45°; \quad \frac{\pi}{6} \text{ rad} = 30°; \quad \frac{\pi}{12} \text{ rad} = 15°;$$

and so on.

Worked problems on radian measure

Problem 1. Convert:

(a) 119° to radians   (b) 73° 33′ to radians

(c) 2.681 radians to degrees   (d) $\frac{3\pi}{7}$ radians to degrees.

(a)   If $180° = \pi \text{ rad}$

then $1° = \dfrac{\pi \text{ rad}}{180}$

Therefore $119° = 119\left[\dfrac{\pi}{180}\right] \text{ rad} = \textbf{2.077 rad}$

(b)   73° 33′ is a mixture of degrees and minutes and it is first necessary to convert to degrees and decimals of a degree.

$$33' = \frac{33°}{60} = 0.55°$$

Therefore 73° 33′ = 73.55°

If $180° = \pi \text{ rad}$

then $73.55° = \dfrac{73.55(\pi) \text{ rad}}{180}$

$$= 1.284 \text{ rad}$$

Therefore 73° 33′ = **1.284 radians**

(c) π rad $= 180°$

Therefore 1 rad $= \dfrac{180°}{\pi}$

2.681 rad $= 2.681 \left[\dfrac{180}{\pi} \right]° = 153.6°$

$0.6° = (0.6 \times 60)' = 36'$

$153.6° = 153° \, 36'$

Therefore 2.681 radians $= \mathbf{153° \, 36'}$

(d) π rad $= 180°$

Therefore 1 rad $= \left[\dfrac{180}{\pi} \right]°$

$\dfrac{3\pi \text{ rad}}{7} = \left[\dfrac{3\pi}{7} \right] \left[\dfrac{180}{\pi} \right]° = \dfrac{3}{7}(180)°$

$= 77.14°$

Now $0.14° = (0.14 \times 60)' = 8.4'$

and $0.4' = (0.4 \times 60)''$

$= 24''$

So $77.14° = 77° \, 8' \, 24''$

Therefore $\dfrac{3\pi}{7}$ radians $= \mathbf{77° \, 8' \, 24''}$

Problem 2. Express the following angles in radians in terms of π:
(a) 150°   (b) 270°   (c) 37.5°   (d) 383° 17' 23"

(a) $180° = \pi$ rad

Therefore $150° = \dfrac{150}{180} \pi$ rad

$= \dfrac{5\pi \text{ rad}}{6}$

Hence $150° = \dfrac{5\pi}{6}$ **radians (or 0.833 3π radians)**

(b) $180° = \pi$ rad

Therefore $270° = \dfrac{270}{180} \pi$ rad

$= \dfrac{3\pi \text{ rad}}{2}$

Hence $270° = \dfrac{3\pi}{2}$ **radians (or 1.5π radians)**

(c) $180° = \pi^c$

Therefore $37.5° = \left(\dfrac{37.5}{180}\right)\pi \text{ rad} = \left(\dfrac{75}{360}\right)\pi \text{ rad} = \left(\dfrac{5\pi}{24}\right)\text{ rad}$

Hence $37.5° = \dfrac{5\pi}{24}$ **radians (or 0.208 3π radians)**

(d) $23'' = \dfrac{23'}{60} = 0.38'$

$17.38' = \dfrac{17.38°}{60} = 0.289\,7° = 0.29°$

Therefore $383° \, 17' \, 23'' = 383.29° = 383.3°$ (to 4 significant figures)

Now $180° = \pi \text{ rad}$

Therefore $383.3° = \dfrac{383.3}{180}\pi \text{ rad} = 2.129\pi \text{ rad}$

Hence $383° \, 17' \, 23'' = \mathbf{2.129\pi}$ **radians**

Problem 3. Find the length of arc of a circle of radius 4.23 cm when the angle subtended at the centre is 1.46 radians.

Length of arc, $s = r\theta^c$

Therefore $s = (4.23)(1.46)$

$= 6.176$

Hence the length of arc is **6.176 cm**

Problem 4. Find the diameter and circumference of a circle if an arc of length 5.67 cm subtends an angle of 2.15 radians.

If $s = r\theta$

then $r = \dfrac{s}{\theta}$

Therefore $r = \dfrac{5.67}{2.15}$

$= 2.637 \text{ cm}$

Diameter $= 2 \times \text{radius}$

$= (2)(2.637)$

$= \mathbf{5.274 \text{ cm}}$

Circumference $= \pi d$

$$= (\pi)(5.274)$$

$$= 16.57 \text{ cm}$$

Hence the diameter is **5.274 cm** and the circumference is **16.57 cm**

Problem 5. In a single swing a pendulum moves through an angle of 9°. Determine the length of the arc traced by the pendulum bob (correct to the nearest centimetre) if the length of the pendulum is 1.4 m.

Since $180° = \pi$ rad, then $9° = \dfrac{9}{180} \pi$ rad

Length of arc, $s = r\theta = (1.4) \left(\dfrac{9}{180} \pi \right)$ m

$$= 0.2199 \text{ m} = \textbf{22 cm},$$

correct to the nearest centimetre

Problem 6. Find the angle, in radians and in degrees, subtended at the centre of a circle of diameter 23.0 mm by an arc of length 31.0 mm

Length of arc, $s = r\theta$

Therefore $\theta = \dfrac{s}{r}$

$s = 31.0$ mm and $r = \dfrac{23.0}{2} = 11.5$ mm

Therefore $\theta = \dfrac{31.0}{11.5} = 2.696$ radians

Now $\pi^c = 180°$

Therefore $2.696^c = 2.696 \left(\dfrac{180°}{\pi} \right)$

$$= 154.47°$$

Now $0.47° = (0.47 \times 60)' = 28'$

Therefore $154.47° = 154° \ 28'$

Hence the angle subtended at the centre is **154° 28'**

Problem 7. If an arc of length 4.627 cm subtends an angle of 231.5° at the centre of a circle find its radius in mm.

Now $180° = \pi^c$

Therefore $231.5° = \dfrac{231.5}{180}\,\pi$ rad

$$= 4.040 \text{ rad}$$

Now $s = r\theta$

Therefore $r = \dfrac{s}{\theta} = \dfrac{4.627}{4.040}$

$$= 1.145 \text{ cm}$$

$$= 11.45 \text{ mm}$$

Hence the radius of the circle is **11.45 mm**

Further problems on radian measure may be found in Section 12.3 (Problems 1–11), page 209.

12.2 Area of a sector of a circle

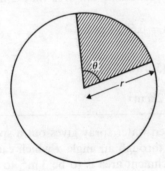

Fig. 12.2

The shaded portion of Fig. 12.2 shows a sector of a circle of radius r with an angle subtended at the centre of θ. This angle may be in degrees or radians.

If the area of a complete circle is πr^2, then the area of a sector is a proportion of πr^2.

If the angle θ is in degrees, then

$$\textbf{Area of sector} = \dfrac{\theta}{360}\,(\pi r^2)$$

For example, if $\theta = 180°$ then the area of the sector is $\frac{180}{360}(\pi r^2)$ or $\frac{1}{2}\pi r^2$. This particular sector is therefore a semicircle.

If the angle θ is in radians, then

$$\text{Area of sector} = \frac{\theta}{2\pi}(\pi r^2)$$
$$= \tfrac{1}{2}r^2\theta$$

Worked problems on the area of a sector of a circle

Problem 1. Determine the area of the sector of a circle of diameter 20 cm which has an angle subtended at the centre of 1.6 radians.

$$\text{Area of sector} = \frac{1}{2}r^2\theta$$

$$= \frac{1}{2}\left(\frac{20}{2}\right)^2(1.6) = \textbf{80 cm}^2$$

Problem 2. A football ground floodlight can spread its illumination over an angle of 40° to a distance of 50 m. Determine (a) the angle in radians and (b) the maximum area that is floodlit.

(a) $40° = \left(40 \times \dfrac{\pi}{180}\right)$ radians $= 0.698\,1$ rad

(b)   Area floodlit $= \dfrac{1}{2}r^2\theta = \dfrac{1}{2}(50)^2(0.698\,1)$

$$= \textbf{872.6 m}^2$$

Problem 3. An automatic garden water spray gives out a spray to a distance of 2 m and revolves through an angle α which can be varied. If the desired spray catchment area is to be 3 m^2, to what should angle α be set (correct to the nearest degree)?

$\text{Area} = \dfrac{1}{2}r^2\alpha$ Hence $3 = \dfrac{1}{2}(2)^2\alpha$

from which, $\alpha = \dfrac{3 \times 2}{2^2} = 1.5$ rad

1.5 rad $= \left(1.5 \times \dfrac{180}{\pi}\right)° = 85.94°$

Hence **angle $\alpha = 86°$**, correct to the nearest degree.

Further problems on the area of sectors of circles may be found in the following Section (12.3) (Problems 12–16), page 210.

12.3   Further problems

Radian measure

1. Convert the following angles from degrees to radians in
    terms of π: (a) 90°   (b) 225°   (c) 75°   (d) 60°   (e) 420°
    (f) 1 110°   (g) 263° 15′ 25″

    (a) $\left[\dfrac{\pi}{2} \text{ or } 0.5\pi\right]$ (b) $\left[\dfrac{5\pi}{4} \text{ or } 1.25\pi\right]$ (c) $\left[\dfrac{5\pi}{12} \text{ or } 0.416\,7\pi\right]$

    (d) $\left[\dfrac{\pi}{3} \text{ or } 0.333\,3\pi\right]$ (e) $\left[\dfrac{7\pi}{3} \text{ or } 2.333\pi\right]$

    (f) $\left[\dfrac{37\pi}{6} \text{ or } 6.167\pi\right]$ (g) $[1.463\pi]$

2. Convert the following angles from degrees to radians correct
    to 4 significant figures: (a) 65°   (b) 34° 17′   (c) 111° 5′
    (d) 251° 52′   (e) 192° 15′   (f) 323° 47′ 15″

    (a) [1.134]   (b) [0.598 4]   (c) [1.939]   (d) [4.396]
    (e) [3.355]   (f) [5.651]

3. Convert the following angles from radians to degrees:

    (a) 2π (b) $\dfrac{\pi}{2}$ (c) $\dfrac{5\pi}{6}$ (d) $\dfrac{5\pi}{3}$ (e) $\dfrac{9\pi}{8}$ (f) $\dfrac{5\pi}{9}$

    (a) [360°]   (b) [90°]   (c) [150°]   (d) [300°]
    (e) [202½°]   (f) [100°]

4. Convert the following angles from radians to degrees, minutes
    and seconds: (a) 2.48   (b) 3.912   (c) 0.013 2   (d) 17.82
    (e) 5.75   (f) 7.325
    (a) [142° 5′ 37″]   (b) [224° 8′ 28″]   (c) [0° 45′ 23″]
    (d) [1 021° 0′ 39″]   (e) [329° 27′ 23″]   (f) [419° 41′ 30″]

5. Find the length of an arc of a circle of radius 11.46 cm when
    the angle subtended at the centre is 2.69 radians.
    [30.83 cm]

6. If the angle subtended at the centre of a circle of diameter
    9.80 cm is 1.79 radians find the lengths of the minor and major
    arcs.   [8.771 cm, 22.02 cm]

7. Find the length of the radius and circumference of a circle if
    an arc length of 21.62 cm subtends an angle of 4.86 radians.
    [4.449 cm, 27.95 cm]

8. An arc subtends an angle of 1.92 radians at the centre of a
    circle of radius 13.2 cm. Find the length of the arc.
    [25.34 cm]

9. If an arc of length 3.25 m subtends an angle of 142° 15′ at the centre of a circle find the radius and circumference of the circle.
    [1.309 m, 8.226 m]

10. Find the angle in (a) radians, (b) degrees, subtended at the centre of a circle of diameter 6.20 cm by an arc of length.
    10.8 cm   (a) [3.484°]   (b) [199.6°]

11. 150 mm of a belt drive is in contact with a pulley of diameter 200 mm. Determine the angle of lap.
    [1.5 rad or 85° 57′]

Area of sector of a circle

12. Calculate the areas of the following sectors of circles:
    (a) radius 42.0 mm, angle subtended at centre 42°
    (b) radius 15.20 cm, angle subtended at centre 102° 17′
    (c) radius 4.80 cm, angle subtended at centre 1.21 radians.
    (a) [646.6 mm²]   (b) [206.3 cm²]   (c) [13.94 cm²]

13. The area of a sector of a circle is 530 mm². If the radius of the circle is 32 mm, determine the angle, in radians and in degrees and minutes, subtended at the centre of the circle by the sector.   [1.035 rad; 59° 19′]

14. Determine the diameter and circumference of a circle if the area of a sector of subtended angle 1.25 rad is 400 mm².
    [50.60 mm; 159.0 mm]

15. Determine the value of the shaded area in Fig. 12.3. Find also the percentage of the whole sector that the shaded area represents.   [280 mm²; 43.75%]

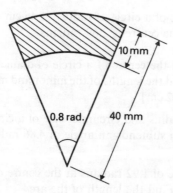

10 mm

0.8 rad.   40 mm

Fig. 12.3

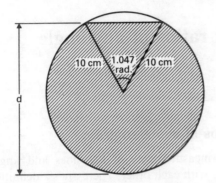

Fig. 12.4

16. Figure 12.4 shows a cross-section through a circular water
     container. The shaded area represents the water in the
     container. Determine (a) the depth d, (b) the area of the
     shaded portion and (c) the area of the unshaded portion.
     (d) Express the unshaded area as a percentage of the whole
     area. (e) If the container is 50 cm long how many litres of water
     are in the container? (1 litre = 1 000 cm³.)
     (a) [18.66 cm]   (b) [305.1 cm²]   (c) [9.06 cm²]
     (d) [2.88%]   (e) [15.26 l]

Chapter 13

Trigonometric ratios of any angle

13.1 Trigonometric ratios of acute angles

Trigonometry deals with the measurements of the sides and angles of triangles and their relationships with each other. There are six definitions of trigonometric ratios which apply to any right-angled triangle. Consider a right-angled triangle ABC shown in Fig. 13.1, and let the angle CAB be denoted by θ. With reference to angle θ, the side BC is called the opposite side, the side AB the adjacent side and the side AC the hypotenuse.

Then by definition, natural sine θ $\quad = \dfrac{\text{opposite side}}{\text{hypotenuse}}$

natural cosine θ $\quad = \dfrac{\text{adjacent side}}{\text{hypotenuse}}$

natural tangent θ $\quad = \dfrac{\text{opposite side}}{\text{adjacent side}}$

natural secant θ $\quad = \dfrac{\text{hypotenuse}}{\text{adjacent side}}$

natural cosecant θ $\quad = \dfrac{\text{hypotenuse}}{\text{opposite side}}$

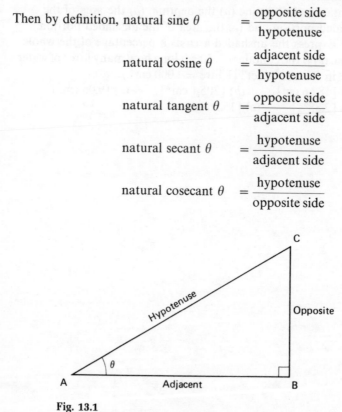

Fig. 13.1

and $\qquad$ natural cotangent $\theta = \dfrac{\text{adjacent side}}{\text{opposite side}}$

Using accepted abbreviations and referring to Fig. 13.1:

$$\sin \theta = \frac{BC}{AC} \qquad \sec \theta = \frac{AC}{AB}$$

$$\cos \theta = \frac{AB}{AC} \qquad \text{cosec } \theta = \frac{AC}{BC}$$

$$\tan \theta = \frac{BC}{AB} \qquad \cot \theta = \frac{AB}{BC}$$

Now $\qquad \dfrac{\sin \theta}{\cos \theta} = \dfrac{\dfrac{BC}{AC}}{\dfrac{AB}{AC}} = \dfrac{BC}{AC} \cdot \dfrac{AC}{AB} = \dfrac{BC}{AB}$

But $\qquad \tan \theta = \dfrac{BC}{AB} \qquad$ **Hence** $\tan \theta = \dfrac{\sin \theta}{\cos \theta}$

Also $\qquad \dfrac{\cos \theta}{\sin \theta} = \dfrac{\dfrac{AB}{AC}}{\dfrac{BC}{AC}} = \dfrac{AB}{AC} \cdot \dfrac{AC}{BC} = \dfrac{AB}{BC}$

But $\qquad \cot \theta = \dfrac{AB}{BC} \qquad$ **Hence** $\cot \theta = \dfrac{\cos \theta}{\sin \theta}$

As $\qquad \sin \theta = \dfrac{BC}{AC}$ and $\text{cosec } \theta = \dfrac{AC}{BC}$

then $\qquad$ **cosec** $\theta = \dfrac{1}{\sin \theta} \left(\text{or } \sin \theta = \dfrac{1}{\text{cosec } \theta} \right)$

Similarly, as $\cos \theta = \dfrac{AB}{AC}$ and $\sec \theta = \dfrac{AC}{AB}$

then $\qquad$ **sec** $\theta = \dfrac{1}{\cos \theta} \left(\text{or } \cos \theta = \dfrac{1}{\sec \theta} \right)$

Similarly, as $\tan \theta = \dfrac{BC}{AB}$ and $\cot \theta = \dfrac{AB}{BC}$

then $\qquad$ **cot** $\theta = \dfrac{1}{\tan \theta} \left(\text{or } \tan \theta = \dfrac{1}{\cot \theta} \right)$

Thus cosecant, secant and cotangents are often called the **'reciprocal trigonometric ratios'**.

Calculators provide a quick method for determining the trigonometric ratios, as shown in *Technician Mathematics 1*. Many calculators contain only the sine, cosine and tangent functions. Thus, for example, to evaluate sec 42°, initially cos 42° is found (i.e. 0.743 144 83) and then the reciprocal taken, giving 1.345 6, correct to 4 decimal places.

Use your calculator to check the following worked problems.

Worked problems on trigonometric ratios of acute angles

Problem 1. For the triangle ABC shown in Fig. 13.2 find the values of the six trigonometric ratios — sin θ, cos θ, tan θ, sec θ, cosec θ and cot θ, each correct to 4 decimal places.

By definition: $\sin \theta = \dfrac{\text{opposite side}}{\text{hypotenuse}} = \dfrac{AB}{AC} = \dfrac{8}{17} = \mathbf{0.470\,6}$

$\cos \theta = \dfrac{\text{adjacent side}}{\text{hypotenuse}} = \dfrac{BC}{AC} = \dfrac{15}{17} = \mathbf{0.882\,4}$

$\tan \theta = \dfrac{\text{opposite side}}{\text{adjacent side}} = \dfrac{AB}{BC} = \dfrac{8}{15} = \mathbf{0.533\,3}$

$\sec \theta = \dfrac{\text{hypotenuse}}{\text{adjacent side}} = \dfrac{AC}{BC} = \dfrac{17}{15} = \mathbf{1.133\,3}$

$\operatorname{cosec} \theta = \dfrac{\text{hypotenuse}}{\text{opposite side}} = \dfrac{AC}{AB} = \dfrac{17}{8} = \mathbf{2.125\,0}$

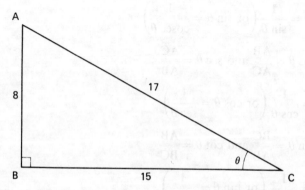

Fig. 13.2

$$\cot \theta = \frac{\text{adjacent side}}{\text{opposite side}} = \frac{BC}{AB} = \frac{15}{8} = 1.8750$$

Problem 2. If $\sin \theta = \dfrac{9}{41}$ find the value of the other five trigonometric ratios.

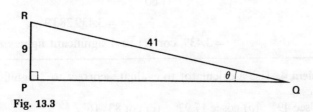

Fig. 13.3

Figure 13.3 shows a triangle PQR with angle PQR equal to θ. If $\sin \theta = \dfrac{9}{41}$ then the length PR = 9 units and the hypotenuse RQ = 41 units. To find the other trigonometric ratios it is necessary to find the value of the third side of the triangle, PQ

From the theorem of Pythagoras: $RQ^2 = PR^2 + PQ^2$

$$41^2 = 9^2 + PQ^2$$
$$PQ^2 = 41^2 - 9^2$$
$$PQ = \sqrt{(41^2 - 9^2)}$$
$$= \sqrt{(1681 - 81)}$$
$$= \sqrt{1600}$$
$$= 40 \text{ units}$$

Hence $\cos \theta = \dfrac{40}{41}$

$$\tan \theta = \frac{9}{40} \qquad\qquad \csc \theta = \frac{1}{\sin \theta} = \frac{41}{9} = 4\tfrac{5}{9}$$

$$\sec \theta = \frac{1}{\cos \theta} = \frac{41}{40} = 1\tfrac{1}{40} \qquad \cot \theta = \frac{1}{\tan \theta} = \frac{40}{9} = 4\tfrac{4}{9}$$

Problem 3. Evaluate, correct to 4 significant figures:
(a) sin 51.42°   (b) cos 12° 52′   (c) tan 73° 47′ 16″

(a) sin 51.42° = 0.781 738 19 . . . = **0.7817**, correct to 4 significant figures

(b) $\cos 12° 52' = \cos 12\dfrac{52°}{60} = \cos 12.86\dot{6} = 0.974\,890\,91\ldots$

= **0.9749**, correct to 4 significant figures

(c) $\tan 73° 47' 16'' = \tan 73° 47\dfrac{16'}{60} = \tan 73° 47.26\dot{6}'$

$$= \tan\left(73\dfrac{47.26\dot{6}}{60}\right)° = \tan 73.787\,\dot{7}$$

$$= 3.439\,283\,9\ldots$$

$$= \textbf{3.439}, \text{ correct to 4 significant figures}$$

Problem 4. Use a calculator to evaluate, correct to 5 significant figures:

(a)   sec 49°   (b) cosec 17.92°   (c) cot 83° 16'

(a) $\sec 49° = \dfrac{1}{\cos 49°} = 1.524\,253\,08\ldots$

$$= \textbf{1.524\,3}, \text{ correct to 5 significant figures}$$

(b) $\text{cosec } 17.92° = \dfrac{1}{\sin 17.92°} = 3.250\,037\,38\ldots$

$$= \textbf{3.250\,0}, \text{ correct to 5 significant figures}$$

(c) $\cos 83° 16' = \dfrac{1}{\tan 83\dfrac{16°}{60}} = 0.118\,062\,84\ldots$

$$= \textbf{0.118\,06}, \text{ correct to 5 significant figures}$$

Problem 5. Evaluate, correct to 4 significant figures:

(a) sin 1.374   (b) $\cos\dfrac{\pi}{6}$ (c) tan 1.19

(a)   'sin 1.374' means 'the sine of 1.374 radians'. Hence a calculator needs to be set on the radian function. Hence sin 1.374 = **0.9807**, correct to 4 significant figures.

(b) $\cos\dfrac{\pi}{6} = \cos 0.523\,598\,77\ldots = \textbf{0.8660}$, correct to 4 significant figures

(Since 2π radians = 360° and π rad = 180°, then $\dfrac{\pi}{6}$ rad = 30° and cos 30° = 0.8660 as above)

(c)   tan 1.19 = 2.497 899 38 ... = **2.498**, correct to 4 significant figures

Problem 6. Evaluate, correct to 4 decimal places:

(a) sec 0.789 (b) cosec $\dfrac{3\pi}{8}$ (c) cot $\dfrac{7\pi}{18}$

(a) Again, with no degrees sign, it is assumed that 0.789 means 0.789 radians.

$$\sec 0.789 = \frac{1}{\cos 0.789} = 1.419\,334\,97\ldots$$

$$= \mathbf{1.419\,3}, \text{ correct to 4 decimal places}$$

(b) $\cosec \dfrac{3\pi}{8} = \dfrac{1}{\sin \dfrac{3\pi}{8}} = \dfrac{1}{\sin 1.178\,097\,24\ldots}$

$$= \mathbf{1.082\,4}, \text{ correct to 4 decimal places}$$

(c) $\cot \dfrac{7\pi}{18} = \dfrac{1}{\tan \dfrac{7\pi}{18}} = \dfrac{1}{\tan 1.221\,730\,47\ldots}$

$$= \mathbf{0.364\,0}, \text{ correct to 4 decimal places}$$

Problem 7. Determine the acute angles:
(a) arcsin 0.452 9 (b) arccos 0.739 2 (c) arctan 1.539 2

(a) 'arcsin θ' is an abbreviation for 'the angle whose sine is equal to θ'. 0.452 9 is entered into the calculator and then the inverse sine (or $\sin^{-1}$) key is pressed.
Hence arcsin 0.4529 = 26.929 897 7 ...

Subtracting 26 leaves 0.929 897 7 ...°

Multiplying by 60 gives 56', correct to the nearest minute. (Many calculators possess a function which changes from decimal to degrees, minutes and seconds.)
Hence arcsin 0.452 9 = **26.93°** or **26° 56′**

Alternatively, in radians,

arcsin 0.452 9 = **0.470 radians**

(b) arccos 0.739 2 = **42.34°** or **45° 20′** or **0.739 radians**

(c) arctan 1.539 2 = **56.99°** or **56° 59′** or **0.995 radians**

Problem 8. Determine the acute angles:
(a) arcsec 2.174 8 (b) arccosec 1.291 6 (c) arccot 3.176 2

(a) arcsec 2.174 8 = $\arccos\left(\dfrac{1}{2.174\,8}\right)$ = arccos 0.459 812 ...

$$= \mathbf{62.62°} \text{ or } \mathbf{62° 37′} \text{ or } \mathbf{1.093 \text{ radians}}$$

(b)   arccosec $1.2916 = \arcsin\left(\dfrac{1}{1.2916}\right) = \arcsin 0.7742335\ldots$

$$= \mathbf{50.74°} \text{ or } \mathbf{50°\ 44'} \text{ or } \mathbf{0.886\ radians}$$

(c)   arccot $3.1762 = \arctan\left(\dfrac{1}{3.1762}\right) = \arctan 0.3148416\ldots$

$$= \mathbf{17.48°} \text{ or } \mathbf{17°\ 29'} \text{ or } \mathbf{0.305\ radians}$$

Problem 9. Given sec $\theta = 1.4723$, where θ is an acute angle, determine cosec θ and cot θ.

If sec $\theta = 1.4723$ then $\theta = $ arcsec $1.4723 = \arccos\left(\dfrac{1}{1.4723}\right)$

i.e. $\theta = 47.218\,106\ldots°$

Hence cosec $\theta = $ cosec $47.218\,106\ldots° = \dfrac{1}{\sin 47.218\,106\ldots°}$

$$= \mathbf{1.3625},$$
correct to 4 decimal places

and cot $\theta = $ cot $47.218\,106\ldots° = \dfrac{1}{\tan 47.218\,106\ldots°}$

$$= \mathbf{0.9254},$$
correct to 4 decimal places

Problem 10. Evaluate, correct to 3 significant figures:

$$\frac{5 \sec 29°\ 10' - 3 \cot 14°\ 21'}{2 \operatorname{cosec} 64°\ 8' \tan 23°\ 17'}$$

By calculator, sec $29°\ 10' = 1.1452\ldots$, cot $14°\ 21' = 3.9089\ldots$, cosec $64°\ 8' = 1.1113\ldots$, tan $23°\ 17' = 0.4303\ldots$

Hence $\dfrac{5 \sec 29°\ 10' - 3 \cot 14°\ 21'}{2 \operatorname{cosec} 64'\ 8' \tan 23'\ 17'} = \dfrac{5(1.1452) - 3(3.9089)}{2(1.1113)(0.4303)}$

$$= \mathbf{-6.26},$$
correct to 3 significant figures

Problem 11. The angle of elevation from a given point, of the top of a tower which stands on horizontal ground is 22°.

From a point 120 m nearer to the tower the angle of elevation is 44°. Find the height of the tower.

The tower BC and the angles of elevation are shown in Fig. 13.4.

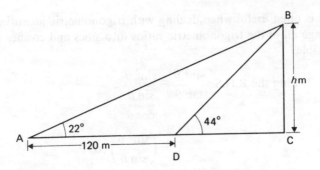

Fig. 13.4

Let $BC = h$

Then $AC = h \cot 22°$

  $DC = h \cot 44°$

and $AD = AC - DC$

  $120 = h \cot 22° - h \cot 44°$

  $= h(\cot 22° - \cot 44°)$

  $= h(2.475\,1 - 1.035\,5)$

  $= 1.439\,6h$

  $h = \dfrac{120}{1.439\,6} = 83.36$

Hence the height of the tower is 83.36 m

Problem 12. Prove the following trigonometric identities:

(a) $\sin \theta \cos \theta = \dfrac{\sin^2 \theta}{\tan \theta}$ (b) $\sin \theta \sec \theta = \tan \theta$

(c) $\dfrac{(\operatorname{cosec} \theta + \cot \theta) \tan \theta}{(\tan \theta + \sec \theta)} = \dfrac{\cos \theta + 1}{\sin \theta + 1}$

A trigonometric 'identity' is an expression that is true for all values of the unknown variable. This is different to an 'equation' which is true only for a certain value or values of the variable. Hence expressions such as

$\operatorname{cosec} \theta = \dfrac{1}{\sin \theta}$ or $\sec \theta = \dfrac{1}{\cos \theta}$ or $\tan \theta = \dfrac{\sin \theta}{\cos \theta}$

are examples of trigonometric identities.

  With trigonometric identities such as in (a), (b) and (c) above it is necessary to start with the left-hand side (L.H.S.) and attempt to make it equal to the right-hand side (R.H.S.) or vice versa.

It is often useful when dealing with trigonometric identities to change all of the trigonometric ratios into sines and cosines where possible.

(a)   Taking the R.H.S.: $\dfrac{\sin^2 \theta}{\tan \theta} = \dfrac{\sin^2 \theta}{\dfrac{\sin \theta}{\cos \theta}}$

$$= \left(\dfrac{\sin^2 \theta}{\sin \theta}\right) \cos \theta$$

$$= \sin \theta \cos \theta$$

$$= \text{L.H.S.}$$

(b)   Taking the L.H.S.: $\sin \theta \sec \theta = (\sin \theta)\left(\dfrac{1}{\cos \theta}\right)$

$$= \dfrac{\sin \theta}{\cos \theta} = \tan \theta$$

$$= \text{R.H.S.}$$

(c) $\operatorname{cosec} \theta = \dfrac{1}{\sin \theta}$, $\cot \theta = \dfrac{\cos \theta}{\sin \theta}$, $\tan \theta = \dfrac{\sin \theta}{\cos \theta}$ and $\sec \theta = \dfrac{1}{\cos \theta}$

Taking the L.H.S.:

$$\dfrac{(\operatorname{cosec} \theta + \cot \theta)}{\tan \theta + \sec \theta} \tan \theta = \dfrac{\left(\dfrac{1}{\sin \theta} + \dfrac{\cos \theta}{\sin \theta}\right) \dfrac{\sin \theta}{\cos \theta}}{\left(\dfrac{\sin \theta}{\cos \theta} + \dfrac{1}{\cos \theta}\right)}$$

Adding the fraction contained in the brackets gives:

$$\text{L.H.S.} = \dfrac{\left(\dfrac{\cos \theta + 1}{\sin \theta}\right) \dfrac{\sin \theta}{\cos \theta}}{\left(\dfrac{\sin \theta + 1}{\cos \theta}\right)}$$

$$= \left(\dfrac{\cos \theta + 1}{\sin \theta}\right)\left(\dfrac{\cos \theta}{\sin \theta + 1}\right) \dfrac{\sin \theta}{\cos \theta}$$

Cancelling gives: $\text{L.H.S.} = \dfrac{\cos \theta + 1}{\sin \theta + 1} = \text{R.H.S.}$

Further problems on trigonometric ratios of acute angles may be found in Section 13.3 (Problems 1–26), page 233.

13.2 Values of the six trigonometric ratios for angles of any magnitude

So far only acute angles (that is angles $<90°$) have been considered. However, it is often necessary to find the value of trigonometric ratios of angles greater than 90°, such as $\sin 261°$ or $\tan 302°$ or $\sec 169°$. Figure 13.5 shows rectangular axes XX' and YY' intersecting at origin O. From a knowledge of graphs it is the adopted convention that measurements made to the right of the origin (i.e. to the right of YY') are positive and measurements to the left negative. Similarly measurements made upwards from the origin (i.e. above XX') are positive and measurements made below are negative.

Let the line OA be free to rotate about O. If movement is anticlockwise angular measurement is considered as positive and if movement is clockwise it is considered as negative. The axis OX represents 0°. In one revolution anticlockwise there are four quadrants encompassed and these are numbered 1 to 4 as shown in Fig. 13.5. Hence the first quadrant spans 0° to 90°, the second quadrant 90° to 180° and so on.

Let OA be rotated such that angle θ_1 is any angle in the first quadrant as shown in Fig. 13.6. If the line AB is constructed perpendicular to OX a right-angled triangle OAB is produced. The length OB and the length AB are both positive values and therefore **each of the six trigonometric ratios is positive in the first quadrant.** (Note that the rotating arm OA is always considered as a positive value.)

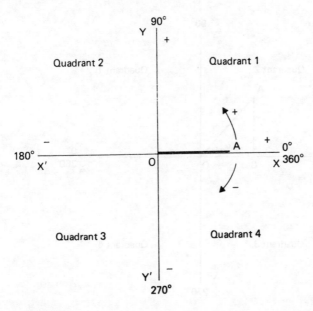

Fig. 13.5

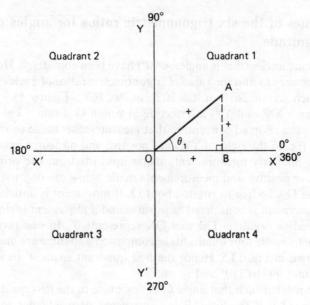

Fig. 13.6

In the three cases which follow, the trigonometric ratios of $\angle AOX$ are defined as those of the related acute angles. For example in Fig. 13.7, $\sin \angle AOX = \sin(180 - \angle AOX') = \sin \theta_2$

Let the line OA now be rotated so that the angle θ_2 is any angle in the

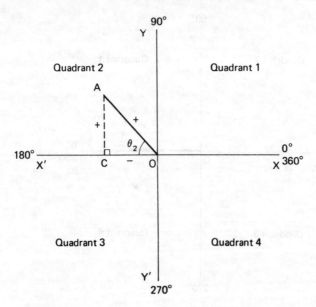

Fig. 13.7

second quadrant (see Fig. 13.7). If the line AC is constructed perpendicular to OX' a right-angled triangle OAC is produced. The length OC is negative and the length AC positive.

Hence, in the second quadrant, and considering only the sign of the trigonometric ratio:

$$\sin \theta_2 \quad = \frac{AC}{AO} = \frac{+}{+} = \text{positive value}$$

$$\cos \theta_2 \quad = \frac{OC}{AO} = \frac{-}{+} = \text{negative value}$$

$$\tan \theta_2 \quad = \frac{AC}{OC} = \frac{+}{-} = \text{negative value}$$

$$\sec \theta_2 \quad = \frac{AO}{OC} = \frac{+}{-} = \text{negative value}$$

$$\operatorname{cosec} \theta_2 = \frac{AO}{AC} = \frac{+}{+} = \text{positive value}$$

$$\cot \theta_2 \quad = \frac{OC}{AC} = \frac{-}{+} = \text{negative value}$$

The last three results may be obtained from the reciprocal ratios. That is, if $\sin \theta_2$ is positive then the reciprocal ratio, $\operatorname{cosec} \theta_2$, is also positive. Similarly if $\cos \theta_2$ is negative it follows that $\sec \theta_2$ is negative and if $\tan \theta_2$ is negative then $\cot \theta_2$ is negative. From here on this will be assumed.

For any angle lying in the second quadrant, for example 150°, $\sin 150°$ is the same as $+\sin 30°$ and $\cos 120°$ is the same as $-\cos 60°$ and so on.

Let the line OA now be rotated so that the angle θ_3 is any angle in the third quadrant (see Fig. 13.8).

If the line AD is constructed perpendicular to OX' a right-angled triangle OAD is produced. The length of OD is negative and the length of AD negative.

Hence, in the third quadrant:

$$\sin \theta_3 \quad = \frac{AD}{AO} = \frac{-}{+} = \text{negative value}$$

$$\therefore \quad \operatorname{cosec} \theta_3 = \text{negative value}$$

$$\cos \theta_3 \quad = \frac{OD}{AO} = \frac{-}{+} = \text{negative value}$$

$$\therefore \quad \sec \theta_3 \quad = \text{negative value}$$

$$\tan \theta_3 \quad = \frac{AD}{OD} = \frac{-}{-} = \text{positive value}$$

$$\therefore \quad \cot \theta_3 \quad = \text{positive value}$$

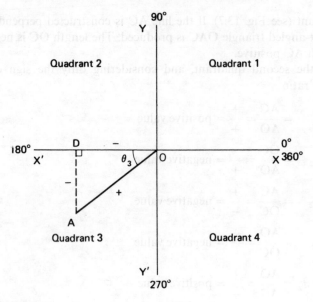

Fig. 13.8

Thus $\tan 240°$ is the same as $+\tan 60°$ and $\operatorname{cosec} 210°$ is the same as $-\operatorname{cosec} 30°$.

Let the line OA now be rotated so that the angle θ_4 is any angle in the fourth quadrant (see Fig. 13.9). If the line AE is constructed perpendicular

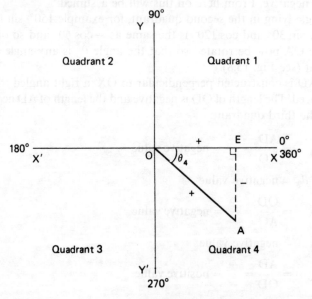

Fig. 13.9

to OX a right-angled triangle OAE is produced. The length OE is positive
and the length AE is negative.

Hence in the fourth quadrant:

$$\sin \theta_4 \quad = \frac{AE}{OA} = \frac{-}{+} = \text{negative value}$$

$\therefore \quad \operatorname{cosec} \theta_4 = \text{negative value}$

$$\cos \theta_4 \quad = \frac{OE}{OA} = \frac{+}{+} = \text{positive value}$$

$\therefore \quad \sec \theta_4 \quad = \text{positive value}$

$$\tan \theta_4 \quad = \frac{AE}{OE} = \frac{-}{+} = \text{negative value}$$

$\therefore \quad \cot \theta_4 \quad = \text{negative value}$

Considering only positive values of sine, cosine and tangent the above
results may be summarised as shown in Fig. 13.10. If the first letter of *A*ll,
*S*ine, *T*angent and *C*osine is used the word CAST is produced starting in
the fourth quadrant and moving in an anticlockwise direction. This provides
a quick check as to the sign of a trigonometric ratio in any quadrant.

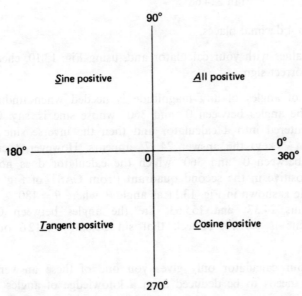

Fig. 13.10

Naturally, a calculator is able to determine the value of the trigonometric ratio of any angle directly. For example:

$$\sin 179° = 0.0175 \qquad\qquad \sin 249° \quad = -0.9336$$

$$\cos 112° = -0.3746 \qquad\qquad \cos 305.21° = 0.5766$$

$$\tan 164° = -0.2867 \qquad\qquad \tan 279.42° = -6.0275$$

$$\sec 98° \qquad = \frac{1}{\cos 98°} \quad = -7.1853$$

$$\sec 205.43° \quad = \frac{1}{\cos 205.43°} = -1.1073$$

$$\text{cosec } 125° \quad = \frac{1}{\sin 125°} = 1.2208$$

$$\text{cosec } 341.92° = \frac{1}{\sin 341.92°} = -3.2222$$

$$\cot 147° \qquad = \frac{1}{\tan 147°} = 1.5399$$

and

$$\cot 224.68° \quad = \frac{1}{\tan 224.68°} = -1.0112$$

each correct to 4 decimal places.

Check these values with your calculator and, using Fig. 13.10, check that each has the correct sign.

A knowledge of angles of any magnitude is needed when finding, for example, all the angles between $0°$ and $360°$ whose sine is, say, 0.4126. If 0.4126 is entered into a calculator and then the inverse sine key is pressed (or $\sin^{-1}$ key) the answer $24.37°$ appears. However, there is a second angle between $0°$ and $360°$ which the calculator does not give. Sine is also positive in the second quadrant (from CAST of Fig. 13.10). The other angle is shown in Fig. 13.11 as angle θ, where $\theta = 180° - 24.37°$ $= 155.63°$. Thus $24.37°$ **and** $155.63°$ are the angles between $0°$ and $360°$ whose sine is 0.4126. (Check that $\sin 155.63° = 0.4126$ on your calculator.)

Be careful! Your calculator only gives you one of these answers. The second answer needs to be deduced from a knowledge of angles of any magnitude.

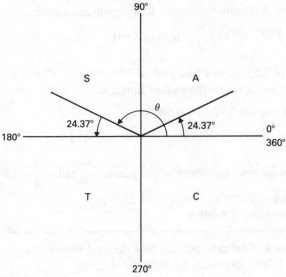

Fig. 13.11

Worked problems on angles of any magnitude

Problem 1. Use a calculator to evaluate, correct to 4 significant figures:
(a) sin 247°   (b) cos 141.37°   (c) tan 311° 17′

(a)   sin 247° = **−0.920 5**
(b)   cos 141.37° = **−0.781 2**

(c) $\tan 311° 17' = \tan 311\dfrac{17°}{60} = \tan 311.283\dot{3} = \mathbf{-1.139}$

Problem 2. Use a calculator to evaluate, correct to 5 significant figures:
(a) sec 163°   (b) cosec 129.73°   (c) cot 321° 43′

(a) $\sec 163° = \dfrac{1}{\cos 163°} = \mathbf{-1.045\,7}$

(b) $\operatorname{cosec} 129.73° = \dfrac{1}{\sin 129.73°} = \mathbf{1.300\,3}$

(c) $\cot 321° 43' = \dfrac{1}{\tan 321\dfrac{43°}{60}} = \dfrac{1}{\tan 321.716\dot{6}}$

$= \mathbf{-1.267\,0}$

Problem 3. Evaluate, correct to 4 significant figures:

(a) sin 3.72   (b) $\cos\dfrac{5\pi}{4}$ (c) tan 3.681

(a) 'sin 3.72' means 'the sine of 3.72 radians'. Hence a calculator needs to be on the radian function.

Hence sin 3.72 = **−0.546 7**

(b) $\cos\dfrac{5\pi}{4} = \cos 3.926\,990\,81\ldots = $ **−0.707 1**

(Note that π radians = 180°, hence $\dfrac{5\pi}{4}$ rad $= \dfrac{5}{4} \times 180° = 225°$

and cos 225° = −0.707 1 as above)

(c) tan 3.681 = **0.598 6**

Problem 4. Evaluate, correct to 4 decimal places:
(a) sec 1.26π (b) cosec 2.189   (c) cot 2.971

(a) sec 1.26π = sec 3.958 406 74 … $= \dfrac{1}{\cos 3.958\,406\,74\ldots}$

$= $ **−1.460 8**

(b) cosec 2.189 $= \dfrac{1}{\sin 1.289} = $ **1.227 1**

(c) cot 2.971 $= \dfrac{1}{\tan 2.971} = $ **−5.804 9**

Problem 5. Evaluate sec 483° 46′.

The angle 483° 46′ is greater than one revolution. In fact 483° 46′ = 360° + 123° 46′. Then, referring to Fig. 13.12, the angle 483° 46′ is equivalent to 123° 46′.

By calculator, sec 123° 46′ $= \dfrac{1}{\cos 123°\ 46′} = $ −1.799 2

Hence sec 483° 46′ = **−1.799 2**

Problem 6. Evaluate, correct to 5 significant figures:
(a) sin(−114°)   (b) cosec(−185° 10′)   (c) cot(−72.37°)

(a) Positive angles are shown on cartesian axes anticlockwise and negative angles are shown clockwise. From Fig. 13.13, −114° is actually the same as +246° (i.e. 360 − 114°). Hence by calculator, sin(−114°) = sin 246° = **−0.913 55**

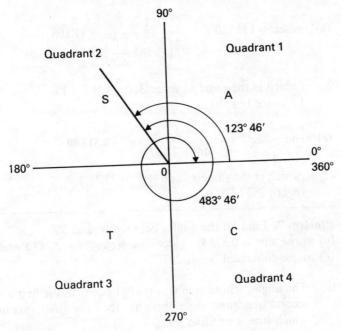

Fig. 13.12

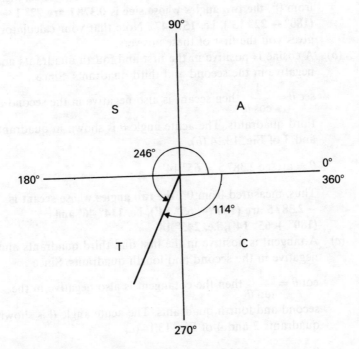

Fig. 13.13

(b) $\cosec(-185° \, 10') = \dfrac{1}{\sin\left(-185\dfrac{10°}{60}\right)} = \mathbf{11.105}$

(which is the same as $\cosec(360° - 185° \, 10')$,
i.e. $\cosec 174° \, 50'$)

(c) $\cot(-72.37°) = \dfrac{1}{\tan(-72.37°)} = \mathbf{-0.317\,80}$

(which is the same as $\cot(360° - 72.37°)$,
i.e. $\cot 287.63°$)

Problem 7. Find all the angles between 0° and 360°
(a) whose sine is 0.378 1,   (b) whose secant is $-2.387\,5$ and
(c) whose cotangent is $-1.237\,7$

(a)   The angles whose sine is $+0.378\,1$ occur in the first and
second quadrants since these are the only quadrants in
which sine is positive.

In Fig. 13.14 (a), $\theta = \arcsin 0.378\,1 = 22° \, 13'$. Thus, measured
from 0°, **the two angles whose sine is 0.378 1 are 22° 13′ and**
(180° − 22° 13′), i.e. 157° 47′. Note that your calculator only
gives you the first of these answers.

(b)   A cosine is positive in the first and fourth quadrants and
negative in the second and third quadrants. Since

$\sec \theta = \dfrac{1}{\cos \theta}$ then secant is also negative in the second and

third quadrants. The acute angle θ is shown in quadrants 2
and 3 of Fig. 13.14 (b).

$\theta = \arcsec 2.387\,5 = 65° \, 14'$

Thus, measured from 0°, **the two angles whose secant is**
−2.387 5 are (180° − 65° 14′), i.e. 114° 46′ and
(180° + 65° 14′), i.e. 245° 14′

(c)   A tangent is positive in the first and third quadrants and
negative in the second and fourth quadrants. Since

$\cot \theta = \dfrac{1}{\tan \theta}$ then the cotangent is also negative in the

second and fourth quadrants. The acute angle θ is shown in
quadrants 2 and 4 of Fig. 13.14 (c)

$\theta = \arccot 1.237\,7 = 38° \, 56'$

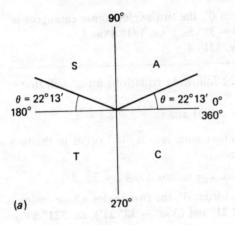

(a)

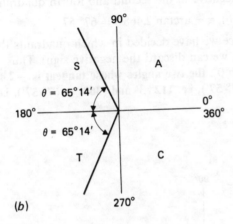

(b)

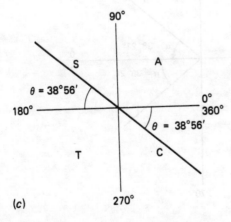

(c)

Fig. 13.14

Thus, measured from 0°, **the two angles whose cotangent is −1.237 7 are (180° − 38° 56′), i.e. 141° 4′ and (360° − 38° 56′), i.e. 321° 4′**

Problem 8. Solve the following equations for α such that $0° < \alpha < 360°$:

(a) arccos 0.784 2 = α (b) arctan −2.468 3 = α

(a)   The angles whose cosine is +0.784 2 occur in the first and fourth quadrants.

In Fig. 13.15 (a), α = arccos 0.784 2 = 38° 21′

Thus, measured from 0°, **the two angles whose cosine is 0.784 2 are 38° 21′ and (360° − 38° 21′), i.e. 321° 39′**

(b)   A tangent is negative in the second and fourth quadrants.

In Fig. 13.15 (b), α = arctan 2.468 3 = 67° 57′

(Note that once we have decided in which quadrants the angle appears, we can discard the negative sign). Thus, measured from 0°, **the two angles whose tangent is −2.468 3 are (180° − 67° 57′), i.e. 112° 3′ and (360° − 67° 57′), i.e. 292° 3′**

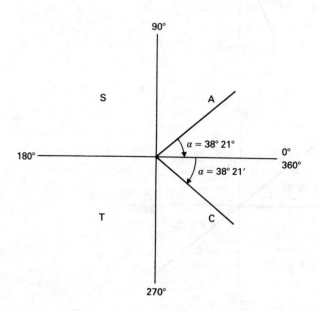

Fig. 13.15 (a)

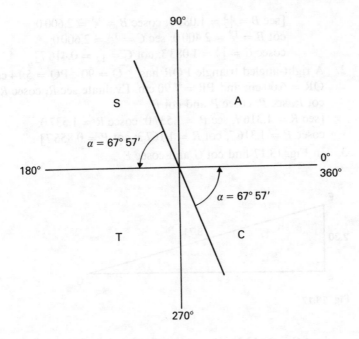

Fig. 13.15 (b)

It may be seen from the last two problems that there are always **two** angles between 0° and 360° that satisfy a particular value of a trigonometric ratio.

Further problems on angles of any magnitude may be found in the following Section (13.3) (Problems 27–39), page 237.

13.3 Further problems

Trigonometric ratios of acute angles

1. In the triangle ABC shown in Fig. 13.16 find sec B, cosec B, cot B, sec C, cosec C and cot C

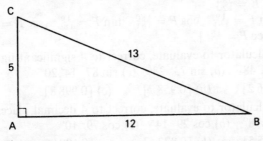

Fig. 13.16

$[\sec B = \frac{13}{12} = 1.0833; \operatorname{cosec} B = \frac{13}{5} = 2.6000;$
$\cot B = \frac{12}{5} = 2.4000;\ \sec C = \frac{13}{5} = 2.6000;$
$\operatorname{cosec} C = \frac{13}{12} = 1.0833; \cot C = \frac{5}{12} = 0.4167]$

2. A right-angled triangle PQR has $\angle Q = 90°$, PQ = 5.14 cm, QR = 6.00 cm and PR = 7.90 cm. Evaluate sec R, cosec R, cot R, sec P, cosec P and cot P
    $[\sec R = 1.3167;\ \sec P = 1.5370;\ \operatorname{cosec} R = 1.5370;$
    $\operatorname{cosec} P = 1.3167;\ \cot R = 1.1673;\ \cot P = 0.8567]$

3. In Fig. 13.17 find cot G and cosec E

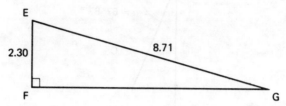

E

2.30   8.71

F   G

Fig. 13.17

$[\cot G = 3.6525; \operatorname{cosec} E = 1.0368]$

4. Find the other five trigonometric ratios, in fraction form, given the following:
    (a) $\sin A = \frac{7}{25}$ (b) $\cos B = \frac{112}{113}$ (c) $\tan C = \frac{12}{5}$
    (d) $\sec D = \frac{85}{84}$ (e) $\operatorname{cosec} E = \frac{37}{12}$ (f) $\cot F = \frac{180}{19}$
    (a) $[\cos A = \frac{24}{25}\quad \tan A = \frac{7}{24}\quad \sec A = \frac{25}{24}\quad \operatorname{cosec} A = \frac{25}{7}$
         $\cot A = \frac{24}{7}]$
    (b) $[\sin B = \frac{15}{113}\quad \tan B = \frac{15}{112}\quad \sec B = \frac{113}{112}\quad \operatorname{cosec} B = \frac{113}{15}$
         $\cot B = \frac{112}{15}]$
    (c) $[\sin C = \frac{12}{13}\quad \cos C = \frac{5}{13}\quad \sec C = \frac{13}{5}\quad \operatorname{cosec} C = \frac{13}{12}$
         $\cot C = \frac{5}{12}]$
    (d) $[\sin D = \frac{13}{85}\quad \cos D = \frac{84}{85}\quad \tan D = \frac{13}{84}\quad \operatorname{cosec} D = \frac{85}{13}$
         $\cot D = \frac{84}{13}]$
    (e) $[\sin E = \frac{12}{37}\quad \cos E = \frac{35}{37}\quad \tan E = \frac{12}{35}\quad \sec E = \frac{37}{35}$
         $\cot E = \frac{35}{12}]$
    (f) $[\sin F = \frac{19}{181}\quad \cos F = \frac{180}{181}\quad \tan F = \frac{19}{180}\quad \sec F = \frac{181}{180}$
         $\operatorname{cosec} F = \frac{181}{19}]$

5. Use a calculator to evaluate, correct to 4 significant figures:
    (a) $\sin 21.48°$ (b) $\sin 73° 28'$ (c) $\sin 87° 14' 20''$
    (a) $[0.3662]$ (b) $[0.9587]$ (c) $[0.9988]$

6. Use a calculator to evaluate, correct to 4 decimal places:
    (a) $\cos 9.78°$ (b) $\cos 29° 14'$ (c) $\cos 79° 10'$
    (a) $[0.9855]$ (b) $[0.8726]$ (c) $[0.1880]$

7. Use a calculator to evaluate, correct to 5 significant figures:
   (a) tan 15.78°   (b) tan 49° 11′   (c) tan 84° 52′
   (a) [0.282 59]   (b) [1.157 8]   (c) [11.132]

8. Use a calculator to evaluate the following, correct to 4 decimal places:
   (a) sec 8° 0′   (b) sec 12° 18′   (c) sec 26° 28′   (d) sec 41° 49′
   (e) sec 68° 13′   (f) sec 76° 15′   (g) sec 87° 4′
   (a) [1.009 8]   (b) [1.023 5]   (c) [1.117 1]   (d) [1.341 8]
   (e) [2.694 7]   (f) [4.207 2]   (g) [19.541 2]

9. Use a calculator to evaluate the following, correct to 5 significant figures:
   (a) cosec 43°   (b) cosec 26° 48′   (c) cosec 35° 40′
   (d) cosec 59° 51′   (e) cosec 83° 9′   (f) cosec 7° 52′
   (g) cosec 14° 11′
   (a) [1.466 3]   (b) [2.217 9]   (c) [1.715 1]
   (d) [1.156 5]   (e) [1.007 2]   (f) [7.306 3]
   (g) [4.081 2]

10. Use a calculator to evaluate the following, correct to 4 significant figures:
    (a) cot 32°   (b) cot 45° 36′   (c) cot 23° 19′   (d) cot 57° 52′
    (e) cot 82° 43′   (f) cot 2° 5′   (g) cot 11° 46′
    (a) [1.600]   (b) [0.979 3]   (c) [2.320]
    (d) [0.628 1]   (e) [0.127 8]   (f) [27.49]
    (g) [4.801]

11. Evaluate, correct to 4 significant figures:
    (a) $\sin \dfrac{\pi}{4}$ (b) cos 0.832   (c) tan 1.293
    (a) [0.707 1]   (b) [0.673 4]   (c) [35.07]

12. Evaluate, correct to 4 decimal places:
    (a) sec 1.121 5   (b) cosec 0.412   (c) $\tan \dfrac{5\pi}{24}$
    (a) [2.302 4]   (b) [2.497 2]   (c) [0.767 3]

13. Determine the acute angles:
    (a) arcsin 0.317 6   (b) arcsin 0.984 2   (c) arcsin 0.711 1
    (a) [18.52° or 18° 31′ or 0.323 rad]
    (b) [79.80° or 79° 48′ or 1.393 rad]
    (c) [45.32° or 45° 19′ or 0.791 rad]

14. Determine the acute angles which have a cosine of:
    (a) 0.129 3   (b) 0.549 2   (c) 0.987 4
    (a) [82.57° or 82° 34′ or 1.441 rad]
    (b) [56.69° or 56° 41′ or 0.989 rad]
    (c) [9.11° or 9° 6′ or 0.159 rad]

15. Determine the acute angles:
   (a) arctan 0.482 9   (b) arctan 0.892 1   (c) arctan 2.461 3
   (a) [25.78° or 25° 47' or 0.450 rad]
   (b) [41.74° or 41° 44' or 0.728 rad]
   (c) [67.89° or 67° 53' or 1.185 rad]

16. Determine the acute angles which have a secant of:
   (a) 1.148 2   (b) 2.683 2   (c) 16.084
   (a) [29.43° or 29° 26' or 0.514 rad]
   (b) [68.12° or 68° 7' or 1.189 rad]
   (c) [86.44° or 86° 26' or 1.509 rad]

17. Determine the acute angles:
   (a) arccosec 2.483 4   (b) arccosec 1.233 5   (c) arccosec 9.864 8
   (a) [23.75° or 23° 45' or 0.414 rad]
   (b) [54.16° or 54° 10' or 0.945 rad]
   (c) [5.82° or 5° 49' or 0.102 rad]

18. Determine the acute angles which have a cotangent of:
   (a) 3.684 1   (b) 1.117 2   (c) 0.016 2
   (a) [15.19° or 15° 11' or 0.265 rad]
   (b) [41.83° or 41° 50' or 0.730 rad]
   (c) [89.07° or 89° 4' or 1.555 rad]

19. If $R = 31° 51'$ and $S = 59° 46'$ evaluate the following correct to 4 significant figures:

   (a) $2 \sec R \cot S$ (b) $\dfrac{\operatorname{cosec} R + \sec S}{1 - \tan R \cos S}$ (c) $\dfrac{5 \cot S}{4 \sin R \operatorname{cosec} S}$

   (a) [1.372]   (b) [5.647]   (c) [1.193]

20. Evaluate the following correct to 4 significant figures:

   (a) $\dfrac{5 \sec 57° + 2 \operatorname{cosec} 41°}{\cot 82°}$

   (b) $\dfrac{(12 \cot 41° - 3 \sin 17°) \operatorname{cosec} 19°}{5 \tan 36° \cot 49°}$

   (a) [87.04]   (b) [12.57]

21. Given $\tan \theta = 1.468 3$ find $\sec \theta$, $\operatorname{cosec} \theta$ and $\cot \theta$
   [$\sec \theta = 1.776 5$; $\operatorname{cosec} \theta = 1.209 9$; $\cot \theta = 0.681 1$]

22. If $\sin \theta = 0.535 8$ and $\cos \theta = 0.844 3$ find the values of $\tan \theta$ and $\cot \theta$ [$\tan \theta = 0.634 6$; $\cot \theta = 1.575 8$]

23. Given that $\sin 40° = 0.642 8$ and $\cos 40° = 0.766 0$ find the values of $\sec 40°$, $\operatorname{cosec} 40°$, $\tan 40°$ and $\cot 40°$
   [$\sec 40° = 1.305 5$; $\operatorname{cosec} 40° = 1.555 7$; $\tan 40° = 0.839 2$; $\cot 40° = 1.191 7$]

24. A surveyor measures the angle of elevation of the top of a building as 19° 17'. He moves 62 m nearer to the building and measures the angle of elevation as 36° 10′. Find the perpendicular height of the building   [41.6 m]

25. From the top of a vertical cliff 75 m high the angles of depression of two boats lying due north of the cliff are 24° and 17°. How far are the boats apart?   [76.9 m]

26. Prove the following trigonometric identities:
    (a) $\sin^2 A \cot A = \sin A \cos A$
    (b) $\cos \theta = \sin \theta \cot \theta$
    (c) $\cos B(\operatorname{cosec} B + \sec B) \tan B = (1 + \tan B)$
    (d) $\sin C \tan C = \dfrac{\sin^2 C}{\cos C}$
    (e) $\dfrac{\cos \theta(\tan \theta + \sec \theta)}{\left(1 + \dfrac{\tan \theta}{\sec \theta}\right)} = 1$

Angles of any magnitude

27. Evaluate the following, correct to 4 decimal places:
    (a) $\sin 47°$ (b) $\sin 132°$ (c) $\sin 241°$ (d) $\sin 347°$
    (a) [0.731 4]   (b) [0.743 1]   (c) [−0.874 6]
    (d) [−0.225 0]

28. Evaluate the following, correct to 5 significant figures:
    (a) $\cos 17°$ (b) $\cos 97°$ (c) $\cos 214°$ (d) $\cos 302°$
    (a) [0.956 30]   (b) [−0.121 87]   (c) [−0.829 04]
    (d) [0.529 92]

29. Evaluate the following, correct to 4 decimal places:
    (a) $\tan 31°$ (b) $\tan 123°$ (c) $\tan 197°$ (d) $\tan 288°$
    (a) [0.600 9]   (b) [−1.539 9]   (c) [0.305 7]
    (d) [−3.077 7]

30. Evaluate the following, correct to 5 significant figures:
    (a) $\sec 72°$ (b) $\sec 111°$ (c) $\sec 242°$ (d) $\sec 291°$
    (a) [3.236 1]   (b) [−2.790 4]   (c) [−2.130 1]
    (d) [2.790 4]

31. Evaluate the following, correct to 6 significant figures:
    (a) $\operatorname{cosec} 80°$ (b) $\operatorname{cosec} 142°$ (c) $\operatorname{cosec} 224°$
    (d) $\operatorname{cosec} 348°$
    (a) [1.015 43]   (b) [1.624 27]   (c) [−1.439 56]
    (d) [−4.809 73]

32. Evaluate the following, correct to 5 significant figures:
    (a) $\cot 54°$ (b) $\cot 152°$ (c) $\cot 236°$ (d) $\cot 312°$
    (a) [0.726 54]   (b) [−1.880 73]   (c) [0.674 51]
    (d) [−0.900 40]

33. Evaluate the following, correct to 5 significant figures:
(a) sin 261° 19' (b) cos 172° 49' (c) tan 123° 8'
(d) sec 301° 57' (e) cosec 191° 5' (f) cot 342° 26'
(a) [−0.988 54]   (b) [−0.992 15]   (c) [−1.532 05]
(d) [1.889 7]   (e) [−5.201 9]   (f) [−3.158 8]

34. Evaluate the following, correct to 4 decimal places:
(a) sin 452° 13' (b) cos 548° 3' (c) tan 702° 58'
(d) sec 602° 41' (e) cosec 1 047° 8' (f) cot 936° 15'
(a) [0.999 3]   (b) [−0.990 1] (c) [−0.306 4]
(d) [−2.179 1]   (e) [−1.842 7]   (f) [1.363 8]

35. Evaluate the following, correct to 4 decimal places:
(a) sin(−79° 8') (b) cos(−143° 3') (c) tan(−62° 15')
(d) sec(−236° 32') (e) cosec(−93° 14') (f) cot(−136° 19')
(a) [−0.982 1]   (b) [−0.799 2]   (c) [−1.900 7]
(d) [−1.813 4]   (e) [−1.001 6]   (f) [1.047 0]

36. Evaluate the following, correct to 4 decimal places:
(a) $\sin \frac{5\pi}{6}$ (b) $\cos \frac{3\pi}{8}$ (c) $\tan \frac{5\pi}{16}$ (d) $\sec \frac{7\pi}{12}$ (e) cosec 4.72π
(f) $\cot \frac{4\pi}{9}$
(a) [0.500 0]   (b) [0.382 7]   (c) [1.496 6]
(d) [−3.863 7]   (e) [1.297 8]   (f) [0.176 3]

37. Find all the angles between 0° and 360°:
(a) whose sine is −0.478 0 (b) whose cosine is 0.863 1
(c) whose tangent is −1.232 8 (d) whose secant is 1.347 3
(e) whose cosecant is −2.468 3
(f) whose cotangent is 0.728 8
(a) [208° 33', 331° 27']   (b) [30° 20', 329° 40']
(c) [129° 3', 309° 3']   (d) [42° 5', 317° 55']
(e) [203° 54', 336° 6']   (f) [53° 55', 233° 55']

38. Solve the following equations for θ such that 0° < θ < 360°:
(a) arcsin 0.317 9 = θ (b) arccos −0.432 1 = θ
(c) arctan 0.462 5 = θ (d) arcsec −2.159 6 = θ
(e) arccosec 1.476 9 = θ (f) arccot −1.329 4 = θ
(a) [18° 32', 161° 28']   (b) [115° 36', 244° 24']
(c) [24° 49', 204° 49']   (d) [117° 35', 242° 25']
(e) [42° 37', 137° 23']   (f) [143° 3', 323° 3']

39. Evaluate correct to 4 significant figures:
4.31 sec 286° 5' − 3.26 cosec 146° 43' + 9.0 cot 312° 15'
[1.442]

Some properties of trigonometric functions

14.1 Relationships between trigonometric ratios

Consider a right-angled triangle ABC, shown in Fig. 14.1, and let angle CAB be denoted by θ. By definition

$$\sin \theta = \frac{BC}{AC} \qquad \cos \theta = \frac{AB}{AC} \qquad \tan \theta = \frac{BC}{AB}$$

$$\sec \theta = \frac{AC}{AB} \qquad \operatorname{cosec} \theta = \frac{AC}{BC} \qquad \cot \theta = \frac{AB}{BC}$$

Thus, $\quad \sec \theta = \dfrac{1}{\cos \theta} \left(\text{or } \cos \theta = \dfrac{1}{\sec \theta} \right)$ \hfill (1)

$$\operatorname{cosec} \theta = \frac{1}{\sin \theta} \left(\text{or } \sin \theta = \frac{1}{\operatorname{cosec} \theta} \right) \hfill (2)$$

$$\cot \theta = \frac{1}{\tan \theta} \left(\text{or } \tan \theta = \frac{1}{\cot \theta} \right) \hfill (3)$$

Using the theorem of Pythagoras on triangle ABC of Fig. 14.1 gives:

$$AB^2 + BC^2 = AC^2$$

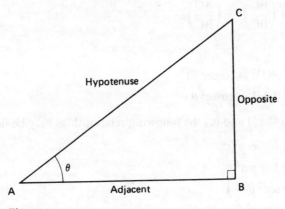

Fig. 14.1

If each term in this equation is divided by AC^2 then

$$\frac{AB^2}{AC^2} + \frac{BC^2}{AC^2} = \frac{AC^2}{AC^2}$$

$$\left(\frac{AB}{AC}\right)^2 + \left(\frac{BC}{AC}\right)^2 = \left(\frac{AC}{AC}\right)^2$$

Therefore

$$(\cos\theta)^2 + (\sin\theta)^2 = (1)^2$$

i.e. $\cos^2\theta + \sin^2\theta = 1$ (4)

If each term in the equation $AB^2 + BC^2 = AC^2$ is divided by AB^2 then

$$\frac{AB^2}{AB^2} + \frac{BC^2}{AB^2} = \frac{AC^2}{AB^2}$$

$$\left(\frac{AB}{AB}\right)^2 + \left(\frac{BC}{AB}\right)^2 = \left(\frac{AC}{AB}\right)^2$$

Therefore

$$(1)^2 + (\tan\theta)^2 = (\sec\theta)^2$$

i.e. $1 + \tan^2\theta = \sec^2\theta$ (5)

If each term in the equation $AB^2 + BC^2 = AC^2$ is divided by BC^2 then

$$\frac{AB^2}{BC^2} + \frac{BC^2}{BC^2} = \frac{AC^2}{BC^2}$$

$$\left(\frac{AB}{BC}\right)^2 + \left(\frac{BC}{BC}\right)^2 = \left(\frac{AC}{BC}\right)^2$$

Therefore

$$(\cot\theta)^2 + (1)^2 = (\operatorname{cosec}\theta)^2$$

i.e. $\cot^2\theta + 1 = \operatorname{cosec}^2\theta$ (6)

From equations (4), (5) and (6) the following relationships may be deduced:

$$\sin^2\theta = 1 - \cos^2\theta$$
$$\cos^2\theta = 1 - \sin^2\theta$$
$$\tan^2\theta = \sec^2\theta - 1$$
$$\cot^2\theta = \operatorname{cosec}^2\theta - 1$$

The relationships shown in equations (1) to (6) are not only true for values of θ between $0°$ and $90°$ but are true for all values of angle θ.

Worked problems on relationships between trigonometric ratios

---

Problem 1. Show that the relationships $\cos^2 \theta + \sin^2 \theta = 1$
$$1 + \tan^2 \theta = \sec^2 \theta$$
$$\text{and } \cot^2 \theta + 1 = \text{cosec}^2 \theta$$
are valid for the following values of θ: (a) $120°$, (b) $210°$, (c) $315°$

(a)   When $\theta = 120°$, $\sin 120° = \sin 60° = \dfrac{\sqrt{3}}{2}$

$$\cos 120° = -\cos 60° = -\frac{1}{2}$$

$$\tan 120° = -\tan 60° = -\sqrt{3}$$
$$\sec 120° = -\sec 60° = -2$$

$$\text{cosec } 120° = \text{cosec } 60° = \frac{2}{\sqrt{3}}$$

$$\cot 120° = -\cot 60° = -\frac{1}{\sqrt{3}}$$

$$\cos^2 \theta + \sin^2 \theta = (\cos 120°)^2 + (\sin 120°)^2$$

$$= \left(-\frac{1}{2}\right)^2 + \left(\frac{\sqrt{3}}{2}\right)^2$$

$$= \frac{1}{4} + \frac{3}{4} = 1$$

Thus, in the second quadrant, when $\theta = 120°$:
$$\cos^2 \theta + \sin^2 \theta = 1$$

$$1 + \tan^2 \theta = 1 + (\tan 120°)^2$$
$$= 1 + (-\sqrt{3})^2$$
$$= 1 + 3 = 4$$
$$\sec^2 \theta = (\sec 120°)^2$$
$$= (-2)^2 = 4$$

Thus, in the second quadrant, when $\theta = 120°$:
$$1 + \tan^2 \theta = \sec^2 \theta$$

$$\cot^2 \theta + 1 = (\cot 120°)^2 + 1$$

$$= \left(-\frac{1}{\sqrt{3}}\right)^2 + 1$$

$$= \frac{1}{3} + 1 = \frac{4}{3}$$

$$\text{cosec}^2 \theta = (\text{cosec } 120°)^2$$

$$= \left(\frac{2}{\sqrt{3}}\right)^2 = \frac{4}{3}$$

Thus, in the second quadrant, when $\theta = 120°$:
$$\cot^2 \theta + 1 = \text{cosec}^2 \theta$$

(b)   When $\theta = 210°$, $\sin 210° = -\sin 30° = -\dfrac{1}{2}$

$$\cos 210° = -\cos 30° = -\frac{\sqrt{3}}{2}$$

$$\tan 210° = \tan 30° = \frac{1}{\sqrt{3}}$$

$$\sec 210° = -\sec 30° = -\frac{2}{\sqrt{3}}$$

$$\text{cosec } 210° = -\text{cosec } 30° = -2$$

$$\cot 210° = \cot 30° = \sqrt{3}$$

$$\cos^2 \theta + \sin^2 \theta = (\cos 210°)^2 + (\sin 210°)^2$$

$$= \left(-\frac{\sqrt{3}}{2}\right)^2 + \left(-\frac{1}{2}\right)^2$$

$$= \quad \frac{3}{4} \quad + \quad \frac{1}{4} \quad = 1$$

Thus, in the third quadrant, when $\theta = 210°$:
$$\cos^2 \theta + \sin^2 \theta = 1$$

$$1 + \tan^2 \theta = 1 + (\tan 210°)^2$$

$$= 1 + \left(\frac{1}{\sqrt{3}}\right)^2$$

$$= 1 + \frac{1}{3} = \frac{4}{3}$$

$$\sec^2 \theta = (\sec 210°)^2$$

$$= \left(-\frac{2}{\sqrt{3}}\right)^2 = \frac{4}{3}$$

Thus, in the third quadrant, when $\theta = 210°$:
$$1 + \tan^2 \theta = \sec^2 \theta$$

$$\cot^2 \theta + 1 = 1 + (\cot 210°)^2$$

$$= 1 + (\sqrt{3})^2$$

$$= 1 + 3 = 4$$

$$\text{cosec}^2 \theta = (-2)^2 = 4$$

Thus, in the third quadrant, when $\theta = 210°$:
$$\cot^2 \theta + 1 = \text{cosec}^2 \theta$$

(c)   When $\theta = 315°$, $\sin 315° = -\sin 45° = -\dfrac{1}{\sqrt{2}}$

$$\cos 315° = \cos 45° = \frac{1}{\sqrt{2}}$$

$$\tan 315° = -\tan 45° = -1$$
$$\sec 315° = \sec 45° = \sqrt{2}$$
$$\operatorname{cosec} 315° = -\operatorname{cosec} 45° = -\sqrt{2}$$
$$\cot 315° = -\cot 45° = -1$$
$$\cos^2 \theta + \sin^2 \theta = (\cos 315°)^2 + (\sin 315°)^2$$

$$= \left(\frac{1}{\sqrt{2}}\right)^2 + \left(-\frac{1}{\sqrt{2}}\right)^2$$

$$= \frac{1}{2} + \frac{1}{2} = 1$$

Thus, in the fourth quadrant, when $\theta = 315°$:
$$\cos^2 \theta + \sin^2 = 1$$

$$1 + \tan^2 \theta = 1 + (\tan 315°)^2$$
$$= 1 + (-1)^2 = 2$$
$$\sec^2 \theta = (\sec 315°)^2$$
$$= (\sqrt{2})^2 = 2$$

Thus, in the fourth quadrant, when $\theta = 315°$:
$$1 + \tan^2 \theta = \sec^2 \theta$$

$$\cot^2 \theta + 1 = (\cot 315°)^2 + 1$$
$$= (-1)^2 + 1$$
$$= 1 + 1 = 2$$
$$\operatorname{cosec}^2 \theta = (\operatorname{cosec} 315°)^2$$
$$= (-\sqrt{2})^2 = 2$$

Thus, in the fourth quadrant, when $\theta = 315°$:
$$\cot^2 \theta + 1 = \operatorname{cosec}^2 \theta$$

Problem 2. Prove the following identities:

(a) $\cos^2 A - \sin^2 A = 2\cos^2 A - 1$

(b) $\dfrac{1 + \tan^2 B}{1 + \cot^2 B} = \tan^2 B$

(c) $\sqrt{\left[\dfrac{1 - \cos C}{1 + \cos C}\right]} = \operatorname{cosec} C - \cot C$

(a) To prove $\cos^2 A - \sin^2 A = 2\cos^2 A - 1$:
Left-hand side (L.H.S.) $= \cos^2 A - \sin^2 A$
Since $\cos^2 A + \sin^2 A = 1$, then $\sin^2 A = 1 - \cos^2 A$
$$\begin{aligned}\text{L.H.S.} &= \cos^2 A - (1 - \cos^2 A)\\ &= \cos^2 A - 1 + \cos^2 A\\ &= 2\cos^2 A - 1 = \text{right-hand side (R.H.S.)}\end{aligned}$$

Thus $\cos^2 A - \sin^2 A = 2\cos^2 A - 1$

(b) To prove $\dfrac{1 + \tan^2 B}{1 + \cot^2 B} = \tan^2 B$:

$1 + \tan^2 B = \sec^2 B$ and $1 + \cot^2 B = \operatorname{cosec}^2 B$

$$\text{L.H.S.} = \frac{1 + \tan^2 B}{1 + \cot^2 B} = \frac{\sec^2 B}{\operatorname{cosec}^2 B}$$

Since $\sec B = \dfrac{1}{\cos B}$ and $\operatorname{cosec} B = \dfrac{1}{\sin B}$

$$\text{L.H.S.} = \frac{\sec^2 B}{\operatorname{cosec}^2 B} = \frac{\dfrac{1}{\cos^2 B}}{\dfrac{1}{\sin^2 B}}$$

$$= \frac{\sin^2 B}{\cos^2 B} = \tan^2 B \left(\text{since } \frac{\sin B}{\cos B} = \tan B\right)$$

Thus $\dfrac{1 + \tan^2 B}{1 + \cot^2 B} = \tan^2 B$

(c) To prove $\sqrt{\left[\dfrac{1 - \cos C}{1 + \cos C}\right]} = \operatorname{cosec} C - \cot C$

$$\text{L.H.S.} = \sqrt{\left[\frac{1 - \cos C}{1 + \cos C}\right]} = \sqrt{\left[\frac{(1 - \cos C)}{(1 + \cos C)}\frac{(1 - \cos C)}{(1 - \cos C)}\right]}$$

$$= \sqrt{\left[\frac{(1 - \cos C)^2}{(1 - \cos^2 C)}\right]}$$

Since $\cos^2 C + \sin^2 C = 1$ then $1 - \cos^2 C = \sin^2 C$

$$\text{L.H.S.} = \sqrt{\left[\frac{(1 - \cos C)^2}{(1 - \cos^2 C)}\right]} = \sqrt{\left[\frac{(1 - \cos C)^2}{\sin^2 C}\right]}$$

$$= \frac{(1 - \cos C)}{\sin C} = \frac{1}{\sin C} - \frac{\cos C}{\sin C}$$

$$= \operatorname{cosec} C - \cot C = \text{R.H.S.}$$

$$\text{Thus } \sqrt{\left[\frac{1 - \cos C}{1 + \cos C}\right]} = \text{cosec } C - \cot C$$

Further problems on relationships between trigonometrical ratios may be found in Section 14.4 (Problems 1–15), page 249.

14.2   Sine and cosine curves

Consider a circle, drawn on rectangular axes Ox and Oy, of centre O and radius 1 unit. The circle is divided into 360° (in 15° sectors), as shown in Fig. 14.2. The horizontal radius arm OA is assumed to represent 0° and 360°, and angles are labelled anticlockwise from this radius arm.

Each radius arm can be considered to have a horizontal component x and a vertical component y. For example, for the 30° radius arm in Fig. 14.2 the horizontal component is shown by the length OC and the vertical component is shown by BC.

At **0°**:   the horizontal component, $x = $ OA $= 1$ unit
the vertical component, $y = 0$

At **15°**: the right-angled triangle shown in Fig. 14.3 shows the horizontal component x and the vertical component y.

The horizontal component x is calculated using the cosine:

$$\cos 15° = \frac{x}{1}$$

Therefore $x = \cos 15° = 0.965\ 9$

The vertical component y is calculated using the sine:

$$\sin 15° = \frac{y}{1}$$

Therefore $y = \sin 15° = 0.258\ 8$

Similarly, at **30°**: $x = \cos 30° = 0.866\ 0$
$y = \sin 30° = 0.500\ 0$

and at **45°**: $x = \cos 45° = 0.707\ 1$
$y = \sin 45° = 0.707\ 1$

and so on.

At 105°, the x and y values are numerically the same as for 75°, except that the horizontal component (x) is negative because it is to the left of the origin.

The x and y values for each of the 15° sectors may thus be found.

Each of the horizontal and vertical component values from 0° to 360° may be projected onto a graph as explained below.

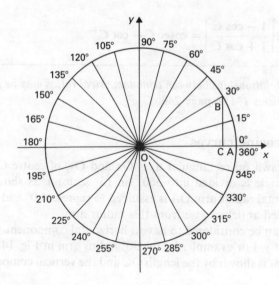

Fig. 14.2

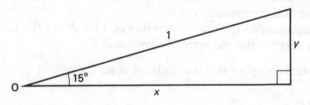

Fig. 14.3

1. Sine curve

The vertical components (that is, the y values from above) are each projected onto a graph of y against angle in degrees, as shown in Fig. 14.4.

The vertical component EF for 15° is projected across to E′F′, the corresponding value of 15° on the graph. A similar projection is made for each of the other 15° increments.

The graph produced shows a complete cycle of a **sine curve**. If values of angles greater than 360° are taken then the graph merely repeats itself indefinitely. If increments smaller than 15° are taken a more accurate sine curve is produced since more points are obtained.

2. Cosine curve

The circle is now redrawn with the radius arm OA in a vertical position and the 360° labelled once again in an anticlockwise direction.

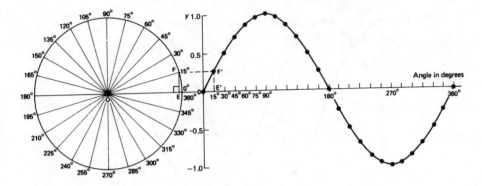

Fig. 14.4

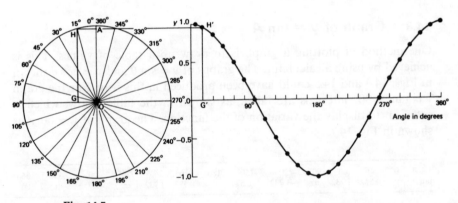

Fig. 14.5

The vertical components are each projected onto the graph of y against angle in degrees, as shown in Fig. 14.5.

The component GH, shown in Fig. 14.5, for 15° is projected across to G′H′, the corresponding value of 15° on the graph. A similar projection is made for each of the other 15° increments.

The graph produced shows a complete cycle of a **cosine curve**.

If a sine curve, $y = \sin \theta$, and a cosine curve, $y = \cos \theta$, were sketched on the same axes over one complete cycle the result would be as in Fig. 14.6. It may be seen that the cosine curve is of the same form as the sine curve but displaced by 90°.

The sine curve representation has many important applications, including depicting the values obtained in simple harmonic motion, wave motion and in electrical alternating currents and voltages.

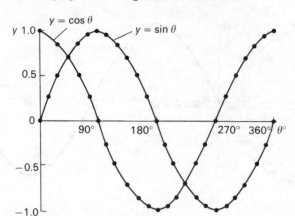

Fig. 14.6

14.3   Graph of $y = \tan A$

One method of plotting a graph is to draw up a table of values. This is achieved by using a calculator. The graphs of $y = \sin \theta$ and $y = \cos \theta$ shown in Figs. 14.4 and 14.5 could have been produced in this way. For the graph $y = \tan A$, 30° intervals are taken, as shown in the table below, which is sufficient to display the variation of the function. The graph of $y = \tan A$ is shown in Fig. 14.7.

A	0°	30°	60°	90°	120°	150°	180°	210°	240°	270°	300°	330°	360°
$\tan A$	0	0.577	1.732	∞	-1.732	-0.577	0	0.577	1.732	∞	-1.732	-0.577	0

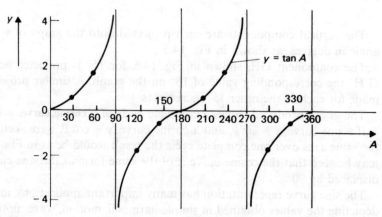

Fig. 14.7

From the graphs of the sine, cosine and tangent curves, the following points are noted:

(i) The sine and cosine graphs oscillate between peak values of ± 1.

(ii) The cosine graph is the same shape as the sine graph but displaced $90°$ to the left.

(iii) In the first quadrant, all three graphs have positive values. In the second quadrant, only the sine graph has positive values. In the third quadrant, only the tangent graph has positive values. In the fourth quadrant, only the cosine graph has positive values (see Section 13.2).

(iv) The sine and cosine graphs are continuous and they repeat at intervals of $360°$. The tangent graphs appear to be discontinuous and repeat at intervals of $180°$.

(v) Since the graphs of sine, cosine and tangent repeat themselves at regular intervals they are called **periodic functions**. Sine and cosine curves have a period of $360°$ or 2π radians and the tangent curve has a period of $180°$ or π radians.

14.4 Further problems

1. Show that the relationships $\cos^2 \theta + \sin^2 \theta = 1$, $1 + \tan^2 \theta = \sec^2 \theta$ and $\cot^2 \theta + 1 = \csc^2 \theta$ are valid for the following values of θ:
   (a) $27°$ (b) $130°$ (c) $242°$ (d) $328° \, 15'$

In Problems 2–15, prove the identities.

2. $\dfrac{1}{\sqrt{(1 - \sin^2 \theta)}} = \sec \theta$

3. $\tan^2 B(\csc^2 B - 1) = 1$

4. $\cos \theta \sqrt{(\sec^2 \theta - 1)} = \sin \theta$

5. $\tan x = \sqrt{\left[\dfrac{(1 - \cos^2 x)}{(\cos^2 x)}\right]}$

6. $\dfrac{\csc X}{\sec X} - \dfrac{\sec X}{\csc X} = (\cos^2 X - \sin^2 X)(\sec X \csc X)$

7. $(1 + \tan \theta)^2 + (1 - \tan \theta)^2 = 2 \sec^2 \theta$

8. $\dfrac{\sec \theta + \csc \theta}{\tan \theta + \cot \theta} = \sin \theta + \cos \theta$

9. $\sin C - \sin^3 C = \dfrac{\sin C}{\sec^2 C}$

10. $\dfrac{1 + \cot y}{1 + \tan y} = \cot y$

11. $(\sin D + \cos D)^2 + (\sin D - \cos D)^2 - 2 = 0$

12. $\dfrac{1 + \sin^4 x - \cos^4 x}{\sin^2 x} = 2$

13. $(1 + \cot E)^2 + (1 - \cot E)^2 = 2 \operatorname{cosec}^2 E$

14. $\cot^2 A - \tan^2 A - \sec^2 A - \operatorname{cosec}^2 A$
$\qquad = \sec^2 A \operatorname{cosec}^2 A(\cos^2 A - \sin^2 A - 1)$

15. $\sqrt{\left[\dfrac{\operatorname{cosec} b + 1}{\operatorname{cosec} b - 1}\right]} = \tan b(\operatorname{cosec} b + 1)$

Chapter 15

The solution and areas of triangles

15.1 Introduction to the solution of triangles

All triangles have 3 sides and 3 angles; that is, 6 facts define a triangle completely. In a right-angled triangle, use of the trigonometric ratios and the theorem of Pythagoras enable the triangle to be solved (that is, all unknown sides and angles found) provided at least 3 facts are known.

In a triangle which is not right-angled, trigonometric ratios and the theorem of Pythagoras **cannot** be used. Unknown sides and angles may be calculated using the **sine and cosine rules**, provided, once again, that at least 3 facts are given. The 3 facts given may be 1 side and any 2 angles, 2 sides and an angle (not the angle between the sides) or 2 sides and the angle between the 2 sides, or 3 sides. There are thus four possible cases. The sine rule is used for the first two cases and the cosine rule is used for the last two.

The sine and cosine rules can be applied to any triangle (this includes right-angled triangles as well, although it would be a rather laborious method of solution in comparison with the more straightforward method using trigonometric ratios and Pythagoras' theorem).

15.2 The sine rule

Let ABC be any triangle with sides a, b and c, as shown in Fig. 15.1.

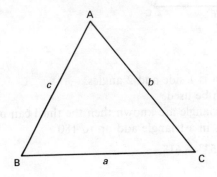

Fig. 15.1

The sine rule states

$$\frac{a}{\sin A} = \frac{b}{\sin B} = \frac{c}{\sin C}$$

The sine rule may be used only when the following information is given:

Either (a) 1 side and any 2 angles

or (b) 2 sides and an angle (not the angle between them).

Worked problems on the sine rule

Problem 1. In a triangle ABC, A = 53°, B = 61° and the length a = 12.60 cm. Find the unknown sides and angle.

The triangle ABC is shown in Fig. 15.2.

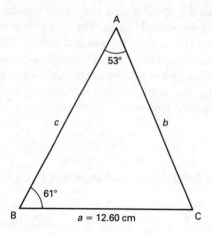

Fig. 15.2

The given information is 1 side and 2 angles.
Thus the sine rule can be used.

If two angles in a triangle are known then the third can be found since the angles in a triangle add up to 180°.

Hence $\angle C = 180° - 53° - 61°$
i.e. $\angle C = 66°$

From the sine rule:

$$\frac{12.60}{\sin 53°} = \frac{b}{\sin 61°} = \frac{c}{\sin 66°}$$

Using $\dfrac{12.60}{\sin 53°} = \dfrac{b}{\sin 61°}$ and transposing gives

$$b = \frac{12.60 \sin 61°}{\sin 53°} = \textbf{13.80 cm}$$

Using $\dfrac{12.60}{\sin 53°} = \dfrac{c}{\sin 66°}$ and transposing gives

$$c = \frac{12.60 \sin 66°}{\sin 53°} = \textbf{14.41 cm}$$

Hence $\angle C = 66°$, side $b = 13.80$ cm and side $c = 14.41$ cm

It is always beneficial to check to see if the solution obtained is sensible; that is, if the longest side is opposite the largest angle and the shortest side is opposite the smallest angle, and so on.

In the above problem the largest angle is 66° (i.e. $\angle C$) and the longest side is 14.41 cm (i.e. side c). The smallest angle is 53° (i.e. $\angle A$) and the shortest side is 12.60 cm (i.e. side a). Hence the solution is feasible.

Problem 2. In a triangle PQR, $\angle R = 41°$, $r = 7.21$ cm and $q = 9.30$ cm. Solve the triangle.

The triangle PQR is shown in Fig. 15.3. In this case 'to solve the triangle PQR' means 'to find $\angle P$ and $\angle Q$ and the length of side p'.

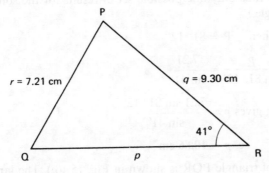

Fig. 15.3

The given information is 2 sides and an angle (not the included one). Thus the sine rule can be used.

Hence $\dfrac{p}{\sin P} = \dfrac{9.30}{\sin Q} = \dfrac{7.21}{\sin 41°}$

Using $\dfrac{9.30}{\sin Q} = \dfrac{7.21}{\sin 41°}$ and transposing gives

$$\sin Q = \frac{9.30 \sin 41°}{7.21} = 0.846\,2$$

Hence $Q = \arcsin 0.846\,2$

It is very important at this stage to realise that there are **TWO** possible values of Q since sine is positive in the first and second quadrants (see Section 13.2). Hence an acute and an obtuse angle can be found. (Note that a calculator usually gives only the acute angle.)

$\angle Q = 57° \, 48'$ or $180° - 57° \, 48'$

i.e. $\angle Q = \mathbf{57° \, 48'}$ **or** $\mathbf{122° \, 12'}$

The next step is to consider whether both of the values for Q are possible in this particular problem.

If $\angle Q = 57° \, 48'$ then $\angle P = 180° - 57° \, 48' - 41°$

i.e. $\angle P = \mathbf{81° \, 12'}$

If $\angle Q = 122° \, 12'$ then $\angle P = 180° - 122° \, 12' - 41°$

i.e. $\angle P = \mathbf{16° \, 48'}$

Hence in this problem there are going to be two separate sets of results and both are correct. (In some problems (see Problem 3), when substituting the obtuse angle to find the third angle, a negative answer results. In such cases the obtuse angle is ignored and there is thus only one possible set of results for the solution of the triangle.)

1st case: when $\angle P = 81° \, 12'$

then $\dfrac{p}{\sin 81° \, 12'} = \dfrac{7.21}{\sin 41°}$

Transposing gives $p = \dfrac{7.21 \sin 81° \, 12'}{\sin 41°}$

i.e. $p = \mathbf{10.86 \; cm}$

A solution of triangle PQR is shown in Fig. 15.4(a). The largest angle (i.e. $\angle P$) is opposite the longest side (i.e. p). The smallest angle (i.e. $\angle R$) is opposite the shortest side (i.e. r). Hence the results are feasible.

2nd case: when $\angle P = 16° \, 48'$

then $\dfrac{p}{\sin 16° \, 48'} = \dfrac{7.21}{\sin 41°}$

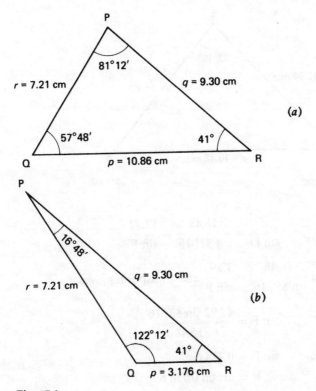

Fig. 15.4

Transposing gives $p = \dfrac{7.21 \sin 16° \, 48'}{\sin 41°}$

i.e. $p = \mathbf{3.176 \ cm}$

A solution of triangle PQR is shown in Fig. 15.4(*b*).

The largest angle (i.e. ∠Q) is opposite the longest side (i.e. *q*).
The smallest angle (i.e. ∠P) is opposite the shortest side (i.e. *p*).
Hence the result is feasible.

Such a problem as this, where there are two possible sets of results, is known as the **ambiguous case**.

Problem 3. Solve the triangle DEF given ∠E = 83° 16′,
e = 16.48 mm and *f* = 12.92 mm.

The triangle DEF is shown in Fig. 15.5. In this case, 'to solve triangle DEF' means 'to find the angles D and F and the length of side *d*'.

The given information is 2 sides and an angle (not the included angle). Thus the sine rule can be used.

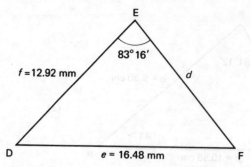

Fig. 15.5

Thus $\qquad \dfrac{d}{\sin D} = \dfrac{16.48}{\sin 83° 16'} = \dfrac{12.92}{\sin F}$

Using $\dfrac{16.48}{\sin 83° 16'} = \dfrac{12.92}{\sin F}$ and transposing gives

$$\sin F = \frac{12.92 \sin 83° 16'}{16.48}$$

Hence $\qquad \sin F = 0.778\,6$

i.e. $\angle F = \arcsin 0.778\,6$

Hence $\qquad \angle F = 51° 8'$ or $128° 52'$

If $\quad \angle F = 51° 8'$ then $\angle D = 180° - 51° 8' - 83° 16'$
i.e. $\angle D = 45° 36'$

If $\angle F = 128° 52'$ then $\angle D = 180° - 128° 52' - 83° 16'$
i.e. $\angle D = -32° 8'$

This value of $\angle D$ is meaningless in the context of triangles and the value $\angle F = 128° 52'$ is thus ignored.

There is therefore only one set of results for the solution of triangle DEF.

If $\angle F = 51° 8'$, $\angle E = 83° 16'$ and $\angle D = 45° 36'$

then $\dfrac{d}{\sin 45° 36'} = \dfrac{16.48}{\sin 83° 16'} \left(= \dfrac{12.92}{\sin 51° 8'} \right)$

Transposing gives $d = \dfrac{16.48 \sin 45° 36'}{\sin 83° 16'}$

i.e. $d = \mathbf{11.86\ mm}$

Hence $\angle D = \mathbf{45° 36'}$, $\angle F = \mathbf{51° 8'}$ and $d = \mathbf{11.86\ mm}$

The largest angle (i.e. $\angle E = 83° \, 16'$) is opposite the longest side (i.e. $e = 16.48$ mm). The smallest angle (i.e. $\angle D = 45° \, 36'$) is opposite the shortest side (i.e. $d = 11.86$ mm). Hence the result is feasible.

Problem 4. A man leaves a point walking at $6.0 \, \text{km h}^{-1}$ in a direction S 30° W. Another man leaves the same point at the same time cycling at a constant speed in a direction S 23° E. After 4.0 hours the two men are 100 km apart. Find the speed of the cyclist.

Let R be the starting point, P be the point that the walker reaches after 4.0 hours (i.e. $4.0 \times 6 = 24$ km from the starting point) and Q be the point the cyclist reaches after 4.0 hours.

Hence, from Fig. 15.6, a triangle is formed with PQ = 100 km and $\angle R = 30° + 23° = 53°$.

The given information is 2 sides and an angle (not the included one). Thus the sine rule can be used.

Hence $\dfrac{p}{\sin P} = \dfrac{24}{\sin Q} = \dfrac{100}{\sin 53°}$

Using $\dfrac{24}{\sin Q} = \dfrac{100}{\sin 53°}$ and transposing gives

$\sin Q = \dfrac{24 \sin 53°}{100}$

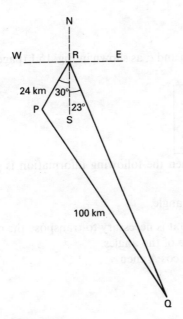

Fig. 15.6

Hence sin Q = 0.191 7

i.e. ∠Q = arcsin 0.191 7

Hence ∠Q = 11° 3′ or 168° 57′

As ∠R = 53°, ∠Q = 168° 57′ is not possible since there are only 180° in a triangle.

Thus there is only one possible solution for this triangle.

If ∠R = 53° and ∠Q = 11° 3′ then ∠P = 180° − 53° − 11° 3′, i.e. ∠P = 115° 57′.

Now $\dfrac{p}{\sin 115° 57'} = \dfrac{100}{\sin 53°}$

Transposing gives $p = \dfrac{100 \sin 115° 57'}{\sin 53°}$

∴ **p = 112.6 km**

Hence the cyclist travelled 112.6 km in 4.0 hours.

Average speed of cyclist $= \dfrac{112.6 \text{ km}}{4.0 \text{ h}}$

$\qquad\qquad\qquad\qquad\qquad\quad = \textbf{28.2 km h}^{-1}$

Further problems on the sine rule may be found in Section 15.5, page 269. Further practical trigonometric problems, which involve the sine rule, may be found in Chapter 16, page 291.

15.3   The cosine rule

Let ABC be any triangle with sides *a*, *b* and *c*, as shown in Fig. 15.1 (page 251). The cosine rule states:

$$
\begin{array}{l}
a^2 = b^2 + c^2 - 2bc \cos A \\
\text{or } b^2 = a^2 + c^2 - 2ac \cos B \\
\text{or } c^2 = a^2 + b^2 - 3ab \cos C.
\end{array}
$$

The cosine rule can be used only when the following information is given:

Either (*a*) 3 sides

  or (*b*) 2 sides and the included angle.

When 3 sides of a triangle are given it is necessary to transpose the cosine rule formula in order to calculate one of the angles.

For example, if $a^2 = b^2 + c^2 - 2bc \cos A$ then

$2bc \cos A = b^2 + c^2 - a^2$

$\cos A = \dfrac{b^2 + c^2 - a^2}{2bc}$

A similar transposition can be obtained for cos B and cos C.

When given 3 sides of a triangle it is good practice to determine firstly whether the triangle is acute or obtuse angled. This is achieved by finding first the largest angle in the triangle (that is, the angle lying opposite the longest side). This step is not essential but it does avoid considering the ambiguous case if the sine rule is later used.

For example, if a triangle has sides 3 cm, 7 cm and 8 cm the largest angle is opposite the 8 cm side.

Hence, if $a = 3$ cm, $b = 7$ cm and $c = 8$ cm then

$$\cos C = \frac{a^2 + b^2 - c^2}{2ab}$$

$$= \frac{3^2 + 7^2 - 8^2}{2(3)(7)}$$

$$= -\frac{6}{42} = -\frac{1}{7}$$

The important thing is that the cosine is negative. That is, the angle C is greater than $90°$ since cosine is negative in the second quadrant. Hence the triangle is obtuse angled.

Similarly, if in a triangle ABC $a = 6$ cm, $b = 4$ cm and $c = 5$ cm then

$$\cos A = \frac{4^2 + 5^2 - 6^2}{2(4)(5)}$$

$$= \frac{5}{40} = \frac{1}{8}$$

The cosine is positive, which means that the angle A is less than $90°$, i.e. the triangle is acute angled.

Worked problems on cosine rule

Problem 1. A triangular template has sides of 8 cm, 7 cm and 5 cm. Find its three angles.

Let the template be labelled ABC with $a = 8$ cm, $b = 7$ cm and $c = 5$ cm, as shown in Fig. 15.7. The lengths of the three sides are given. Thus the cosine rule can be used.

It is usual to start by calculating the largest angle (this will show whether the triangle is acute or obtuse angled). The largest angle is opposite the longest side. Thus the largest angle is A.

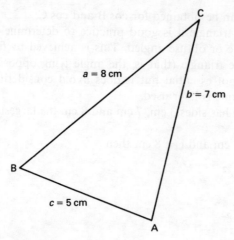

Fig. 15.7

$$\cos A = \frac{b^2 + c^2 - a^2}{2bc}$$

$$= \frac{7^2 + 5^2 - 8^2}{2(7)(5)}$$

$$= \frac{10}{70} = \frac{1}{7} = 0.142\,9$$

Hence $\angle A = \arccos 0.142\,9$

i.e. $\angle A = \mathbf{81°\,47'}$

Hence the triangle is acute angled since cos A is positive. (Theoretically, if cos A = 0.142 9 then angle A is 81° 47′ or 278° 13′. Since any triangle can have only 180° the latter value is impossible and is neglected.)

Using the sine rule:

$$\frac{8}{\sin 81°\,47'} = \frac{7}{\sin B}$$

Transposing gives

$$\sin B = \frac{7 \sin 81°\,47'}{8}$$

$$\angle B = \mathbf{60°}$$

If $\angle A = 81°\,47'$ and $\angle B = 60°$ then $\angle C = 180° - 81°\,47' - 60°$, i.e. $\angle C = \mathbf{38°\,13'}$

Problem 2. Solve the triangle ABC, given $\angle C = 67°$, $a = 16.40$ cm and $b = 11.80$ cm.

The triangle ABC is shown in Fig. 15.8. The given information is 2 sides and the included angle. Thus the cosine rule can be used.

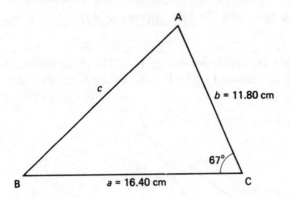

Fig. 15.8

$$c^2 = a^2 + b^2 - 2ab \cos C$$
$$= (16.40)^2 + (11.80)^2 - 2(16.40)(11.80) \cos 67°$$
$$= 269.0 + 139.2 - 151.2$$
$$= 257$$
$$c = \sqrt{257}$$

i.e. $c = 16.03$ cm

Now, using the sine rule:

$$\frac{16.03}{\sin 67°} = \frac{11.80}{\sin B}$$

Transposing gives: $\sin B = \dfrac{11.80 \sin 67°}{16.03}$

Hence $\sin B = 0.677\ 6$

i.e. $\angle B = \arcsin 0.677\ 6$

Hence $\angle B = 42° 39'$ or $137° 21'$

As $\angle C = 67°$, $\angle B = 137° 21'$ is not possible since there are only $180°$ in a triangle. Thus only the value of $\angle \mathbf{B} = \mathbf{42° 39'}$ is valid.

If $\angle C = 67°$ and $\angle B = 42° 39'$ then $\angle A - 180° - 67° - 42° 39'$, i.e. $\angle \mathbf{A} = \mathbf{70° 21'}$.

The largest angle (i.e. $\angle A = 70° 21'$) is opposite the longest side (i.e. $a = 16.40$ cm). The smallest angle (i.e. $\angle B = 42° 39'$) is opposite the shortest side (i.e. $b = 11.80$ cm). Hence the solution is feasible.

Note that in any problem involving the solution of a triangle the cosine rule need only be used once, and then the sine rule used. (Calculations involving the use of the sine rule tend to be shorter than those involving the cosine rule.)

Problem 3. Solve triangle JKL, given $\angle J = 123° \, 17'$, JK = 72 mm and JL = 43 mm.

The triangle JKL is shown in Fig. 15.9. The given information is 2 sides and the included angle. Therefore the cosine rule can be used.

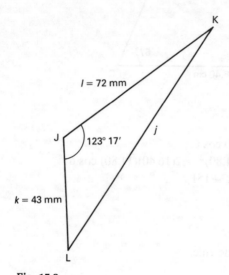

Fig. 15.9

Thus $j^2 = l^2 + k^2 - 2lk \cos J$

$$= 72^2 + 43^2 - 2(72)(43) \cos 123° \, 17'$$

$$j^2 = 5\,184 + 1\,849 + 3\,398 = 10\,341$$

$$j = \sqrt{10\,341}$$

i.e. $j = \mathbf{102 \; mm}$

Using the sine rule:

$$\frac{102}{\sin 123° \, 17'} = \frac{43}{\sin K}$$

Transposing gives $\sin K = \dfrac{43 \sin 123° \ 17'}{102}$

$$= 0.352 \ 4$$

i.e. $\angle K = \arcsin 0.352 \ 4$

Hence $\angle K = 20° \ 38'$ or $159° \ 22'$

As $\angle J = 123° \ 17'$, $\angle K = 159° \ 22'$ is not possible since there are only 180° in a triangle. Thus only the value of $\angle K = 20° \ 38'$ is valid.
    If $\angle J = 123° \ 17'$ and $\angle K = 20° \ 38'$ then
$\angle L = 180° - 123° \ 17' - 20° \ 38'$, i.e. $\angle L = 36° \ 05'$

Problem 4. Two ships P and Q set sail at the same time from the same port. P sails at a steady speed of 30 km h⁻¹, N 41° W and Q at 24 km h⁻¹, N 32° E. Find their distance apart after 3 hours.

Let point A represent the port from which the two ships sail. After 3 hours P has travelled $3 \times 30 = 90$ km, N 41° W. This distance is represented by AB on Fig. 15.10. After 3 hours Q has travelled $3 \times 24 = 72$ km, N 32° E. This distance is represented by AC.

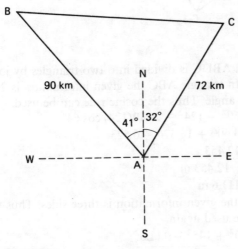

Fig. 15.10

The given information is 2 sides and the included angle. Thus the cosine rule can be used.

Hence $a^2 = 90^2 + 72^2 - 2(90)(72) \cos 73°$
$= 8\,100 + 5\,184 - 3\,789$
$= 9\,495$
$a = \sqrt{9\,495}$
i.e. $a = 97.4$ km

Hence the distance apart after 3 hours is 97.4 km

Problem 5. A quadrilateral plot of ground ABDC has the dimensions AB = 60 m, BD = 130 m, DC = 145 m and CA = 124 m. The angle BAC is 64°. Determine BC and the angle BDC.

The plot of ground is shown in Fig. 15.11.

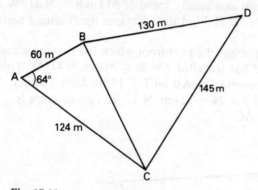

Fig. 15.11

The quadrilateral ABDC is divided into two triangles by joining BC (Fig. 15.11). In triangle ABC the given information is 2 sides and the included angle. Thus the cosine rule can be used.

Hence $BC^2 = 60^2 + 124^2 - 2(60)(124) \cos 64°$
$= 3\,600 + 15\,376 - 6\,523$
$= 12\,453$
$BC = \sqrt{12\,453}$ m
BC = 111.6 m

In triangle BCD the given information is three sides. Thus the cosine rule can be used again.

$$\cos \angle BDC = \frac{130^2 + 145^2 - 111.6^2}{2 \times 130 \times 145}$$
$$= \frac{16\,900 + 21\,025 - 12\,455}{37\,700} = 0.675\,6$$

Hence $\angle BDC = 47.50° = 47° \, 30'$

Further problems on the cosine rule may be found in Section 15.5, page 269.
Further practical trigonometric problems, which involve the cosine rule, may
be found in Chapter 16, page 291.

15.4   The area of any triangle

The area of any triangle = $\frac{1}{2}$ × base × perpendicular height

In Fig. 15.12, ABC is any triangle with AD constructed such that AD is perpendicular to BC. Hence area of triangle ABC = $\frac{1}{2}$ × BC × AD. However,
in the right-angled triangle ADC of Fig. 15.12

$$AD = AC \sin C$$

$$\therefore \text{ Area of triangle ABC} = \tfrac{1}{2} \times BC \times AC \sin C$$

$$= \tfrac{1}{2}ab \sin C$$

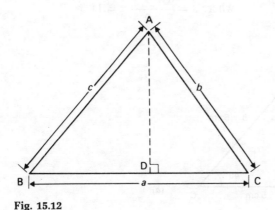

Fig. 15.12

By similar reasoning the area of triangle ABC may also be shown to equal
$\frac{1}{2}ac \sin B$ and $\frac{1}{2}bc \sin A$.

Thus, if the information given is two sides and the included angle, the area
can be found.

The following formula for the area of a triangle ABC can be used for the
case when 3 sides and no angles are given:

$$\textbf{Area of triangle ABC} = \sqrt{[s(s-a)(s-b)(s-c)]}$$

$$\text{where } s = \frac{a+b+c}{2}$$

Summary of formula for finding the area of any triangle

1. $\frac{1}{2} \times$ base $\times$ height
2. $\frac{1}{2}ab \sin C$ or $\frac{1}{2}ac \sin B$ or $\frac{1}{2}bc \sin A$
3. $\sqrt{[s(s-a)(s-b)(s-c)]}$, where $s = \dfrac{a+b+c}{2}$

Worked problems on the areas of triangles

Problem 1. Calculate the areas of the triangles shown in Fig. 15.13(*a*) and (*b*).

(*a*)   Area of triangle ABC $= \frac{1}{2}ab \sin C$

$$= \frac{1}{2}(5)(6) \sin 49° = \textbf{11.3 cm}$$

(*b*)   Area of triangle DEF $= \sqrt{[s(s-d)(s-e)(s-f)]}$

$$\text{where } s = \frac{6+9+7}{2} = 11$$

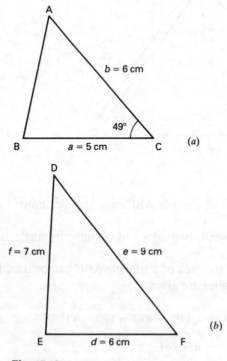

(*a*)

(*b*)

Fig. 15.13

$$\text{Area of triangle DEF} = \sqrt{[11(11-6)(11-9)(11-7)]}$$
$$= \sqrt{[11(5)(2)(4)]}$$
$$= \sqrt{440}$$
$$= \mathbf{21.0\ cm^2}$$

Problem 2. Two sides of an acute angled triangular plot of ground are 48.0 m and 26.0 m respectively. If the area of the plot is 550 m², find the length of the third side and the angles of the triangular plot.

Let the plot of ground be represented by the triangle ABC shown in Fig. 15.14.

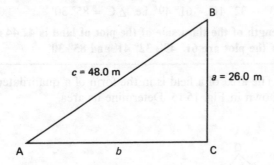

Fig. 15.14

Let AB = 48.0 m and BC = 26.0 m

$$\text{Area of triangle ABC} = \tfrac{1}{2}ac \sin B$$
$$550 = \tfrac{1}{2}(26.0)(48.0) \sin B$$
$$\sin B = \frac{2(550)}{(26.0)(48.0)}$$
$$\sin B = 0.881\ 4$$
$$\angle B = \arcsin 0.881\ 4$$

i.e. **∠B = 61° 49′** (the other possible value of 118° 11′ being ignored since the triangular plot is acute angled)

Using the cosine rule:
$$b^2 = a^2 + c^2 - 2ac \cos B$$
$$= (26.0)^2 + (48.0)^2 - 2(26.0)(48.0) \cos 61° 49'$$
$$= 676 + 2\ 304 - 1\ 179$$
$$= 1\ 801$$
$$\therefore b = \sqrt{1\ 801}$$

i.e. $b = \mathbf{42.44\ m}$

Using the sine rule:

$$\frac{26.0}{\sin A} = \frac{42.44}{\sin 61° \, 49'}$$

$$\sin A = \frac{26.0 \sin 61° \, 49'}{42.44}$$

$$= 0.540 \, 0$$

Hence $\angle A = \arcsin 0.540 \, 0$

    i.e. $\angle A = $ **32° 41′** (the other possible value of 147° 19′ being
                      ignored since the triangular plot is
                      acute angled)

If $\angle A = 32° \, 41'$ and $\angle B = 61° \, 49'$ then
    $\angle C = 180° - 32° \, 41' - 61° \, 49'$, i.e. $\angle C = 85° \, 30'$

**Hence the length of the third side of the plot of land is 42.44 m and
the angles of the plot are 61° 49′, 32° 41′ and 85° 30′**

Problem 3. The area of a field is in the form of a quadrilateral
PQRS, as shown in Fig. 15.15. Determine its area.

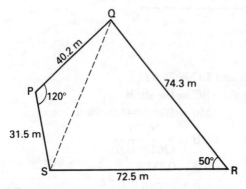

Fig. 15.15

If a diagonal is drawn from Q to S the quadrilateral is divided
into two triangles.

Area of quadrilateral PQRS = area of triangle PQS
                                   + area of triangle QRS
                                   $= \frac{1}{2}(40.3)(31.5) \sin 120°$
                                   $+ \frac{1}{2}(74.3)(72.5) \sin 50°$
                                   $= 548.3 + 2 \, 063.2$
                                   $= \mathbf{2 \, 612 \ m^2}$

Further problems on areas of triangles may be found in the following Section (15.5).

15.5   Further problems

1. Use the sine rule to solve the following triangles ABC:
   (a) $A = 37°$, $B = 73°$, $b = 4.30$ m
   (b) $A = 58°$, $B = 69°$, $b = 7.40$ cm
   (c) $A = 71°$, $B = 36°$, $a = 23.7$ mm
   (d) $B = 68° 15'$, $C = 57° 47'$, $b = 1.47$ m
   (e) $A = 114° 8'$, $C = 21° 5'$, $a = 14.6$ cm
   (f) $B = 111° 49'$, $C = 15° 19'$, $c = 43.81$ cm
   (a) $[a = 2.71$ m, $c = 4.23$ m, $C = 70°]$
   (b) $[a = 6.72$ cm, $c = 6.33$ cm, $C = 53°]$
   (c) $[b = 14.73$ mm, $c = 23.97$ mm, $C = 73°]$
   (d) $[a = 1.28$ m, $c = 1.339$ m, $A = 53° 58']$
   (e) $[b = 11.27$ cm, $c = 5.755$ cm, $B = 44° 47']$
   (f) $[a = 132.2$ cm, $b = 154.0$ cm, $A = 52° 52']$

2. Use the sine rule to solve the following triangles PQR:
   (a) $p = 8.52$ cm, $r = 12.7$ cm, $R = 24° 9'$
   (b) $q = 17.86$ m, $r = 12.67$ m, $Q = 83° 46'$
   (c) $p = 45$ mm, $r = 35$ mm, $P = 51° 19'$
   (d) $p = 19.53$ cm, $q = 15.96$ cm, $P = 102° 57'$
   (a) $[P = 15° 56', Q = 139° 55', q = 19.99$ cm$]$
   (b) $[P = 51° 23', R = 44° 51', p = 14.04$ m$]$
   (c) $[R = 37° 23', Q = 91° 18', q = 57.6$ mm$]$
   (d) $[Q = 52° 47', R = 24° 16', r = 8.237$ cm$]$

3. Use the sine rule to solve the following triangles DEF:
   (a) $d = 37.2$ cm, $e = 31.6$ cm, $E = 37°$
   (b) $e = 45.3$ mm, $f = 35.7$ mm, $F = 36° 47'$
   (c) $d = 1.46$ m, $f = 1.62$ m, $D = 26°$
   (d) $e = 43.36$ cm, $f = 30.2$ cm, $F = 29° 32'$
   (a) $[D = 45° 7', F = 97° 53', f = 52.01$ cm$]$
   or $[D = 134° 53', F = 8° 7', f = 7.414$ cm$]$
   (b) $[E = 49° 21', D = 93° 52', d = 59.57$ mm$]$
   or $[E = 130° 39', D = 12° 34', d = 12.97$ mm$]$
   (c) $[F = 29° 6', E = 124° 54', e = 2.732$ m$]$
   or $[F = 150° 54', E = 3° 6', e = 0.180\,1$ m$]$
   (d) $[E = 45° 3', D = 105° 25', d = 59.06$ cm$]$
   or $[E = 134° 57', D = 15° 31', d = 16.39$ cm$]$

4. Use the cosine rule to solve the following triangles XYZ:
   (a) $y = 11$ cm, $z = 15$ cm, $X = 55°$
   (b) $x = 62.8$ mm, $y = 41.2$ mm, $Z = 62°\ 11'$
   (c) $x = 19.47$ cm, $z = 12.83$ cm, $Y = 38°\ 29'$
   (d) $y = 32.8$ mm, $z = 43.7$ mm, $X = 102°\ 8'$
   (a) $[x = 12.5$ cm, $Y = 46°\ 8',\ Z = 78°\ 52']$
   (b) $[z = 56.80$ mm, $X = 77°\ 55',\ Y = 39°\ 54']$
   (c) $[y = 12.35$ cm, $X = 101°\ 10',\ Z = 40°\ 21']$
   (d) $[x = 59.90$ mm, $Y = 32°\ 22',\ Z = 45°\ 30']$

5. Use the cosine rule to solve the following triangles ABC:
   (a) $a = 9$ cm, $b = 7$ cm, $c = 6$ cm
   (b) $a = 12.9$ cm, $b = 15.8$ cm, $c = 5.90$ cm
   (c) $a = 25$ mm, $b = 32$ mm, $c = 39$ mm
   (d) $a = 8.983$ cm, $b = 12.46$ cm, $c = 15.91$ cm
   (a) $[A = 87°\ 16',\ B = 50°\ 59',\ C = 41°\ 45']$
   (b) $[A = 50°\ 43',\ B = 108°\ 33',\ C = 20°\ 44']$
   (c) $[A = 39°\ 43',\ B = 54°\ 52',\ C = 85°\ 25']$
   (d) $[A = 34°\ 13',\ B = 51°\ 20',\ C = 94°\ 27']$

6. Find the areas of the triangles ABC of question 1.
   (a) $[5.473$ m$^2]$ (b) $[19.86$ cm$^2]$ (c) $[166.96$ mm$^2]$
   (d) $[0.796\ 0$ m^2 (or $7\ 960$ cm^2)]$
   (e) $[29.59$ cm$^2]$ (f) $[2\ 689$ cm^2 (or $0.268\ 9$ m^2)]$

7. Find the areas of the triangles PQR of question 2.
   (a) $[34.84$ cm$^2]$ (b) $[88.42$ m$^2]$
   (c) $[787$ mm$^2]$ (d) $[64.05$ cm$^2]$

8. Find the areas of the triangles DEF of question 3.
   (a) $[582.2$ cm^2 or 82.99 cm$^2]$
   (b) $[806.8$ mm^2 or 175.9 mm$^2]$
   (c) $[0.969\ 9$ m^2 or 639.5 cm$^2]$
   (d) $[631.2$ cm^2 or 175.2 cm$^2]$

9. Find the areas of the triangles XYZ of question 4.
   (a) $[67.6$ cm$^2]$ (b) $[1\ 144$ mm$^2]$
   (c) $[77.72$ cm$^2]$ (d) $[700.7$ mm$^2]$

10. Find the areas of the triangles ABC of question 5.
    (a) $[21$ cm$^2]$ (b) $[36.08$ cm$^2]$
    (c) $[399$ mm$^2]$ (d) $[55.80$ cm$^2]$

11. The sides of a triangle are in the ratio 4:5:6.
    Find the angles of the triangle.
    $[41°\ 25',\ 55°\ 46',\ 82°\ 49']$

12. Two voltage phasors are shown in Fig. 15.16. If $v_1 = 60$ V and $v_2 = 90$ V, calculate the value of their resultant (i.e. the length AB) and the angle the resultant makes with v_1.
    [145 V, 18° 4′]

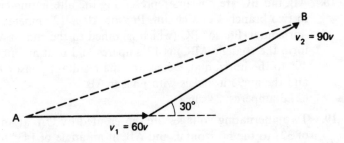

Fig. 15.16

13. Two ships, A and B, leave a port at the same time. A sails at a steady speed of 52 km h^{-1}, S 32° W and B at 38 km h^{-1}, S 24° E. Find their distance apart after $2\frac{1}{2}$ hours.
    [110 km]

14. Two sides of an acute angled triangular plot of land are 46 m and 25 m respectively. If the area of the plot is 526 m^2, find the length of fencing required to enclose the plot of land.
    [114 m]

15. A room 9 m wide has a span roof which slopes at 32° on one side and 41° on the other. Find the length of the roof slopes, to the nearest centimetre.   [6.17 m, 4.99 m]

16. A jib crane consists of a vertical stanchion AB, 5.2 m in length, the inclined jib BC, 12.8 m in length, and a tie AC. Angle BAC is 122°. Calculate (a) the length of the tie and (b) the inclination of the jib to the vertical.
    (a) [9.26 m]   (b) [37° 51′]

17. A reciprocating engine mechanism is shown in Fig. 15.17, where XY represents the rotating crank, YZ is the connecting rod and Z is the piston which moves vertically along the broken line XZ.

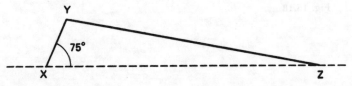

Fig. 15.17

If the rotating crank is 1.24 cm in length and the connecting rod is 6.48 cm, calculate for the position shown (a) the inclination of the connecting rod to the vertical and (b) the distance XZ.   (a) [10° 39′]   (b) [6.689 cm]

18. AB and BC are phasors representing the alternating currents in two branches of a circuit. Phasor AB is 12 amperes and is horizontal. Phasor BC (which is joined to the end of AB to form the triangle ABC) is 17 amperes and is at an angle of 40° to the horizontal. Determine the resultant phasor AC and the angle it makes with phasor AB.
[27.3 amperes; 23° 36′]

19. Two alternating voltages are represented by OA, at an angle of 32° to the horizontal, and AB, at an angle of 143° to the horizontal. (Voltage phasor AB is joined to the end of voltage phasor OA to form the triangle OAB.) If OA = 23.0 volts and AB = 19.0 volts, find the resultant voltage OB and its angle measured to the horizontal.
[24.02 volts; 79° 36′]

20. A park is in the form of a quadrilateral ABCD, as shown in Fig. 15.18, and its area is 2 791 m². Determine the length of (a) the perimeter fencing and (b) the short-cut BD across the park.   (a) [219.4 m]   (b) [70.37 m]

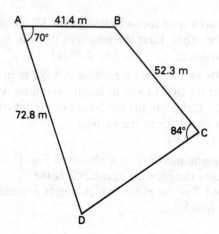

Fig. 15.18

Chapter 16

Practical trigonometric problems

16.1 Introduction

There are many practical examples where the use of trigonometry is essential to find unknown sides and angles of triangles. By using trigonometric ratios and Pythagoras's theorem for right-angled triangles, and the sine and cosine rules for other triangles, any problem involving triangles may be solved provided sufficient information is given.

It is very important that any diagram associated with a problem can be visualised and hence, a clearly labelled sketch should always be made. If the diagram is 'three-dimensional' then a number of constituent triangles will be evident. It is often useful to redraw the constituent triangles separately together with all relevant dimensional information.

16.2 The angle between a line and a plane

The angle between a line and a plane is defined as the angle between the line and its projection on the plane.

In Fig. 16.1 the line AB is shown making an angle with the plane DEFG.

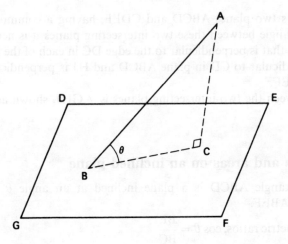

Fig. 16.1

273

If the line AC is constructed perpendicular to plane DEFG then the projection of the line AB on the plane DEFG is given by the length BC (where BC lies in the plane DEFG and is thus perpendicular to AC). In Fig. 16.1 angle θ is the angle between the line AB and the plane DEFG.

16.3   The angle between two intersecting planes

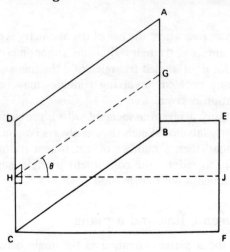

Fig. 16.2

Figure 16.2 shows two planes ABCD and CDEF, having a common edge CD. To find the angle between these two intersecting planes it is necessary to establish a line that is perpendicular to the edge DC in each of the planes, i.e. GH is perpendicular to CD in plane ABCD and HJ is perpendicular to CD in plane CDEF.

The angle between the two intersecting planes is $\angle$ GHJ, shown as angle θ in Fig. 16.2.

16.4   Lengths and areas on an inclined plane

In Fig. 16.3, rectangle ABCD is a plane inclined at an angle θ to the horizontal plane ABEF.

From trigonometric ratios, $\cos \theta = \dfrac{BE}{BC}$

from which $\qquad\qquad BC = \dfrac{BE}{\cos \theta} (= BE \sec \theta)$

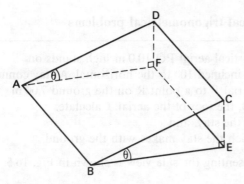

Fig. 16.3

Hence the line of greatest slope on an inclined plane is given by

$$\left(\frac{1}{\cos\theta}\right) \text{(its projection onto the horizontal plane)}$$

Area of ABCD = (AB) (BC)

$$= (AB)\left(\frac{BE}{\cos\theta}\right) = \left(\frac{1}{\cos\theta}\right)\text{(area of horizontal plane)}$$

An example where this latter expression is used is in calculations of roof areas in construction work where the plane area is known (see Problems 9 and 10).

16.5 Angles of elevation and depression

Let AB in Fig. 16.4(a) represent horizontal ground and BC a vertical flagpole. The **angle of elevation** of the top of the flagpole, C, from the point A is the angle that the imaginary straight line AC must be raised (or elevated) from the horizontal AB, i.e. angle θ.

Let PR represent a vertical cliff and Q a ship at sea as shown in Fig. 16.4(b). The **angle of depression** of the ship from point R is the angle that the imaginary straight line RQ must be lowered (or depressed) from the horizontal drawn through R to the ship, i.e. angle α. (Angle PQR is also equal to angle α since they are alternate angles between parallel lines.)

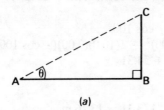

(a)   (b)

Fig. 16.4

Worked practical trigonometrical problems

Problem 1. A vertical aerial PQ 10.0 m high stands on ground which is inclined 10° to the horizontal. A stay connects the top of the aerial P to a point R on the ground 7.00 m downhill from Q, the foot of the aerial. Calculate:
(*a*) the length of the stay, and
(*b*) the angle which the stay makes with the ground.

A diagram representing the side view is shown in Fig. 16.5.

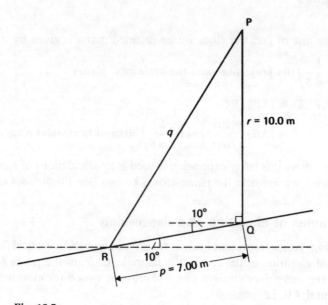

Fig. 16.5

(*a*)   To find the length of stay PR (i.e. side *q*):
         Since PQ = 10.0 m, QR = 7.00 m and
         ∠PQR = 90° + 10° = 100°, the cosine rule is used.

$$\text{Thus } q^2 = r^2 + p^2 - 2\,rp\cos 100°$$
$$= (10.0)^2 + (7.00)^2 - 2(10.0)(7.00)(-\cos 100°)$$
$$= 100.0 + 49.0 + 24.31$$
$$= 173.31$$

Hence length of stay $q = \sqrt{173.31} = 13.16$ **m**
(*b*)   To find the angle which the stay makes with the ground

(i.e. angle PRQ):

Using the sine rule: $\dfrac{r}{\sin R} = \dfrac{q}{\sin Q}$

$$\sin R = \frac{r \sin Q}{q} = \frac{10.0 \sin 100°}{13.16}$$

Therefore sin R = 0.748 3

Hence angle PRQ = 48° 27′

Problem 2. A vertical mast stands on horizontal ground. A man, positioned due west of the mast, finds the elevation of the top to be 51° 5′. He moves due south 20.0 m and finds the elevation to be 47° 19′. Calculate the height of the mast.

In the diagram shown in Fig. 16.6, AB represents the vertical mast (hence ∠ABC is a right angle), C is the point where the man is positioned initially and D is his final position. Since he moves due south from C, ∠BCD is a right angle. The relevant constituent triangles are shown in Fig. 16.7

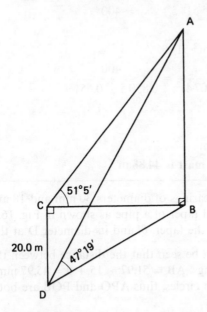

Fig. 16.6

From triangle ABC   :   BC = AB cot 51° 5′
From triangle ABD   :   BD = AB cot 47° 19′
From triangle BCD, using the theorem of Pythagoras:
$$BD^2 = CD^2 + BC^2$$

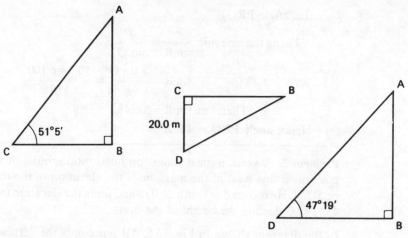

Fig. 16.7

$(AB \cot 47° \ 19')^2 = (20.0)^2 + (AB \cot 51° \ 5')^2$

$(AB \cot 47° \ 19')^2 - (AB \cot 51° \ 5')^2 = (20.0)^2$

$AB^2 (\cot^2 47° \ 19' - \cot^2 51° \ 5') \qquad = 400$

$$AB^2 = \frac{400}{\cot^2 47° \ 19' - \cot^2 51° \ 5'}$$

$$AB^2 = \frac{400}{(0.922\,2)^2 - (0.807\,4)^2} = \frac{400}{0.850\,5 - 0.651\,9}$$

$$= \frac{400}{0.198\,6} = 2\,014$$

$AB \ = \sqrt{2\,014} = 44.88 \text{ m}$

Hence the height of the mast is 44.88 m

Problem 3. Two ball bearings of diameters 30 mm and 18 mm are used to find the internal taper of a pipe as shown in Fig. 16.8. Determine the angle of the taper, θ, and its diameter D at the top.

From Fig. 16.9(*a*) it can be seen that the distance between the centre of the ball bearings, $AB = 51.97 - 15 + 9 = 45.97$ mm. PQ is a tangent to both circles, thus APQ and PQB are both right angles.

Since $BQ = 9$ mm, then $TP = 9$ mm and $AT = 15 - 9 = 6$ mm.

Since the whole taper is angle θ, then angle ABT is $\dfrac{\theta}{2}$

Using trigonometric ratios, $\sin\dfrac{\theta}{2} = \dfrac{AT}{AB} = \dfrac{6}{45.97}$

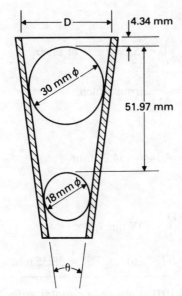

Fig. 16.8

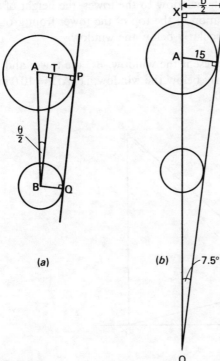

(a)   (b)

Fig. 16.9

Hence $\dfrac{\theta}{2} = \arcsin\!\left(\dfrac{6}{45.97}\right)$ and

$$\theta = 2\arcsin\!\left(\dfrac{6}{45.97}\right) = 15° = \text{angle of taper}$$

From Fig. 16.9(*b*), $\sin 7.5° = \dfrac{15}{\text{AO}}$, from which,

$$\text{AO} = \dfrac{15}{\sin 7.5°} = 114.92 \text{ mm}$$

Length $\text{OX} = 114.92 + 15 + 4.34 = 134.26 \text{ mm}$

Using trigonometric ratios,

$$\tan 7.5° = \dfrac{\dfrac{D}{2}}{\text{OX}}, \text{ from which, } \dfrac{D}{2} = \text{OX}\tan 7.5°$$

Hence top diameter, $D = 2(134.26)\tan 7.5° = \textbf{35.35 mm}$

Problem 4. From a window 10.0 m above horizontal ground the angle of elevation of the top of a vertical tower is 42° and the angle of depression of the bottom of the tower is 13°. Calculate the distance from the window to the tower, the height of the tower and the elevation of the top of the tower from ground level at a point perpendicularly below the window.

Let the point A represent the window, BC the tower and D the point perpendicularly below the window, i.e. AD = 10.0 m, as shown in Fig. 16.10

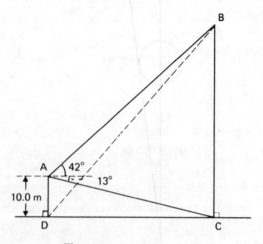

Fig. 16.10

To find the distance from the window to the tower (i.e. length DC):

Since the angle of depression of C from A is 13° then ∠ACD is 13° (alternate angles between parallel lines).

Hence the distance from the window to the tower, DC = 10.0 cot 13°
$$= \textbf{43.31 m}$$

To find the height of the tower (i.e. length BC):

From triangle ADC: AC = 10.0 cosec 13° = 44.45 m
Since ∠ACD = 13°, ∠BCA = 77° because BC is vertical.
Hence ∠ABC = 180° − 42° − 13° − 77° = 48°

Using the sine rule on triangle ABC: $\dfrac{AC}{\sin B} = = \dfrac{BC}{\sin A}$

$$BC = \frac{AC \sin A}{\sin B}$$

$$= \frac{(44.45)(\sin (42° + 13°))}{\sin 48°}$$

$$BC = 49.0 \text{ m}$$

Hence the height of the tower is 49.0 m

To find the elevation of point B from D (i.e. to find ∠BDC):

$$\tan BDC = \frac{BC}{DC} = \frac{49.0}{43.31} = 1.131 \, 4$$

Hence ∠BDC = arctan 1.131 4 = 48° 32′

Hence the angle of elevation of the top of the tower from point D is 48° 32′

Problem 5. A crank mechanism shown in Fig. 16.11(a) comprises arm OP, of length 1.20 m, which rotates anticlockwise about the fixed point O, and connecting rod PQ of length 4.70 m. End Q moves horizontally in a straight line OR. If ∠POR is initially zero how far does end Q travel
    (a) in ¼ revolution of OP?
    (b) in ½ revolution of OP?
Find also the distance Q moves (to the nearest centimetre) when ∠POR changes from 30° to 145°.

(a)   When ∠POR is zero, OQ = 1.20 m + 4.70 m = 5.90 m.
    After ¼ revolution ∠POR is 90° as shown in Fig. 16.11(b).
    Using the theorem of Pythagoras: OQ = $\sqrt{(4.70^2 - 1.20^2)}$
                                              = 4.540 m

Hence the distance moved by Q in ¼ revolution is 5.90 − 4.54, i.e. 1.36 m

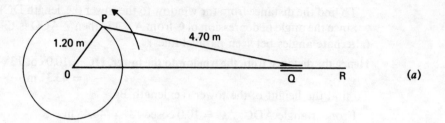

(a)

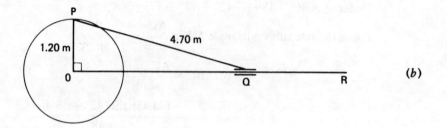

(b)

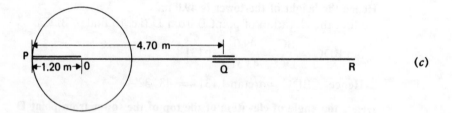

(c)

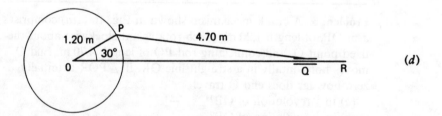

(d)

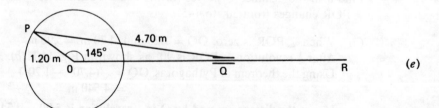

(e)

Fig. 16.11

(b)   After $\frac{1}{2}$ revolution (shown in Fig. 16.11(c)),
OQ = 4.70 − 1.20 = 3.50 m.

Hence Q has moved from 5.90 m in the zero position to 3.50 m at 180°, a distance of 2.40 m

When ∠POR is 30°, as shown in Fig. 16.11(d), ∠PQO and then the length OQ may be found using the sine rule.

Hence $\dfrac{OP}{\sin PQO} = \dfrac{PQ}{\sin POQ}$. Therefore $\dfrac{1.20}{\sin PQO} = \dfrac{4.70}{\sin 30°}$

$$\therefore \sin PQO = \frac{1.20 \sin 30°}{4.70} = 0.127\,7$$

$$\angle PQO = 7°\,20'$$

Therefore ∠OPQ = 180° − 30° − 7° 20′ = 142° 40′

Hence $\dfrac{OQ}{\sin OPQ} = \dfrac{PQ}{\sin POQ}$

$$\frac{OQ}{\sin 142°\,40'} = \frac{4.70}{\sin 30°}$$

Therefore OQ $= \dfrac{4.70 \sin 142°\,40'}{\sin 30°} =$ **5.70 m**

When ∠POR is 145°, as shown in Fig. 16.11(e), ∠PQO and then the length OQ may be found using the sine rule.

Hence $\dfrac{PQ}{\sin POQ} = \dfrac{OP}{\sin PQO}$

$$\frac{4.70}{\sin 145°} = \frac{1.20}{\sin PQO}$$

$$\sin PQO = \frac{1.20 \sin 145°}{4.70} = 0.146\,4$$

$$\angle PQO = 8°\,25'$$

Therefore ∠OPQ = 180° − 145° − 8° 25′ = 26° 35′

Hence $\dfrac{OQ}{\sin OPQ} = \dfrac{PQ}{\sin POQ}$

$$\frac{OQ}{\sin 26°\,35'} = \frac{4.70}{\sin 145°}$$

$$OQ = \frac{4.70 \sin 26°\,35'}{\sin 145°} = \textbf{3.67 m}$$

Hence when angle POR (i.e. ∠POQ) changes from 30° to 145°, Q moves a distance of 5.70 − 3.67 or 2.03 m

Problem 6. A grading hopper is initially in the shape of a right pyramid, of vertex A, and whose base is a rectangle BCDE. F and G are the midpoints of BC and CD. AB is 15.0 cm, BC is 11.0 cm and CD is 8.00 cm. Calculate:

  (i) the perpendicular height of the pyramid,
  (ii) the angle that the edge AB makes with the base,
  (iii) the angles which the faces ABC and ACD make with the base,
and (iv) the angle the plane AFG makes with the base.

The pyramid is shown in Fig. 16.12.

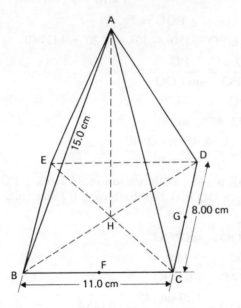

Fig. 16.12

  (i) As the pyramid is regular the perpendicular height is given by AH where H is the intersection of diagonals BD and CE. The length AH is part of the right-angled triangle ABH shown in Fig. 16.13(a).

Diagonal BD = $\sqrt{(11.0^2 + 8.00^2)} = \sqrt{185} = 13.60$ cm
Hence BH $= \frac{1}{2}BD = \frac{1}{2}(13.60)$ $= 6.80$ cm

Using the theorem of Pythagoras on right-angled triangle ABH:

AH $= \sqrt{(15.0^2 - 6.80^2)} = 13.37$ cm

Hence the perpendicular height is 13.37 cm

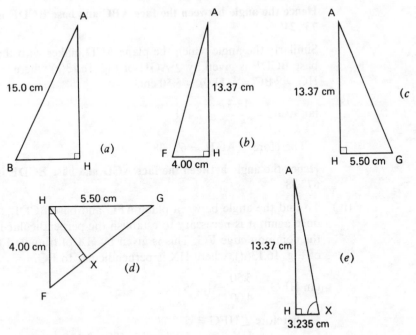

Fig. 16.13

(ii)   The angle which a line makes with a plane is the angle
which it makes with its projection on the plane. In the
pyramid, BH is the projection of AB on to the base BCDE.
Hence the required angle is ∠ABH.

$$\cos ABH = \frac{BH}{AB} = \frac{6.80}{15.0} = 0.4533$$

Therefore ∠ABH = 63° 3′

Hence the edge AB makes an angle of 63° 3′ with the base.

(iii)   To find the angle between two planes it is necessary to
establish a line perpendicular to their common edge. In this
case the line from F to H in Fig. 16.12 is perpendicular to
the common edge BC of the planes ABC and BCDE. Thus
∠AFH is the angle between planes ABC and BCDE.
Right-angled triangle AHF is shown in Fig. 16.13(*b*) with
length FH = ½CD = ½(8.00) = 4.00 cm.

$$\tan AFH = \frac{13.37}{4.00} = 3.343$$

Therefore ∠AFH = 73° 21′

Hence the angle between the face ABC and base BCDE is 73° 21′

Similarly the angle which the plane ACD makes with the base BCDE is given by $\angle$ AGH of Fig. 16.13(c) where HG $= \frac{1}{2}$BC $= \frac{1}{2}(11.0) = 5.50$ cm.

$$\tan AGH = \frac{13.37}{5.50} = 2.431$$

Therefore $\angle$ AGH $= 67°\ 38′$

Hence the angle between the face ACD and base BCDE is 67° 38′

(iv) To find the angle between plane AFG and base BCDE, once again it is necessary to establish the perpendicular to the common edge FG. This is given by HX of triangle FGH of Fig. 16.13(d), where HX is perpendicular to FG.

$$\tan HFG = \frac{5.50}{4.00} = 1.375$$

Therefore $\angle$ HFG $= 53°\ 58′$
    HX $= 4.00 \sin 53°\ 58′ = 3.235$ cm

The angle between the plane AFG and the base BDCE is the angle AXH of the right-angled triangle AHX shown in Fig. 16.13(e).

$$\tan AXH = \frac{13.37}{3.235} = 4.1329$$

Therefore AXH $= 76°\ 24′$

Hence the angle between plane AFG and base BCDE is 76° 24′

Problem 7. An aeroplane is sighted due west from a radar station at an elevation of 38° 0′ and a height of 7 500 m and later at an elevation of 34° 0′ and height 5 000 m in direction W 65° 17′ S. If it is descending uniformly, find the angle of descent. Find also the speed of the aeroplane (in m s^{-1} and in km h^{-1}) if the time between the two observations is 50.0 seconds.

Figure 16.14 shows radar station R, the initial position, A, of the aeroplane and the later position, B, of the aeroplane.

From the right-angled triangle ACR, CR $= 7\,500 \cot 38°\ 0′$
                    CR $= 9\,600$ m
From the right-angled triangle BRD, DR $= 5\,000 \cot 34°\ 0′$
                    DR $= 7\,413$ m

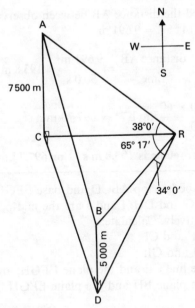

Fig. 16.14

Using the cosine rule on triangle CRD,

$$CD^2 = CR^2 + DR^2 - 2(CR)(DR) \cos 65° \, 17'$$
$$= (9\,600)^2 + (7\,413)^2 - 2(9\,600)(7\,413) \cos 65° \, 17'$$
$$= 87\,600\,000$$

Therefore $CD = \sqrt{87\,600\,000} = 9\,359$ m

The angle of descent is shown as angle ϕ in Fig. 16.15

$$\tan \phi = \frac{AE}{EB} = \frac{2\,500}{9\,359} = 0.267\,1$$

Hence the angle of descent, $\phi = 14° \, 57'$

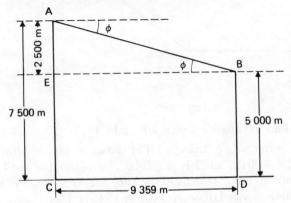

Fig. 16.15

The aeroplane travelled the distance AB between observations where AB = 2 500 cosec 14° 57′ = 9 691 m.

$$\text{Speed of the aeroplane} = \frac{\text{distance AB}}{\text{time}} = \frac{9\,691\text{ m}}{50.0\text{ s}} = 193.8\text{ m s}^{-1}$$

$$193.8\text{ m s}^{-1} = \frac{193.8 \times 60 \times 60}{1\,000}\text{ km h}^{-1} = 697.7\text{ km h}^{-1}$$

Hence the speed of the aeroplane is 193.8 m s^{-1} or 697.7 km h^{-1}

Problem 8. A 4.00-cm cube has top ABCD and base EFGH with vertical edges AE, BF, CG and DH. I and J are the mid-points of sides EH and HG respectively. Calculate:
(*a*) the angle between DF and CF;
(*b*) the angle between DI and GI;
(*c*) the angle between the line CE and the plane EFGH; and
(*d*) the angle between the plane BIJ and the plane EFGH.

The cube is shown in Fig. 16.16.

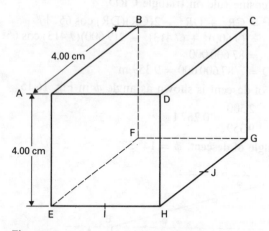

Fig. 16.16

(*a*) To find the angle between DF and CF:

DF is one side of triangle DFH shown in Fig. 16.17(*a*) where
DH = 4.00 cm and FH = $\sqrt{[(4.00)^2 + (4.00)^2]}$ = 5.657 cm.
Hence all the diagonals of the cube are 5.657 cm. Since
∠DHF is 90°, DF = $\sqrt{[(4.00)^2 + (5.657)^2]}$, i.e.
DF = 6.928 cm.

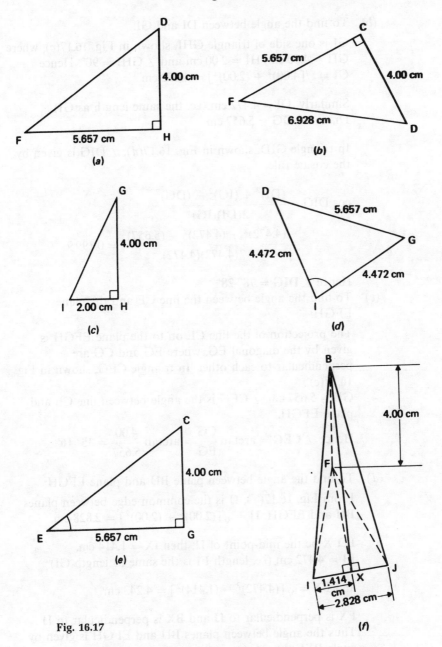

Fig. 16.17

In triangle CFD, shown in Fig. 16.17(*b*), FD = 6.928 cm, CF = 5.657 cm and CD = 4.00 cm.

$$\angle\, \mathbf{CFD} = \arctan \frac{4.00}{5.657} = \mathbf{35^\circ\ 16'}$$

(b) To find the angle between DI and GI:

GI is one side of triangle GHI, shown in Fig. 16.17(c), where
GH = 4.00 cm, HI = 2.00 cm and $\angle$ GHI = 90°. Hence
GI = $\sqrt{[4.00)^2 + (2.00)^2]}$ = 4.472 cm.

Similarly, DI = 4.472 cm (i.e. the same length as GI).
Diagonal DG = 5.657 cm

In triangle GID, shown in Fig. 16.17(d), $\angle$ DIG is given by
the cosine rule.

$$\cos DIG = \frac{(DI)^2 + (IG)^2 - (DG)^2}{2(DI)(IG)}$$

$$= \frac{(4.472)^2 + (4.472)^2 - (5.657)^2}{2(4.472)(4.472)} = 0.199\,9$$

Hence $\angle$ **DIG = 78° 28′**

(c) To find the angle between the line CE and the plane
EFGH:

The projection of the line CE on to the plane EFGH is
given by the diagonal EG, where EG and CG are
perpendicular to each other. In triangle CEG, shown in Fig.
16.17(e),
GE = 5.657 cm, $\angle$ CEG is the angle between line CE and
plane EFGH.

Hence $\angle$ **CEG** $= \arctan \dfrac{CG}{EG} = \arctan \dfrac{4.00}{5.657} = $ **35° 16′**

(d) To find the angle between plane BIJ and plane EFGH:

From Fig. 16.17(f), IJ is the common edge between planes
BIJ and EFGH. IJ = $\sqrt{[(2.00)^2 + (2.00)^2]}$ = 2.828 cm

Let X be the mid-point of IJ, then IX = 1.414 cm,
FI = 4.472 cm (i.e. length FI is the same as length GI)

and FX = $\sqrt{[(4.472)^2 - (1.414)^2]}$ = 4.243 cm

FX is perpendicular to IJ and BX is perpendicular to IJ.
Thus the angle between planes BIJ and EFGH is given by
angle BXF.

Hence $\angle$ **BXF** $= \arctan \dfrac{4.00}{4.243} = $ **43° 19′**

Problem 9. A hipped roof has a rectangular plan 6 m by 4.5 m. Determine the surface area of the roof if it slopes at 40° to the horizontal.

From Section 16.4, area of sloping roof $= \left(\dfrac{1}{\cos \theta}\right)$ (plan area)

$$= \left(\dfrac{1}{\cos 40°}\right)(6 \times 4.5) = 35.25 \text{ m}^2$$

Problem 10. A solid cylindrical bar of steel has a diameter of 20 mm. Determine the area of a section cut across the bar at an angle of 45°.

Area of section (which is elliptical)

$$= \left(\dfrac{1}{\cos 45°}\right) \text{(area of circular cross-section)}$$

$$= \left(\dfrac{1}{\cos 45°}\right)\left(\dfrac{\pi d^2}{4}\right) = \left(\dfrac{1}{\cos 45°}\right)\left(\dfrac{\pi 20^2}{4}\right)$$

$$= 444.3 \text{ mm}^2$$

Further problems on practical trigonometrical problems may be found in the following Section (16.6) (Problems 1–20).

16.6 Further problems

1. The angle of elevation of the top of a hill from a certain point is 6° 12′. After walking 1.42 km directly towards the hill the elevation of the top is 15° 17′. Find the height of the hill.
   [256.1 m]

2. From the top of a vertical cliff 75.0 m high, forming one bank of a river, the angles of depression of the top and bottom of a vertical cliff which forms the opposite bank are 22° and 58° respectively. Determine the height of the second cliff and the width of the river.   [56.06 m, 46.87 m]

3. A vertical aerial MN stands on ground which is inclined at 14° 11′ to the horizontal. A stay 24.3 m long connects the top of the aerial M to a point O on the ground 15.8 m downhill from N the foot of the aerial. Calculate (a) the height of the aerial, and (b) the angle which the stay makes with the ground.
   (a) [14.99 m]   (b) [36° 44′]

4. A crank mechanism comprises arm AB of length 150 cm which rotates anticlockwise about the fixed point A, and connecting rod BC of length 5.10 m. End C is constrained to move in a

straight line AD. Find the distance C travels (to the nearest centimetre) when angle CAB changes from zero to 120°. [2.42 m]

5. A tent is in the form of a regular octagonal pyramid of base side 4.0 m. If the perpendicular height of the tent is 10 m calculate
    (a) the angle between a sloping face and the base, and
    (b) the angle of inclination that an edge of the sloping face makes with the base.   (a) [64° 14′]   (b) [62° 24′]

6. The base of a roof, shown in Fig. 16.18 is a rectangle ABCD. The equal faces ADE and BCF each make an angle of 45° with the base; the equal faces ABFE and CDEF each make an angle of 50° with the base. Calculate:
    (a) the perpendicular height of the roof,
    (b) the length EF,
    (c) the length of the sloping edge AE, and
    (d) the angle AE makes with the base.
    (a) [5.96 m]   (b) [4.08 m]   (c) [9.80 m]   (d) [37° 27′]

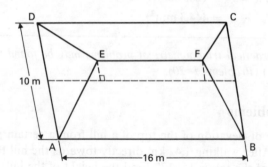

Fig. 16.18

7. The base of a right pyramid of vertex P is a rectangle QRST. A and B are the midpoints of QR and RS respectively. PQ is 9.20 cm, QR is 7.60 cm and RS is 4.20 cm. Calculate:
    (a) the perpenciular height of the pyramid,
    (b) the angle that edge PT makes with the base,
    (c) the angles which faces PQR and PRS make with the base, and
    (d) the angle the plane PAB makes with the base.
    (a) [8.111 cm]   (b) [61° 50′]   (c) [75° 29′, 64° 54′]
    (d) [77° 14′]

8. Two ball bearings of diameter 36 mm and 20 mm are used to find the internal taper of a tube as shown in Fig. 16.19. Determine (a) the angle of the taper, θ, and (b) the diameter, D.
    (a) [17°]   (b) [43.64 mm]

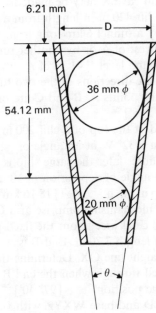

6.21 mm

D

36 mm φ

54.12 mm

20 mm φ

θ

Fig. 16.19

9. A vertical mast stands on horizontal ground. A man standing due south of the mast finds the elevation of the top to be 38° 52'. He walks due west 42.0 m and finds the elevation of the top of the mast to be 33° 10'. Calculate the height of the mast. [46.91 m]

10. An aeroplane is sighted due north from a radar station at an elevation of 49° 4' with a height of 10 000 m and later at an elevation of 37° 19' and height 6 250 m in direction E 24° S. If it is descending uniformly find the angle of descent and the speed of the aeroplane (in m s⁻¹ and km h⁻¹) if the time between the two observations is 1 minute 12 seconds. [14° 50', 203.34 m s⁻¹ or 732.03 km h⁻¹]

11. A ship, situated S 17° W of a vertical lighthouse, measures the angle of elevation of the top of the lighthouse as 13°. Later, when the ship is S 02° E from the lighthouse the angle of elevation is 9°. Find the height of the lighthouse if the distance between the two observations is 146 m. [55.55 m]

12. A surveyor, standing W 30° S of a tower, measures the angle of elevation of the top of the tower as 44° 19'. From a position E 28° S from the tower the elevation of the top is 36° 43'. Calculate the height of the tower if the distance between the two observations is 120 m. [57.86 m]

13. The access to an underground cave X may be made from three entrances; in a vertical lift 420 m in length from a point A at ground level or by a tunnel 6 km long from a point B at ground level due west of A or by another tunnel 4.2 km from a point C at ground level E 41° S of A. Calculate the angle between the directions of the two tunnels (i.e. ∠ BXC) assuming that the points A, B and C are on horizontal ground.   [137° 55′]

14. A coastguard station situated at the top of a cliff 200 m high observes a ship in a direction S 32° W at an angle of depression of 9° 14′. Five minutes later the same ship is seen in a direction W 25° N at an angle of depression of 10° 51′. Calculate the speed of the ship in km h^{-1}.   [18.16 km h^{-1}]

15. A crank and connecting rod mechanism comprise arm OA of length 0.5 m which rotates clockwise about the fixed point O and connecting rod AB of length 3.70 m. End B is constrained to move in a straight line OX. Determine the angle that arm OA has moved through when the end B has moved 60 cm from its extreme position.   [97° 40′]

16. A 300-cm cube has top ABCD and base WXYZ with vertical edges AW, BX, CY and DZ. P and Q are the mid-points of sides WZ and ZY respectively. Calculate: (a) the angle between DX and CX; (b) the angle between DP and YP; (c) the angle between the line CW and the plane WXYZ; and (d) the angle between the plane BPQ and the plane WXYZ.
(a) [35° 16′]   (b) [78° 28′]   (c) [35° 16′]
(d) [43° 19′]

17. A 15.30 cm-high right circular cone of vertex A has a base diameter of 9.80 cm. If point B is on the circumference of the base find the inclination of AB to the base. C is also a point on the circumference of the base, distance 7.60 cm from B. Find the inclination of the plane ABC to the base.
[72° 15′, 78° 34′]

18. A hipped roof has a rectangular plan 5 m by 4 m. Find the surface area of the roof if its slope is 37° to the horizontal. [25.04 m^2]

19. A vertical, cylindrical ventilation shaft of diameter 300 mm has its end at an angle of 24° to the horizontal. Determine the area in cm^2 of a plate required to cover the end.   [773.8 cm^2]

20. A chimney stack has a diameter of 1.2 m and passes through a roof that has a pitch of 35°. Determine the area of the resulting void in the roof covering.   [1.38 m^2]

Chapter 17

Areas and volumes

17.1 Volumes of pyramids, cones and spheres

A solid with a plane end and straight sides meeting in a point is called a **pyramid**. The shape of the plane end may be triangular, rectangular, circular, pentagonal or that of any other polygon.

A pyramid with a circular base is given the special name of a **cone**.

A **sphere** is a solid such that every point on its surface is a constant distance, called the radius, from a fixed point, called the centre.

(a)   The volume V of any **pyramid** of base A and perpendicular height h is given by:

Volume $= \frac{1}{3} \times$ area of base $\times$ perpendicular height.

   i.e. $V = \frac{1}{3} A h$ cubic units.

(b)   The volume V of a **cone** of base radius r and perpendicular height h is given by:

$V = \frac{1}{3}(\pi r^2) h$ **cubic units.**

   since the base is a circle of area πr^2 (i.e. area of base $A = \pi r^2$).

(c)   The volume V of a **sphere** of radius r is given by:

$V = \frac{4}{3}\pi r^3$ **cubic units.**

17.2 Surface areas of pyramids, cones and spheres

(a)   The surface area S of a **pyramid** is given by:

   $S = $ **[Area of base + sum of the triangles forming the sides] square units.**

(b)   Consider a hollow cone of base radius r and slant height ℓ shown in Fig. 17.1. If a perpendicular cut is made from a point on the circumference of the base to the vertex (shown by the broken line in Fig. 17.1(a)), and the cone is opened out, a sector of a circle is produced (Fig. 17.1(b)). The circumference of the base of the cone is $2\pi r$. Thus the arc length of the sector of the circle is $2\pi r$.

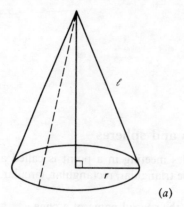

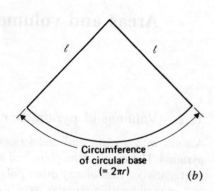

Circumference
of circular base
(= 2πr)

(a)   (b)

Fig. 17.1

The area of a sector of a circle of radius ℓ and arc length $2\pi r$ is given by:

$$\left(\frac{\text{arc length of sector}}{\text{circumference of circle}}\right) \times (\text{area of circle})$$

That is, $\dfrac{2\pi r}{2\pi\ell}(\pi\ell^2) = \pi r\ell$ square units.

Hence the curved surface area S_c of a **cone** is given by:

$S_c = \pi r\ell$ **square units**

The total surface area S_T of a **solid cone** (that is, the area of the curved surface and the area of the base) is given by:

$S_T = (\pi r\ell + \pi r^2)$ **square units**

(c)   The surface area S of a **sphere** of radius r is given by:

$S = 4\pi r^2$ **square units**

Worked problems on volumes and surface areas of pyramids, cones and spheres

In each of the following problems the accuracy of the calculations is consistent with the given data and is taken as one significant figure more than the least significant data.

Problem 1. Find the volume and the total surface area of the square pyramid of perpendicular height 9.41 cm and length of side of base 2.92 cm as shown in Fig. 17.2.

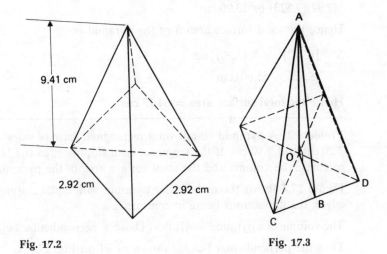

Fig. 17.2   **Fig. 17.3**

The volume of a pyramid, $V = \frac{1}{3}$(area of base × perpendicular height),

$$\therefore V = \tfrac{1}{3}(2.92)^2(9.41)$$

Hence volume, $V = 26.74$ cm^3

The total surface area consists of a square base and four equal triangles. To find the area of one of the triangles, the slant height needs to be calculated (that is, the length AB shown in Fig. 17.3, where B is the mid-point of the side CD).

The perpendicular height of the pyramid, OA, is 9.41 cm and the length OB is $\dfrac{2.92}{2}$ or 1.46 cm

Using the theorem of Pythagoras:

$$AB^2 = AO^2 + OB^2 = (9.41)^2 + (1.46)^2$$

$$= 88.55 + 2.13 = 90.68$$

$$\therefore AB = \sqrt{90.68} \qquad = \pm 9.523 \text{ cm}$$

In practical examples such as this, the negative value of the square root is always neglected since it has no meaning in this context. In all future practical problems of this type the negative value will be ignored.

Hence the slant height, AB = 9.523 cm

Therefore the area of one of the triangles forming the pyramid is
$\frac{1}{2}(2.92)(9.523)$ or $13.90\ \text{cm}^2$

Hence the total surface area S of the pyramid is:

$S = [(2.92)^2 + 4(13.90)]\ \text{cm}^2$

$\quad = (8.526 + 55.60)\ \text{cm}^2$

Hence the total surface area $= 64.13\ \text{cm}^2$

Problem 2. A pyramid stands on a rectangular base of sides
3.20 cm and 5.80 cm. If the length of each sloping edge is 12.60 cm
calculate the volume and the total surface area of the pyramid.

Figure 17.4 shows the rectangular pyramid with equal sloping
edges, the dimensions being in centimetres.

The volume of a pyramid $= \frac{1}{3}$[area of base $\times$ perpendicular height].

Thus the perpendicular height, shown as EF in Fig. 17.4, is
required. The length EF is one side of the triangle EFC. FC is
half the length of the diagonal AC. Using the theorem of Pythagoras:

$(AC)^2 = (AD)^2 + (DC)^2 = (3.20)^2 + (5.80)^2$

$\quad = 10.24 + 33.64 \quad = 43.88$

$\therefore\ AC\ = \sqrt{43.88} \qquad = 6.624\ \text{cm}$

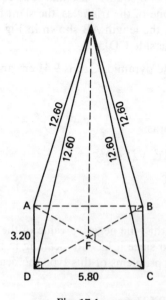

Hence $FC = \dfrac{6.624}{2} = 3.312\ \text{cm}$

Using the theorem on triangle EFC
gives:

$(EC)^2 = (EF)^2 + (FC)^2$

$\therefore\ (EF)^2 = (EC)^2 - (FC)^2$

$\qquad = (12.60)^2 - (3.312)^2$

$\qquad = 158.8 - 10.97$

$\qquad = 147.8$

$\therefore\ EF\ = \sqrt{147.8}$

$\qquad = 12.16\ \text{cm}$

Fig. 17.4

Hence the perpendicular height of the pyramid is 12.16 cm.

Therefore the volume of the pyramid $= \frac{1}{3}(3.20)(5.80)(12.16)$

$$= 75.23 \text{ cm}^3$$

The total surface area consists of four isosceles triangles and a rectangular base.

It is necessary to find the perpendicular height of each of the four triangles forming the sides of the pyramid.

In Fig. 17.5(*a*), EG represents the perpendicular height of triangle ECB.

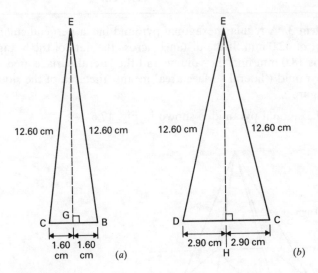

Fig. 17.5

Using the theorem of Pythagoras:

$$(EB)^2 = (EG)^2 + (GB)^2$$

$$\therefore (EG)^2 = (EB)^2 - (GB)^2 = (12.60)^2 - (1.60)^2$$

$$= 158.8 - 2.56 \quad = 156.2$$

$$\therefore EG \quad = \sqrt{156.2} \quad = 12.50 \text{ cm}$$

Hence the area of triangle ECB (which equals the area of triangle EAD) is $\frac{1}{2}(3.20)(12.50)$, i.e. 20.0 cm^2.

In Fig. 17.5(*b*), EH represents the perpendicular height of triangle EDC. Using the theorem of Pythagoras:

$$(EC)^2 = (EH)^2 + (HC)^2$$

$$\therefore \quad (EH)^2 = (EC)^2 - (HC)^2 \quad = (12.60)^2 - (2.90)^2$$

$$= 158.8 - 8.41 \quad = 150.4$$

$$\therefore \quad EH \quad = \sqrt{150.4} \quad = 12.26 \text{ cm}$$

Hence the area of triangle EDC (which equals the area of triangle EAB) is $\frac{1}{2}(5.80)(12.26)$, i.e. 35.55 cm^2

Therefore the total surface area of the pyramid is:

$$2(20.0) + 2(35.55) + (3.20)(5.80) = 40.0 + 71.10 + 18.56$$

$$= \textbf{129.7 cm}^2$$

Problem 3. A regular hexagonal pyramid has a perpendicular height of 42.0 mm. If the distance across the flats of the hexagonal base is 18.0 mm find the volume and the lateral surface area of the pyramid ('lateral surface area' means 'the area of the sides of the figure').

The hexagonal pyramid is shown in Fig. 17.6.

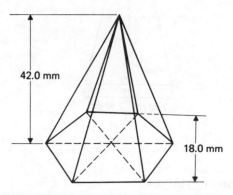

42.0 mm

18.0 mm

Fig. 17.6

The volume of a pyramid $= \frac{1}{3}$[area of base × perpendicular height].

The area of the regular hexagon needs to be calculated. The hexagon consists of 6 equal isosceles triangles as shown in Fig. 17.7. In the triangle ABC, DC is the perpendicular height of $\dfrac{18.0}{2}$ or 9.0 mm

In the triangle DBC, $\tan 30° = \dfrac{DB}{DC} = \dfrac{DB}{9.0}$

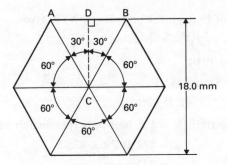

Fig. 17.7

∴ DB = 9.0 tan 30° = 5.196 mm

Hence the base AB of triangle ABC is 2 × 5.196 = 10.39 mm

The area of triangle ABC $= \frac{1}{2}(10.39)(9.0)$

$= 46.76$ mm^2

Area of the hexagon $= 6 × 46.76$

∴ Base of pyramid $= 280.6$ mm^2

∴ **Volume of the pyramid** $= \frac{1}{3}(280.6)(42.0)$

$= \textbf{3 928 mm}^3$

To find the lateral surface area of the pyramid the areas of six identical isosceles triangles forming the sloping sides need to be calculated. The perpendicular height of each triangle is given by the length XY shown in Fig. 17.8.

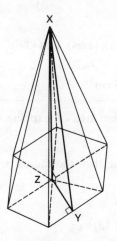

Fig. 17.8

Using the theorem of Pythagoras on triangle XYZ gives:

$$(XY)^2 = (XZ)^2 + (YZ)^2$$
$$= (42.0)^2 + (9.0)^2$$
$$= 1\,764 + 81 = 1\,845$$
$$\therefore XY = \sqrt{1\,845} = 42.95 \text{ mm}$$

Hence the area of one triangle $= \frac{1}{2} \times$ base $\times$ perpendicular height
$$= \frac{1}{2} \times 10.39 \times 42.95$$
$$= 223.1 \text{ mm}^2$$

Hence the lateral surface area is 6 × 223.1 or 1 339 mm²

Problem 4. Find the volume and the total surface area of a cone of radius 4.86 cm and perpendicular height 10.58 cm

The volume of a cone $= \frac{1}{3}[$(area of base)(perpendicular height)$]$
$$= \frac{1}{3}\pi r^2 h$$
$$= \frac{1}{3}\pi(4.86)^2(10.58)$$

Hence the volume of the cone = 261.7 cm²

The total surface area of a cone, $S_T =$ area of curved surface
$$+ \text{ area of base}$$
$$= \pi r \ell + \pi r^2$$

Since the height $h = 10.58$ cm and the radius $r = 4.86$ cm, the slant height ℓ may be found using the theorem of Pythagoras:

$$\ell^2 = (10.58)^2 + (4.86)^2$$
$$= 111.9 + 23.62 = 135.5$$
$$\therefore \ell = \sqrt{135.5} = 11.64 \text{ cm}$$

Therefore the total surface area $= \pi(4.86)(11.64) + \pi(4.86)^2$
$$= 177.7 + 74.20$$
$$= \mathbf{251.9 \text{ cm}^2}$$

Problem 5. Find the volume and surface area of a sphere of diameter 8.24 cm.

If the diameter of the sphere is 8.24 cm, then the radius is $\dfrac{8.24}{2}$ or 4.12 cm

The volume of a sphere $= \frac{4}{3}\pi r^3$
$$= \frac{4}{3}\pi(4.12)^3$$

Hence the volume of the sphere = 292.9 cm³

The surface area of a sphere $= 4\pi r^2$
$$= 4\pi(4.12)^2$$
Hence the surface area of the sphere $= 213.3$ cm^2

Further problems on the volumes and surface areas of pyramids, cones and spheres may be found in Section 17.5 (Problems 1–21), page 315.

17.3   The volumes and surface areas of frusta of pyramids and cones

The frustum of a pyramid or cone is the portion remaining when a part containing the vertex is cut off by a plane parallel to the base (note that the plural of frustum is frusta). The volume denoted by ABCD in Fig. 17.9 is a frustum of the cone ABE.

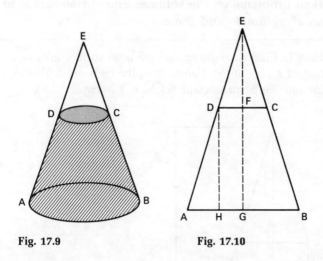

Fig. 17.9   Fig. 17.10

The volume of the frustum ABCD = [the volume of the large cone ABE] − [the volume of the small cone DCE]

Similarly, the curved surface area of the frustum ABCD = [the curved surface area of the large cone ABE] − [the curved surface area of the small cone DCE]

In Fig. 17.10 a section perpendicular to the base through the vertex is shown. $\angle$ EDF $= \angle$ DAH and $\angle$ DEF $= \angle$ ADH (since DC is parallel to AB).

Hence triangles EDF and DAH are similar.

Therefore: $\dfrac{EF}{DF} = \dfrac{DH}{AH}$

The following procedure is adopted when finding the volume or surface area of a frustum of a pyramid or cone.

1. Sketch a section perpendicular to the base, through the vertex of the complete pyramid (such as in Fig. 17.10).
2. Use similar triangles to find unknown lengths.
3. The volume of any frustum of a pyramid or cone is given by the volume of the whole pyramid or cone minus the volume of the small pyramid or cone cut off.
4. The surface area of the sides of any frustum of a pyramid or cone is given by the surface area of the whole pyramid or cone minus the surface area of the small pyramid or cone cut off. This gives the lateral surface area of the frustum. If the complete surface area of the frustum is required then the surface area of the two parallel ends must be added to the lateral surface area.

If the slant height of a pyramid is required use is made of the theorem of Pythagoras.

Worked problems on the volumes and surface areas of frusta of pyramids and cones

Problem 1. Find the volume and the total surface area of a frustum of a cone if the diameters of the ends are 5.0 cm and 3.0 cm and the perpendicular height is 3.20 cm.

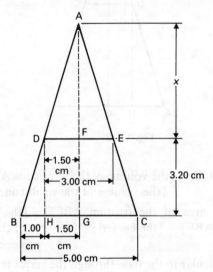

Fig. 17.11

A section through the vertex of the complete cone is shown in Fig. 17.11.

Using similar triangles: $\dfrac{AF}{DF} = \dfrac{DH}{BH}$

$$\therefore \quad \frac{x}{1.50} = \frac{3.20}{1.00}$$

$$\therefore \quad x = (1.50)(3.20) = 4.80$$

Hence the height of the complete cone is $3.20 + 4.80$, i.e. 8.00 cm.

The volume of the frustum of the cone = [the volume of the large cone] − [the volume of the small cone cut off]

Volume of the large cone $= \frac{1}{3}\pi(2.50)^2(8.00)$

$$= 52.36 \text{ cm}^3$$

Volume of the small cone $= \frac{1}{3}\pi(1.50)^2(4.80)$

$$= 11.31 \text{ cm}^3$$

Therefore the volume of the frustum $= 52.36 - 11.31$

$$= \mathbf{41.05 \text{ cm}^3}$$

The curved surface area of the frustum = [the curved surface area of the large cone] − [the curved surface area of the small cone cut off]

To find the curved surface area of a cone the slant height ℓ is required. Hence, referring to Fig. 17.11, the lengths AB (the slant height of the large cone) and AD (the slant height of the small cone) are required.

Using the theorem of Pythagoras:

$$(AB)^2 = (AG)^2 + (BG)^2 = (8.00)^2 + (2.50)^2$$

$$= 64.0 + 6.25 \qquad = 70.25$$

$$\therefore AB \quad = \sqrt{70.25} \qquad = 8.382 \text{ cm}$$

Similarly,

$$(AD)^2 = (AF)^2 + (DF)^2 = (4.80)^2 + (1.50)^2$$

$$= 23.04 + 2.25 \qquad = 25.29$$

$$\therefore AD \quad = \sqrt{25.29} \qquad = 5.029 \text{ cm}$$

The curved surface area of the complete cone $= \pi r \ell = \pi(\text{BG})(\text{AB})$

$$= \pi(2.50)(8.382)$$

$$= 65.83 \text{ cm}^2$$

The curved surface area of the small cone $\quad = \pi r \ell = \pi(\text{DF})(\text{AD})$

$$= \pi(1.50)(5.029)$$

$$= 23.70 \text{ cm}^2$$

Hence the curved surface area of the frustum $= 65.83 - 23.70$

$$= 42.13 \text{ cm}^2$$

The total surface area of the frustum is the curved surface area plus the area of the two circular ends.

i.e. $42.13 + \pi(2.50)^2 + \pi(1.50)^2$ or **68.83 cm²**

Problem 2. Find the volume and total surface area of a frustum of a pyramid, the ends being squares of sides 6.40 m and 3.60 m and the height being 4.00 m.

The frustum is shown in Fig. 17.12

A section perpendicular to the base through the vertex is shown in Fig. 17.13.

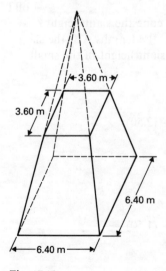

Fig. 17.12

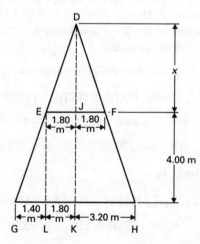

Fig. 17.13

Using similar triangles: $\dfrac{DJ}{EJ} = \dfrac{EL}{GL}$

$$\therefore \quad \frac{x}{1.80} = \frac{4.00}{1.40}$$

$$\therefore \quad x = \frac{(1.80)(4.00)}{(1.40)} = 5.143 \text{ m}$$

Hence the height of the complete pyramid is $5.143 + 4.00$, i.e. 9.143 m.

The volume of the large pyramid $\qquad = \frac{1}{3}(6.40)^2(9.143)$

$$= 124.8 \text{ m}^3$$

The volume of the small pyramid cut off $= \frac{1}{3}(3.60)^2(5.143)$

$$= 22.22 \text{ m}^3$$

Hence the volume of the frustum $= 124.8 - 22.22$

$$= 102.6 \text{ m}^3$$

The lateral surface area of the frustum consists of four equal trapeziums. To find the area of trapezium KLMN shown in Fig. 17.14, the perpendicular distance between the two parallel sides KL and MN, i.e. distance QR, is required.

The thickness of the frustum, PQ $= 4.00$ m

The length PR is the same as the length GL of Fig. 17.13, that is, 1.40 m.

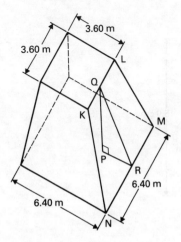

Fig. 17.14

Using the theorem of Pythagoras:

$$(QR)^2 = (QP)^2 + (PR)^2 = 4.00^2 + 1.40^2$$

$$= 16.00 + 1.96 \quad = 17.96$$

$$\therefore \; QR \quad = \sqrt{17.96} \qquad \quad = 4.24 \text{ m}$$

Hence the area of one trapezium $= \frac{1}{2}(3.60 + 6.40)4.24$

$$= 21.20 \text{ m}^2$$

The area of four trapeziums $\quad = 4(21.20)$

$$= 84.80 \text{ m}^2$$

The area of the two square ends $= (3.60)^2 + (6.40)^2$

$$= 12.96 + 40.96$$

$$= 53.92 \text{ m}^2$$

Hence the total surface area of the frustum $= 84.80 + 53.92$

$$= \mathbf{138.7 \ m^2}$$

Problem 3. A cone 15.0 cm high and of base diameter 12.0 cm is cut by a plane parallel to the base and 9.0 cm from the base. Find the ratio of the volume of the two parts thus formed.

The complete cone is shown in Fig. 17.15.

If a plane section, labelled AEH, is taken perpendicular to the base, then using similar triangles: $\dfrac{AC}{BC} = \dfrac{BF}{EF}$

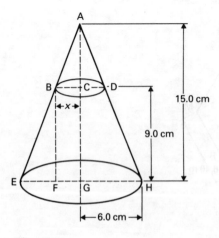

Fig. 17.15

Let $x = BC$, the radius of the top face of the frustum. Hence $EF = (6.0 - x)$ cm.

$$\therefore \frac{6.0}{x} = \frac{9.0}{6.0 - x}$$

$$\therefore 6.0(6.0 - x) = 9.0x$$

$$\therefore 36.0 - 6.0x = 9.0x$$

$$\therefore 36.0 = 15.0x$$

$$\therefore x = \frac{36.0}{15.0} = 2.40 \text{ cm}$$

The volume of the whole cone $= \frac{1}{3}\pi(6.0)^2(15.0)$
$$= 565.5 \text{ cm}^3$$

The volume of the small cone cut off $= \frac{1}{3}\pi(2.40)^2(6.0)$
$$= 36.19 \text{ cm}^3$$

Hence the volume of the frustum $= 565.5 - 36.19$
$$= 529.3 \text{ cm}^3$$

Thus the ratio of the volume of the small cone to the frustum is 36.2 : 529.3, i.e. **1 : 14.62**

Further problems on volumes and surface areas of frusta of pyramids and cones may be found in Section 17.5 (Problems 22–36), page 317.

17.4 Composite figures

A composite figure consists of two or more shapes joined together. Such shapes may be triangles, rectangles, circles, prisms, cones, cylinders, spheres or frusta. In problems involving composite figures, the figure must be broken down into its known component shapes before proceeding to find areas or volumes.

Worked problems on composite figures

Problem 1. The following area, representing a plot of land, is to be marked out on a map: From a given point A, move due north 6.00 cm, then due east 4.00 cm, then due south 2.00 cm, then south-east 1.42 cm, then due south 3.00 cm, then south-west 1.42 cm, then due south 2.00 cm, then S 30° W 5.00 cm, then N 30° W 5.00 cm and then in a straight line back to point A. Draw the plot of land to scale. If the scale of the map is 1.00 cm

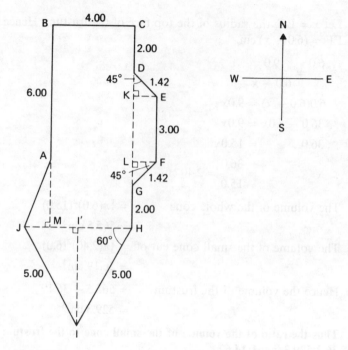

Fig. 17.16

representing 650 m find the length of fencing, in kilometres, required to enclose the land completely, and also the area of the land.

The shape of the plot of land is shown in Fig. 17.16, the dimensions being in centimetres. The component shapes consist of rectangle BCHM, triangle DEK, rectangle KEFL, triangle LFG, triangle HIJ and triangle JAM.

Triangle DEK is isosceles, since $\angle$KDE = $\angle$DEK = 45°

Hence DK = KE = 1.42 sin 45°

Therefore DK = KE = 1.00 cm

Triangle LFG is identical to triangle DEK

Triangle HIJ is equilateral since each of its angles is 60°. Hence each of the sides is 5.00 cm

The perpendicular height of triangle HIJ is given by the length II′ in Fig. 17.16.

Hence sin 60° = $\dfrac{\text{II}'}{5.00}$

Therefore II′ = 5.00 sin 60° = 4.330 cm

Since BC is 4.00 cm then MH is also 4.00 cm

Hence JM is 1.00 cm (since JH = 5.00 cm)

The distance from C to H is CD + DK + KL + LG + GH, i.e.
2.00 + 1.00 + 3.00 + 1.00 + 2.00, i.e. 9.00 cm

But CH = BA + AM

i.e. 9.00 = 6.00 + AM

Hence AM = 3.00 cm

In triangle JAM the length AJ may be found using the theorem of Pythagoras:

$$(AJ)^2 = (AM)^2 + (JM)^2 = (3.00)^2 + (1.00)^2$$

$$= 9.00 + 1.00 \qquad = 10.00$$

$$\therefore \ AJ \quad = \sqrt{10.00} \qquad \qquad = 3.162 \text{ cm}$$

The length of the perimeter of the plot of land on the map is
AB + BC + CD + DE + EF + FG + GH + HI + IJ + JA

That is,
6.00 + 4.00 + 2.00 + 1.42 + 3.00 + 1.42 + 2.00 + 5.00 + 5.00 + 3.162
$$= 33.002 \text{ cm}$$

If 1 cm represents 650 m then 33.002 cm represents
(33.002)(650) m, i.e. 21 450 m or 21.45 km.

Hence 21.45 km of fencing is required for the plot of land

Area of rectangle BCHM = 9.00 × 4.00 = 36.00 cm^2

Area of triangle DEK $= \frac{1}{2}(1.00)(1.00) = 0.50$ cm^2

Area of rectangle KEFL = 3.00 × 1.00 = 3.00 cm^2

Area of triangle LFG $= \frac{1}{2}(1.00)(1.00) = 0.50$ cm^2

Area of triangle HIJ $= \frac{1}{2}(5.00)(4.33) = 10.83$ cm^2

Area of triangle AJM $= \frac{1}{2}(1.00)(3.00) = 1.50$ cm^2

Total area of plot of land on map $= 52.33$ cm^2

If 1 cm represents 650 m then 1 cm^2 represents $(650)^2$ m^2

Hence 52.33 cm^2 represents $(52.33)(650)^2$ m^2

i.e. 22 110 000 m^2 or 22.11 km^2

Hence the area of the land is 22.11 km^2

Problem 2. The construction of a boiler consists of a cylindrical section of length 10.0 m and diameter 7.0 m, on one end of which is surmounted a hemispherical section of diameter 7.0 m and on the other end a conical section of height 6.0 m. Find the surface area of metal required and the volume of the boiler.

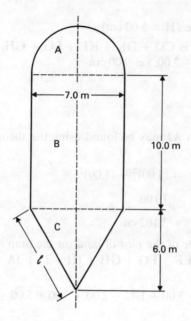

Fig. 17.17

A section through the boiler is shown in Fig. 17.17

The component shapes consist of a hemisphere (A), a cylinder (B) and a cone (C).

Hemisphere (A)

Surface area $= 2\pi r^2 = 2\pi(3.50)^2 = 76.97$ m^2

Volume $\quad = \frac{2}{3}\pi r^3 = \frac{2}{3}\pi(3.50)^3 = 89.80$ m^3

Cylinder (B)

Surface area $= 2\pi rh = 2\pi(3.50)(10.0) = 219.9$ m^2

Volume $\quad = \pi r^2 h = \pi(3.50)^2(10.0) = 384.9$ m^3

Cone (C)

Curved surface area $= \pi r \ell$

The slant height ℓ may be found using the theorem of Pythagoras:

$$\ell^2 \qquad\qquad = (6.0)^2 + (3.50)^2$$
$$= 36.0 + 12.25$$
$$= 48.25$$
$$\therefore \ell \qquad\qquad = \sqrt{48.25} = 6.946 \text{ m}$$

Hence the curved surface area $= \pi(3.50)(6.946) = 76.38$ m^2

Volume $= \frac{1}{3}\pi r^2 h \qquad\qquad = \frac{1}{3}\pi(3.50)^2(6.0) = 76.97$ m^3

The total surface area of the boiler $= 76.97 + 219.9 + 76.38$

$$= 373.3 \text{ m}^2$$

Total volume of the boiler $\qquad = 89.80 + 384.9 + 76.97$

$$= 551.7 \text{ m}^3$$

Problem 3. A cooling tower is in the form of a cylinder of height 15.0 m surmounted by a frustum of a cone. The diameter of the cylinder and the bottom of the frustum is 30.0 m and the diameter at the top of the tower is 18.0 m. The height of the tower is 40.0 m. Calculate the volume of air space in the tower if 35% of the space is used for pipes and other structures.

The cooling tower is shown in Fig. 17.18

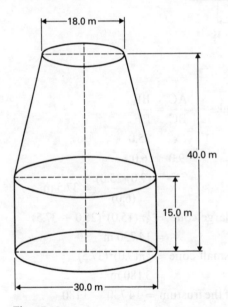

Fig. 17.18

The volume of the cylindrical portion of the cooling tower $= \pi r^2 h = \pi (15.0)^2 (15.0) = 10\,600 \text{ m}^2$

A section through the complete cone from which the frustum is formed is shown in Fig. 17.19

The volume of the frustum BDHE = [the volume of the large cone AEH] − [the volume of the small cone ABD].

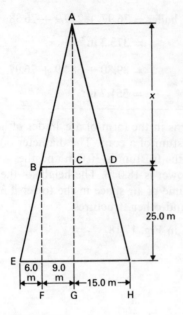

Fig. 17.19

Using similar triangles: $\dfrac{AC}{BC} = \dfrac{BF}{EF}$

$\therefore \qquad \dfrac{x}{9.0} = \dfrac{25.0}{6.0}$

$\therefore \qquad x = \dfrac{(9.0)(25.0)}{(6.0)} = 37.5 \text{ m}$

The volume of the large cone $= \frac{1}{3}\pi(15.0)^2(25.0 + 37.5)$

$= 14\,730 \text{ m}^3$

The volume of the small cone $= \frac{1}{3}\pi(9.0)^2(37.5)$

$= 3\,180 \text{ m}^3$

Hence the volume of the frustum $= 14\,730 - 3\,180$

$= 11\,550 \text{ m}^3$

The total volume of the cooling tower $= 10\,600 + 11\,550$

$= 22\,150 \text{ m}^3$

If 35% of the space is occupied by pipes and other structures then 65% is air space.

Hence air space volume $= 0.65 \times 22\,150$

$= \mathbf{14\,400 \text{ m}^3}$

Further problems on composite figures may be found in the following Section (17.5) (Problems 37–42), page 318.

17.5 Further problems

Volumes and surface areas of pyramids, cones and spheres

1. Find the volume and total surface area of a cone of radius 6.00 cm and perpendicular height 8.00 cm.
   [301.6 cm^3, 301.6 cm^2]

2. A cone, slant height 4 cm and base radius 1.5 cm, is opened out to form a sector of a circle. Calculate the angle at the centre of this circle in radians and degrees. $\left[\dfrac{3\pi^c}{4}, 135°\right]$

3. If a sphere has a diameter of 54.0 mm, calculate its volume and surface area.
   [82 460 mm^3 or 82.46 cm^3; 9 162 mm^2 or 91.62 cm^2]

4. If the diameter of a sphere is 140 mm find its surface area in cm^2 correct to four significant figures.   [615.8 cm^2]

5. If the volume of a sphere is 476 cm^2 find its radius.   [4.843 cm]

6. A hemisphere has a diameter of 3.60 cm. Find its volume and total surface area.   [12.22 cm^3, 30.54 cm^2]

7. A cone has a base diameter of 12.74 cm and a slant height of 12.46 cm. Find the volume and the lateral surface area of the cone.   [455.03 cm^3, 249.35 cm^2]

8. A triangular pyramid has a perpendicular height of 25.00 cm. If the base is an equilateral triangle of side 3.00 cm, find its volume.   [32.48 cm^3]

9. A metal sphere weighing 36.0 kg is melted down and recast into a solid cone of base radius 12.0 cm. If the density of the metal is 8.00 g cm^{-3} find

   (a) the diameter of the metal sphere and
   (b) the perpendicular height of the solid cone, assuming that 12.5% of the metal is lost in the process.

   (a) [20.48 cm]   (b) [26.11 cm]

10. The volume of a pyramid whose base is an equilateral triangle of side 3.25 cm is 42.8 cm^3. Find its perpendicular height.   [28.07 cm]

11. Find the volume of a pyramid of perpendicular height 15.0 cm

and whose base is a triangle of sides 2.60 cm, 3.90 cm and
4.30 cm.   [24.97 cm³]

12. Calculate the volume and total surface area of a pyramid of
vertical height 11.2 cm and square base of side 4.60 cm.
[79.00 cm³, 126.4 cm²]

13. A square pyramid has a perpendicular height of 24.0 mm. If
the side of the base is 3.20 cm find the volume and the total
surface area of the pyramid.   [8.192 cm³, 28.70 cm²]

14. A pyramid stands on a square base of side 15.0 cm. If the
length of each sloping edge is 21.0 cm calculate the volume
and total surface area of the pyramid.
[1 359 cm³, 813.5 cm²]

15. A regular pyramid has a square base and all its edges are
4.00 cm. Calculate its volume and total surface area.
[15.08 cm³, 43.71 cm²]

16. A pyramid with a square base of side 8.4 cm has the same
height as a cone of equal volume. Find the diameter of the
base of the cone.   [9.48 cm]

17. A hollow sphere has inner and outer diameters of 5.48 cm
and 7.84 cm respectively. Calculate

(a) the outer surface area of the sphere,
(b) the volume of the material of the sphere.

The hollow sphere is melted down and recast into a cone of
vertical height 4.20 cm. Find the base radius of the cone
assuming no material is lost in the process.
(a) [193.1 cm²]   (b) [166.2 cm³; 6.147 cm]

18. A pyramid stands on a rectangular base whose sides are
5.00 cm and 4.00 cm. If the length of each sloping edge is
11.00 cm, calculate the volume and the total surface area of
the pyramid.   [70.20 cm³, 116.8 cm²]

19. Find the volume of a regular hexagonal pyramid if the
perpendicular height is 12.00 cm and the side of base is
4.00 cm.   [166.3 cm³]

20. A solid pyramid of vertical height 9.60 cm has a regular
hexagonal base. If the distance across the flats of the
hexagon is 4.20 cm find the volume and total surface area of
the pyramid.   [48.89 cm³, 86.76 cm²]

21. The side of the base of a regular octagonal pyramid is
6.20 cm and the perpendicular height is 14.80 cm. Find the
volume, the lateral surface area and the total surface area of
the pyramid.   [915.6 cm³, 411.3 cm², 596.9 cm²]

Volumes and surface areas of frusta of cones and pyramids

22. The radii of the faces of a frustum of a cone are 3.00 cm and 4.00 cm, and its height is 5.00 cm. Find its volume. [193.7 cm^3]

23. Calculate the total surface area of the frustum of a cone of which the end radii are 15.0 cm and 10.0 cm respectively and the slant height measures 11.0 cm.   [1 885 cm^2]

24. A lampshade is in the shape of a frustum of a cone. The vertical height of the shade is 10.00 cm and the diameters of the ends are 10.00 cm and 6.00 cm respectively. Find the area of material needed to form the lampshade.   [256.4 cm^2]

25. A cooling tower is in the form of a frustum of a square pyramid. The base has sides of length 40.0 m, the top has sides of length 24.0 m and the vertical height is 30.0 m. Calculate the volume of the tower.   [31 360 m^3]

26. A cone of height d is cut by a plane parallel to and distance $\dfrac{d}{3}$ from the base. Show that the volume of the frustum produced is 70.37% of the original cone.

27. A cone 46.0 mm high and of base diameter 40.0 mm is cut by a plane parallel to the base and 20.0 mm from the base. Find the ratio of the volume of the two parts.   [1:4.538]

28. A loudspeaker diaphragm is in the form of a frustum of a cone. If the end diameters are 30.00 cm and 5.00 cm and the vertical height is 30.00 cm find the curved surface area of material needed to cover the speaker in square metres.   [0.178 7 m^2]

29. Find the volume and total surface area of a tumbler in the form of a frustum of a cone, if the diameters of the ends are 6.50 cm and 3.50 cm and the perpendicular height of the tumbler is 7.80 cm.   [157.7 cm^3, 167.6 cm^2]

30. A hole is to be dug in the form of a frustum of a pyramid. The top is to be a square of side 6.40 m and the bottom a square of 3.60 m. If the depth of the hole is to be 4.00 m calculate the volume of earth to be removed. If the hole is now filled with concrete to a depth of 2.00 m find the amount of concrete required.   [102.6 m^3, 43.1 m^3]

31. A pyramid on a square base of side 12.00 cm and of perpendicular height 22.00 cm is cut by a plane 7.00 cm from the base and parallel to it. Find
    (a) the volume of the frustum formed and
    (b) the total surface area of the frustum.
    (a) [721.3 cm^3]   (b) [503.7 cm^2]

32. Find the volume and total surface area of a bucket consisting of an inverted frustum of a cone, of slant height 35.0 cm and end diameters 60.0 cm and 40.0 cm. [66 730 cm³, 6 754 cm²]

33. Calculate the volume and the total surface area of the frustum of a cone of height 16.00 m with diameter of ends 12.00 m and 8.00 m.   [1 273 m³, 669.9 m²]

34. A cylindrical tank 2.0 m in diameter and 2.5 m high is to be replaced by a tank of the same capacity, in the form of a frustum of a cone. If the frustum is to have 1.0 m diameter small end and 2.2 m diameter large end, calculate the vertical height required correct to the nearest centimetre. [3.73 m]

35. A solid cone has a base radius of 12.0 cm and a vertical height of 40.0 cm. A part of the cone is removed by making a cut along a plane 30 cm from the base end parallel to the base. Calculate the percentage of the original mass removed. [1.56%]

36. A rectangular piece of metal with dimensions 5.2 cm by 8.6 cm by 11.3 cm is melted down and recast into a frustum of a square pyramid. If the ends of the frustum are squares of side 4.0 cm and 10.0 cm find the height of the frustum. Assume no loss of metal in the casting process.   [9.72 cm]

Composite figures

37. A buoy consists of a hemisphere surmounted by a cone. The diameter of the hemisphere is equal to the diameter of the base of the cone and is 2.0 m. The slant height of the cone is 3.6 m. Calculate the total surface area of the buoy. [17.6 m²]

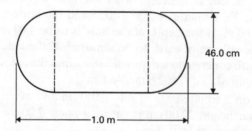

46.0 cm

1.0 m

Fig. 17.20

38. Figure 17.20 shows a water container in the form of a

central cylindrical portion with two hemispherical ends. Calculate its capacity in litres (1 litre = 1 000 cm^3).
[140.7 litres]

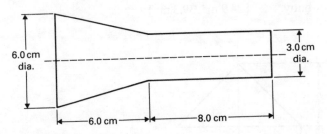

6.0 cm dia.

3.0 cm dia.

6.0 cm

8.0 cm

Fig. 17.21

39. Figure 17.21 shows a section of the glass envelope of a cathode ray tube with the small end of the tube left open. Calculate the total outside surface area of the glass envelope and the volume enclosed by the tube.   [191 cm^2, 156 cm^3]

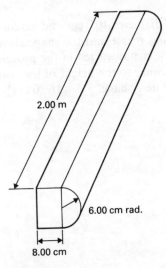

2.00 m

6.00 cm rad.

8.00 cm

Fig. 17.22

40. Figure 17.22 shows a 2.00 m section of metal rod. Calculate
   (a) its volume in cm^3 and m^3, and
   (b) its total surface area in cm^2 and m^2.

   (a) [30 510 cm^3 or 0.030 51 m^3]
   (b) [9 675 cm^2 or 0.967 5 m^2]

41. A buoy consists of a frustum of a cone whose ends have radii 1.0 m and 1.8 m and whose height is 2.0 m, to which is attached, at the larger end, a hemisphere of radius 1.8 m. Calculate the volume and surface area of the buoy.   [24.9 m³, 39.3 m²]

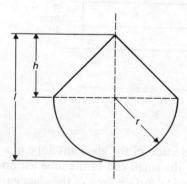

Fig. 17.23

42. A solid metal cylinder of diameter 10.0 cm and height 25.0 cm is melted down and recast into the shape shown in Fig. 17.23, with 15% of the metal wasted in the process. If the radius of the cone is equal to the height of the cone find the overall length *l* of the new shape.   [16.20 cm]

Chapter 18

Irregular areas and volumes

18.1 Areas of irregular figures

There are several practical situations where it is necessary to estimate the area of irregular figures. Examples include estimation of areas of plots of land by surveyors, areas of indicator diagrams of steam engines by engineers and areas of water planes and transverse sections of a ship by naval architects.

There are many methods whereby the area of an irregular plane surface may be found and these include:

(a) Use of a planimeter,
(b) Trapezoidal rule,
(c) Mid-ordinate rule and
(d) Simpson's rule.

(a) The planimeter

A planimeter is an instrument for directly measuring areas bounded by an irregular curve. There are many different types of the instrument but all consist basically of two rods AB and BC, hinged at B (see Fig. 18.1). The end labelled A is fixed, preferably outside of the irregular area being

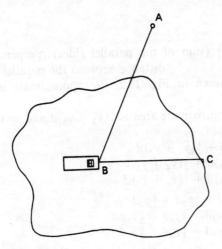

Fig. 18.1

measured. Rod BC carries at B a wheel whose plane is at right angles to the plane formed by ABC. Point C, called the tracer, is guided round the boundary of the figure to be measured. The wheel is geared to a dial which records the area directly. If the length BC is adjustable, the scale can be altered and readings obtained in mm², cm², m² and so on.

(b) Trapezoidal rule

To find the area ABCD of Fig. 18.2, the base AD is divided into a number of equal intervals of width d. This can be any number; the greater the number, the more accurate the result. The ordinates y_1, y_2, y_3, etc. are accurately measured. The approximation used in this rule is to assume that each strip is equal to the area of a trapezium.

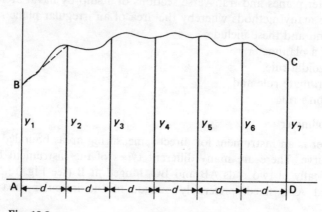

Fig. 18.2

The area of a trapezium = $\frac{1}{2}$ (sum of the parallel sides) (perpendicular distance between the parallel sides).

Hence for the first strip, shown in Fig. 18.2, the approximate area is $\frac{1}{2}(y_1 + y_2)d$

For the second strip, the approximate area is $\frac{1}{2}(y_2 + y_3)d$ and so on. Hence the approximate area of

$$
\begin{aligned}
\text{ABCD} &= \tfrac{1}{2}(y_1 + y_2)d + \tfrac{1}{2}(y_2 + y_3)d \\
&\quad + \tfrac{1}{2}(y_3 + y_4)d + \tfrac{1}{2}(y_4 + y_5)d \\
&\quad + \tfrac{1}{2}(y_5 + y_6)d + \tfrac{1}{2}(y_6 + y_7)d \\
&= \tfrac{1}{2}y_1 d + \tfrac{1}{2}y_2 d + \tfrac{1}{2}y_2 d + \tfrac{1}{2}y_3 d \\
&\quad + \tfrac{1}{2}y_3 d + \tfrac{1}{2}y_4 d + \tfrac{1}{2}y_4 d + \tfrac{1}{2}y_5 d \\
&\quad + \tfrac{1}{2}y_5 d + \tfrac{1}{2}y_6 d + \tfrac{1}{2}y_6 d + \tfrac{1}{2}y_7 d \\
&= \tfrac{1}{2}y_1 d + y_2 d + y_3 d + y_4 d + y_5 d + y_6 d + \tfrac{1}{2}y_7 d \\
&= d\left[\frac{y_1 + y_7}{2} + y_2 + y_3 + y_4 + y_5 + y_6\right]
\end{aligned}
$$

Generally, the trapezoidal rule states that the area of an irregular figure is given by:

> **Area = (width of interval) [½(first + last ordinate)**
> **+ sum of remaining ordinates]**

(c) Mid-ordinate rule

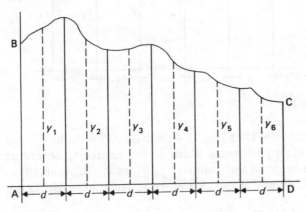

Fig. 18.3

To find the area of ABCD of Fig. 18.3, the base AD is divided into any number of equal strips of width d. (As with the trapezoidal rule, the greater the number of intervals used the more accurate the result.) If each strip is assumed to be a trapezium then the average length of the two parallel sides will be given by the length of a mid-ordinate, i.e. an ordinate erected in the middle of each trapezium. This is the approximation used in the mid-ordinate rule.

The mid-ordinates are labelled y_1, y_2, y_3, etc. as in Fig. 18.3 and each is then accurately measured. Hence the approximate area of ABCD

$$= y_1d + y_2d + y_3d + y_4d + y_5d + y_6d$$
$$= d(y_1 + y_2 + y_3 + y_4 + y_5 + y_6)$$

where $d = \dfrac{\text{length of AD}}{\text{number of mid-ordinates}}$

Generally, the mid-ordinate rule states that the area of an irregular figure is given by:

> **Area = (width of interval) (sum of mid-ordinates)**

(d) Simpson's rule

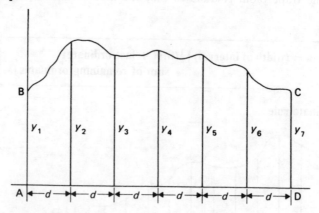

Fig. 18.4

To find the area ABCD of Fig. 18.4 the base AD must be divided into an **even** number of strips of equal width d, thus producing an **odd** number of ordinates. The length of each ordinate, y_1, y_2, y_3, etc., is accurately measured. Simpson's rule states that the area of the irregular area ABCD is given by:

$$\text{Area of ABCD} = \frac{d}{3}[(y_1 + y_7) + 4(y_2 + y_4 + y_6) + 2(y_3 + y_5)]$$

More generally, the area under any irregular figure is given by:

$$\text{Area} = \tfrac{1}{3}(\text{width of interval})\left[\begin{pmatrix}\text{first and last} \\ \text{ordinates}\end{pmatrix} + 4\begin{pmatrix}\text{sum of even} \\ \text{ordinates}\end{pmatrix} + 2\begin{pmatrix}\text{sum of} \\ \text{remaining odd ordinates}\end{pmatrix}\right]$$

When estimating areas of irregular figures Simpson's rule is generally regarded as the most accurate of the approximate methods available.

18.2   Volumes of irregular solids

Simpson's rule may be applied to find the volume of a solid of varying section when bounded by two parallel planes. Figure 18.5 shows a cross-section of an irregular solid, in this case a boat. If the cross-sectional areas A_1, A_2, A_3, etc. are known at equal intervals of width d then Simpson's rule states that the volume of an irregular solid is given by:

$$\text{Volume} = \frac{d}{3}[(A_1 + A_7) + 4(A_2 + A_4 + A_6) + 2(A_3 + A_5)]$$

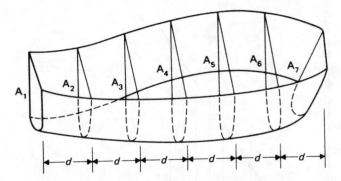

Fig. 18.5

More generally, the volume of an irregular solid is given by:

$$\text{Volume} = \frac{1}{3}(\text{width of interval})\left[\binom{\text{first and last}}{\text{ordinate area}} + 4\binom{\text{sum of even}}{\text{ordinate areas}} + 2\binom{\text{sum of remaining}}{\text{odd ordinate areas}}\right]$$

Worked problems on areas of irregular figures and volumes of irregular solids

Problem 1. The values of the y ordinates of a curve and their distance x from the origin are given in the table below. Plot the graph and find the area under the curve by:
(a) the trapezoidal rule,
(b) the mid-ordinate rule,
(c) Simpson's rule.

x	0	1	2	3	4	5	6
y	2	5	8	11	14	17	20

The graph is shown in Fig. 18.6.

(a) **Trapezoidal rule**

Using 7 ordinates with interval width 1 the area under the curve is given by:

$$\text{Area} = 1[\tfrac{1}{2}(2 + 20) + 5 + 8 + 11 + 14 + 17]$$
$$= [11 + 5 + 8 + 11 + 14 + 17]$$
$$= 66 \text{ square units}$$

(b) **Mid-ordinate rule**

Using 6 intervals of width 1 the mid-ordinates of the 6 strips are measured.

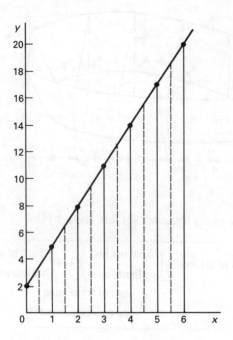

Fig. 18.6 Graph of *y* against *x*

The area under the curve is given by:

Area = 1(3.5 + 6.5 + 9.5 + 12.5 + 15.5 + 18.5)
     = 66 square units

(c) Simpson's rule

Using 7 ordinates gives an even number of strips, i.e. 6, each of width 1, thus the area under the curve is given by:

Area $= \frac{1}{3}[(2 + 20) + 4(5 + 11 + 17) + 2(8 + 14)]$
     $= \frac{1}{3}[22 + 4(33) + 2(22)]$
     $= \frac{1}{3}[22 + 132 + 44]$
     $= \frac{198}{3}$
     = 66 square units

The area under the curve is a trapezium and may be calculated using the formula $\frac{1}{2}(a + b)h$, where a and b are the lengths of the parallel sides and h the perpendicular distance between the parallel sides. Hence area $= \frac{1}{2}(2 + 20)(6) = 66$ square units.

This problem demonstrates the methods for finding areas under curves. Obviously the three 'approximate' methods would not normally be used for an area such as in this problem since it is not 'irregular'.

Problem 2. Sketch a semicircle of radius 10 cm. Erect ordinates at intervals of 2 cm and determine the lengths of ordinates and mid-ordinates. Determine the area of the semicircle using the three approximate methods. Calculate the true area of the semicircle and hence determine the percentage inaccuracy, correct to two decimal places for each of the results.

The semicircle is shown in Fig. 18.7 with the lengths of the ordinates and mid-ordinates marked, the dimensions being centimetres.

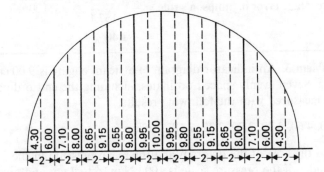

Fig. 18.7

(a) **Trapezoidal rule**

Area $= 2[\frac{1}{2}(0 + 0) + 6.00 + 8.00 + 9.15 + 9.80 + 10.00 + 9.80$
$\qquad\qquad + 9.15 + 8.00 + 6.00]$

$\qquad = 2(75.90)$

$\qquad = $ **151.8 square units**

(b) **Mid-ordinate rule**

Area $= 2[4.30 + 7.10 + 8.65 + 9.55 + 9.95 + 9.95 + 9.55 + 8.65$
$\qquad\qquad + 7.10 + 4.30]$

$\qquad = 2(79.10)$

$\qquad = $ **158.2 square units**

(c) **Simpson's rule**

Area $= \frac{2}{3}[(0 + 0) + 4(6.00 + 9.15 + 10.00 + 9.15 + 6.00)$
$\qquad\qquad + 2(8.00 + 9.80 + 9.80 + 8.00)]$

$\qquad = \frac{2}{3}[0 + 4(40.30) + 2(35.60)]$

$\qquad = \frac{2}{3}(161.2 + 71.2)$

$\qquad = \frac{2}{3}(232.4)$

$\qquad = $ **154.9 square units**

The true area is given by $\dfrac{\pi r^2}{2}$, i.e. $\dfrac{\pi (10)^2}{2} = \textbf{157.1 square units}$

Percentage error in trapezoidal rule $= \dfrac{151.8 - 157.1}{157.1} \times 100\%$

$$= -\textbf{3.37}\%$$

Percentage error in mid-ordinate rule $= \dfrac{158.2 - 157.1}{157.1} \times 100\%$

$$= +\textbf{0.70}\%$$

Percentage error in Simpson's rule $= \dfrac{154.9 - 157.1}{157.1} \times 100\%$

$$= -\textbf{1.40}\%$$

Problem 3. An indicator diagram of a steam engine is 9.00 cm long. Seven evenly spaced ordinates, including the end ordinates, are measured with the following results:

5.10, 4.60, 3.20, 2.70, 2.32, 2.18, 2.06 cm

Find the area of the diagram and the mean pressure in the cylinder, if the pressure scale is 100 kN m^{-2} to 1 cm and given that mean pressure $= \dfrac{\text{Area of diagram}}{\text{base}}$

A sketch of the indicator diagram is shown in Fig. 18.8

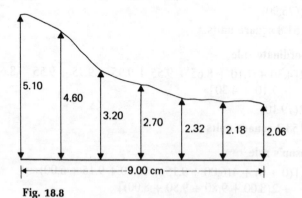

Fig. 18.8

With an odd number of ordinates (i.e. 7) and therefore an even number of strips (i.e. 6) Simpson's rule is used. (In this problem the trapezoidal rule may also be used. However, Simpson's rule is generally considered to be the more accurate.) The width of each strip is $\frac{9}{6}$ or 1.50 cm.

$$
\begin{aligned}
\textbf{Area} &= \frac{1.50}{3}\,[(5.10 + 2.06) + 4(4.60 + 2.70 + 2.18) \\
&\qquad\qquad + 2(3.20 + 2.32)] \\
&= 0.50[7.16 + 4(9.48) + 2(5.52)] \\
&= 0.50(7.16 + 37.92 + 11.04) \\
&= 0.50(56.12) \\
&= \textbf{28.06 cm}^2
\end{aligned}
$$

$$
\begin{aligned}
\textbf{Mean pressure} &= \frac{\text{area of diagram}}{\text{base of diagram}} = \frac{28.06}{9.00}\ \text{cm} \\[2mm]
&= \frac{28.06}{9.00} \times 100\ \text{kN m}^{-2} \\[2mm]
&= \textbf{311.8 kN m}^{-2}
\end{aligned}
$$

Problem 4. A stream of width 30.00 m is measured for depth at intervals of 5.00 m across its width

Distance from bank (m) 0   5.00   10.00   15.00   20.00   25.00   30.00
Depth (m)   0   0.90   1.70   2.70   2.30   1.20   0

Determine the cross-sectional area of the flow of water at this point using Simpson's rule.

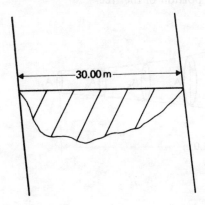

Fig. 18.9

A sketch of the cross-section of the river being measured is shown in Fig. 18.9. It is often more convenient to draw up a table when using Simpson's rule. This is especially true when a large number of ordinates are used. The table is shown below.

Number of the ordinate	Length of ordinate	Simpson's multiplier	Totals
1	0	1	0
2	0.90	4	3.60
3	1.70	2	3.40
4	2.70	4	10.80
5	2.30	2	4.60
6	1.20	4	4.80
7	0	1	0

Sum = 27.20

Since the width of each strip is 5.00 m:

$$\text{Area} = \frac{5.00}{3}(27.20) = 45.33 \text{ m}^2$$

Hence the cross-sectional area of the flow of water is 45.33 m²

Problem 5. A portion of a small tree is 6.0 m in length and has a varying circular section. The cross-sectional areas measured from one end at intervals of 1.0 m are: 240, 191, 152, 102, 79, 62, 49 cm². Find the volume of the portion of the tree.

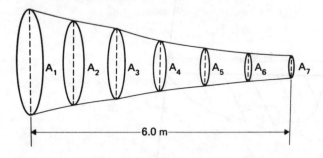

Fig. 18.10

A sketch of the portion of the tree is shown in Fig. 18.10, where $A_1 = 240$ cm², $A_2 = 191$ cm², $A_3 = 152$ cm² and so on. An odd number of ordinates is given (i.e. 7) resulting in an even number of strips or sections (i.e. 6). Hence Simpson's rule for estimating volumes of irregular solids may be used. The width of each interval is 1.0 m or 100 cm.

Volume $= \frac{100}{3}[(240 + 49) + 4(191 + 102 + 62) + 2(152 + 79)]$

$\qquad\quad = \frac{100}{3}[289 + 4(355) + 2(231)]$

$\qquad\quad = \frac{100}{3}(289 + 1\,420 + 462)$

$\qquad\quad = \frac{100}{3}(2\,171)$

$\qquad\quad = 72\,400 \text{ cm}^3 \text{ or } 0.072\,4 \text{ m}^3$

Hence the volume of the portion of the tree is 72 400 cm³ or 0.072 4 m³

Problem 6. The areas of equidistantly spaced sections of the underwater form of a small vessel are as follows: 1.23, 2.49, 2.86, 2.37, 0.92 m² respectively. They are spaced 4.00 m apart. Find the underwater volume.

A sketch of the underwater form of the vessel is shown in Fig. 18.11 where $A_1 = 1.23$ m², $A_2 = 2.49$ m² and so on. An odd

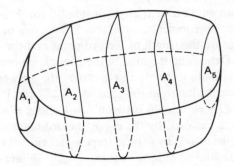

Fig. 18.11

number of ordinates is given (i.e. 5), resulting in an even number of sections. Hence Simpson's rule may be used and is shown in tabular form below.

Number of the ordinate	Length of ordinate	Simpson's multiplier	Totals
1	1.23	1	1.23
2	2.49	4	9.96
3	2.86	2	5.72
4	2.37	4	9.48
5	0.92	1	0.92

Sum = 27.31

Since the width of each section is 4.0 m:

$$\text{Volume} = \frac{4.0}{3}(27.31) = 36.4 \text{ m}^3$$

Hence the underwater volume is 36.4 m³

Further problems on areas of irregular figures and volumes of irregular solids may be found in the following Section (18.3) (Problems 1–20).

18.3   Further problems

1. Draw the graph of $y = 6x - x^2$ by compiling a table of values for y from $x = 0$ to $x = 6$. Find the area enclosed by the curve, the ordinates $x = 1$ and $x = 5$ and the x-axis by (a) the trapezoidal rule, (b) the mid-ordinate rule and (c) Simpson's rule.   [30.67 square units]

2. Draw to scale an equilateral triangle of side 8.0 cm. From one of the sides erect perpendicular ordinates at intervals of 1.0 cm and measure the length of the ordinates and the mid-ordinates. Determine the area of the triangle by each of the three approximate methods and compare the answers with a trigonometric calculation of the area.   [27.7 cm²]

3. Plot the graph of $y = 3x^2 + 6$ between $x = 1$ and $x = 4$. Estimate the area enclosed by the curve, the ordinates $x = 1$ and $x = 4$ and the x-axis by (a) the trapezoidal rule, (b) the mid-ordinate rule and (c) Simpson's rule.   [81 square units]

4. Find the area enclosed by the curve $y = 8x + 20 - x^2$ and the positive x- and y-axis using Simpson's rule. [266.7 square units]

5. The shape of a piece of land is shown in Fig. 18.12.

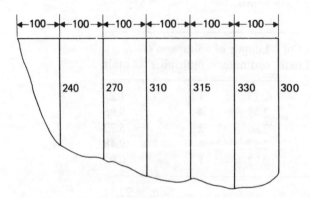

Fig. 18.12

To estimate the area of the land, measurements at intervals
of 100 m are taken perpendicular to the straight portion with
the results shown in Fig. 18.12, the dimensions being in metres.
Estimate the area of the land in hectares (1 hectare = 10^4 m^2).
[16.63 ha]

6. An indicator diagram of a steam engine is 12.0 cm in length.
Seven evenly spaced ordinates, including the end ordinate,
are measured with the following results: 6.30, 5.73, 4.31, 3.74,
3.42, 3.31, 3.17 cm. Make an estimation of the area of the
diagram and from it find the mean pressure in the cylinder, if
the pressure scale is 95.0 kN m^{-2} to 1.00 cm.
[50.70 cm^2, 401.4 kN m^{-2}]

7. A stream of width 18.0 m was measured for depth at intervals
of 3.0 m across its width. The results were:

Distance from bank (m)	0	3.0	6.0	9.0	12.0	15.0	18.0
Depth (m)		0	1.21	2.23	4.14	3.06	1.72	0

Determine the cross-sectional area of the flow of water at
this point.   [38.9 m^2]

8. The deck of a ship is 30 m long. At equal intervals of 6 m
the width is given by the following table:

Width (m) 0 3.2 5.8 7.1 6.0 4

Estimate the area of the deck.   [144.6 m^2]

9. A river is 15.0 m wide. Soundings of the depth are made
at equal intervals across the river and are shown in the
table below:

Depth (m) 0, 2.40, 3.60, 4.70, 4.40, 2.80, 0

Calculate the cross-sectional area of the river at that point.
[46.33 m^2]

10. A vehicle starts from rest and its velocity is measured every
second for 6.0 seconds.

Time, t (seconds)	0	1.0	2.0	3.0	4.0	5.0	6.0
Velocity, v (metres per sec)	0	1.2	2.4	3.7	5.2	6.0	9.2

Using Simpson's rule, calculate the distance travelled in 6.0
seconds (i.e. the area under the v/t graph) and the average
speed over this period (i.e. the area $\div$ time).
[22.7 m, 3.78 m s^{-1}]

11. An alternating current i has the following values at equal
intervals of 2.0 ms.

time (ms)	0	2.0	4.0	6.0	8.0	10.0	12.0
current (A)	0	1.7	3.5	5.0	3.7	2.0	0

Estimate the average current over the 12.0 ms period.
[2.73 A]

12. An indicator diagram for a steam engine is shown in Fig. 18.13. The base line has been divided into six equally spaced strips and the lengths of the seven ordinates measured and the results shown, the dimensions being in centimetres. Find the area of the indicator diagram. If 110 kN m^{-2} represents 1.00 cm find the mean pressure in the cylinder given that the

$$\text{mean pressure} = \frac{\text{area of diagram}}{\text{base}}$$

$[45.24 \text{ cm}^2, 311.0 \text{ kN m}^{-2}]$

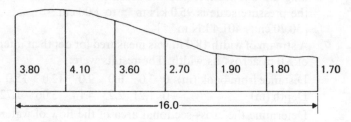

| 3.80 | 4.10 | 3.60 | 2.70 | 1.90 | 1.80 | 1.70 |

←————————————16.0————————————→

Fig. 18.13

13. To estimate the amount of earth to be removed when constructing a cutting, the cross-sectional areas at intervals of 10.0 m were firstly estimated and the results are shown below:

    Cross-sectional area (m²) 0 3.20 4.10 4.90 4.50 2.80 0

    Calculate the volume of earth to be excavated.   [202.7 m³]

14. A tree trunk is 12.0 m in length and has a varying circular cross-section. The cross-sectional areas at intervals of 2.0 m measured from one end are 0.95, 0.87, 0.76, 0.66, 0.54, 0.48, 0.45 m². Estimate the volume of the tree trunk.   [8.03 m³]

15. The areas of equidistantly spaced sections of the underwater form of a small vessel are as follows:

    2.0 3.2 3.8 3.6 2.8 1.8 0.8 m²

    The section is 6.0 m in length. Estimate the underwater volume. [16.8 m³]

16. The areas of seven horizontal cross-sections of a reservoir at invervals of 20.0 m are 330, 380, 440, 460, 350, 270, 220 m². How many litres of water will fill the reservoir? Express the answer in standard form.   [4.38×10^7 litres]

17. A cooling tower is 20.0 m high. The inside diameter of the tower at different heights is given in the following table:

Height (m)	0	5.00	10.00	15.00	20.00
Diameter (m)	16.00	13.30	10.70	8.60	8.00

    Estimate the capacity of the tower in m³.   [2 032 m³]

18. The pressure (p) of a gas at a given temperature varies with the volume (v) as shown below:

v (dm^3)	1.00	1.23	1.43	1.77	2.27	3.58
p (Pa)	100	75	60	45	32	15

Plot pressure against volume and determine the work done in compressing the gas from 3.58 dm^3 to 1.00 dm^3. This is done by measuring the area under the curve.
[94×10^{-3} joules]

19. A piece of timber has a cross-sectional area A sq. units perpendicular to its length at distance d units from its end, as shown by these figures:

d	0	2	4	6	8	10	12
A	2.8	4.7	6.2	7.3	7.1	6.7	6.2

Draw a graph of A against d and use Simpson's rule with 12 intervals to find the volume of the piece of timber.
[73.5 cubic units]

20. The heat capacity (C) of nitrogen varies with the natural logarithm (ln) of the absolute temperature (T) as shown below:

ln T	5.70	5.86	5.99	6.11	6.21	6.31	6.40
C(JK^{-1} mol^{-1})	28.8	29.1	29.4	29.7	30.0	30.3	30.6

Plot C against ln T and determine the entropy change between values of ln T of 5.70 and 6.40 by measuring the area under the curve.   [20.50 JK^{-1}]

Chapter 19

Further areas and volumes

19.1　The ellipse

An **ellipse** is the name given to the regular oval shape ACBD shown in Fig. 19.1.

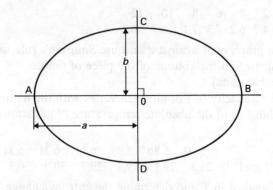

Fig. 19.1

A practical way of drawing an ellipse is to slip a loop of string around two fixed pins and then to slide a pencil along the string in such a way as to keep it taut at all points.

Let a loop of string of length l be used and let the distance between the two fixed pins F_1 and F_2 be x (Fig. 19.2).

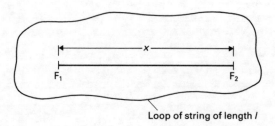

Loop of string of length l

Fig. 19.2

If the string is now pulled taut by sliding a pencil along it then points such as P_1, P_2 and P_3 can be marked, resulting ultimately in the oval shape shown in Fig. 19.3.

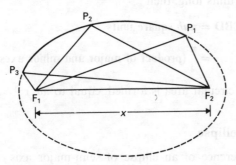

Fig. 19.3

The length $P_1F_1 + P_1F_2 =$ the length $l - x$

Similarly $P_2F_1 + P_2F_2 =$ the length $l - x$

and $P_3F_1 + P_3F_2 =$ the length $l - x$

Since the sum of such distances P_1, P_2, P_3 from the two fixed points, known as foci (i.e. plural of focus), is constant, then a definition of an ellipse emerges:

'An ellipse is a plane closed curve in which the sum of the distances of any point from the two foci is a constant quantity.'

The circle is a special case of an ellipse, occurring when the two foci coincide with each other to form the centre.

An ellipse is often referred to as a 'conic section' and this is because the ellipse is derived from a plane through a circular cone which is not parallel with the base. Such an ellipse is shown in Fig. 19.4.

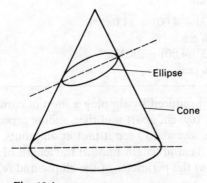

Fig. 19.4

Area of an ellipse

In Fig. 19.1 an ellipse ACBD is shown. AB is called the **major axis** and CD the **minor axis**. If the semi-major axis AO is 'a' units long and if the semi-minor axis CO is 'b' units long, then

Area of ellipse ACBD = πab square units

$$= \frac{\pi}{4} \text{ (product of major and minor axes)}$$

(Note that if $b = a$ then a circular area is formed, equal to πa^2)

Perimeter of an ellipse

The perimeter or circumference of an ellipse of semi-major axis 'a' and semi-minor axis 'b' is given by

Perimeter = $\pi(a + b)$ units

$$= \frac{\pi}{2} \text{ (sum of major and minor axes)}$$

Note that if $b = a$ then the perimeter (or circumference) of the circle formed is given by $2\pi a$.

Worked problems on the area and perimeter of an ellipse

Problem 1. The major axis of an ellipse is 14.00 cm and the minor axis 8.00 cm. Find the perimeter and the area of the ellipse.

If the major axis = 14.0 cm then the semi-major axis
$$= 7.00 \text{ cm}$$
If the minor axis = 8.00 cm then the semi-minor axis
$$= 4.00 \text{ cm}$$

$$\textbf{Perimeter} = \pi(7.00 + 4.00) = 11.0\pi$$
$$= \textbf{34.56 cm}$$
$$\textbf{Area} = \pi(7.00)(4.00) = 28.00\pi$$
$$= \textbf{87.96 cm}^2$$

Problem 2. An ellipse is produced by slipping a loop of cord around two fixed points 10.0 cm apart and then sliding a pencil along the cord in such a way as to keep it taut at all points.
If the loop of cord is 25.0 cm in length find (a) the length of the minor and major axes, (b) the perimeter of the ellipse and (c) the area of the ellipse.

(a)   The two fixed points 10.0 cm apart are shown as F_1 and F_2 in Fig. 19.5. When the pencil is in position P_1 such that P_1O is perpendicular to F_1F_2 then $F_1O = F_2O = 5.00$ cm and $F_1P_1 = 7.50$ cm since the loop of cord is 25.00 cm long.

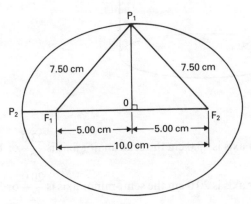

Fig. 19.5

Using the theorem of Pythagoras:
$$OP_1 = \sqrt{(7.50^2 - 5.00^2)} = 5.59 \text{ cm}$$

Thus the semi-minor axis is 5.59 cm and the **minor axis is 2 × 5.59 cm or 11.18 cm**
When the pencil is in position P_2, $P_2F_2 = \dfrac{25.0}{2} = 12.50$ cm and $F_1F_2 = 10.0$ cm. Hence $P_2F_1 = 2.50$ cm

Thus the semi-major axis $(P_2O) = 7.50$ cm and the **major axis is 2 × 7.50 cm or 15.0 cm**

(b)   **Perimeter of ellipse** $= \frac{1}{2}$(sum of major and minor axes)π

$$= \tfrac{1}{2}(15.0 + 11.18)\pi$$

$$= 13.09\pi = \textbf{41.12 cm}$$

(c)   **Area of ellipse** $=$ (product of semi-axes)π

$$= (5.59)(7.50)\pi$$

$$= \textbf{131.7 cm}^2$$

Problem 3. Find the total surface area and the volume of a regular oval closed cylinder, the diameters being 23.0 cm (major axis) and 20.0 cm (minor axis) respectively and the height 10.0 cm.

The cylinder is shown in Fig. 19.6

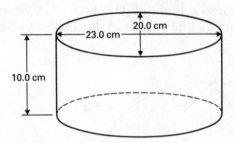

Fig. 19.6

Since the major axis is 23.0 cm the semi-major axis is $\dfrac{23.0}{2}$ or 11.5 cm

Since the minor axis is 20.0 cm the semi-minor axis is $\dfrac{20.0}{2}$ or 10.0 cm

$$\text{Area of elliptical end} = (11.5)(10.0)\pi = 115\pi \text{ cm}^2$$
$$\text{Perimeter of elliptical end} = (11.5 + 10.0)\pi = 21.5\pi \text{ cm}$$
$$\textbf{Total surface area of cylinder} = (21.5\pi)(10.0) + 2(115\pi)$$
$$= 215\pi + 230\pi$$
$$= 445\pi = \textbf{1 398 cm}^2$$
$$\textbf{Volume of elliptical cylinder} = (115\pi)(10.0)$$
$$= 1\,150\pi = \textbf{3 613 cm}^2$$

Problem 4. An elliptical plot of land has an area of 200 m² and a circumference of 60.0 m. Find the maximum length and the maximum width of the plot.

Let the semi-major axis and the semi-minor axis be denoted by a and b respectively, as shown in Fig. 19.1

$$\text{Area of plot} = 200 \text{ m}^2 = \pi ab \tag{1}$$
$$\text{Circumference of plot} = 60.0 \text{ m} = \pi(a + b) \tag{2}$$

From equation (2) $a = \dfrac{60}{\pi} - b$

Substituting $a = \dfrac{60}{\pi} - b$ in equation (1) gives

$$200 = \pi\left(\frac{60}{\pi} - b\right)b = 60b - \pi b^2$$
$$\pi b^2 - 60b + 200 = 0$$

This is a quadratic equation and is solved using the quadratic formula.

Hence $b = \dfrac{60 \pm \sqrt{[(60)^2 - (4)(\pi)(200)]}}{2\pi}$

$\qquad = \dfrac{60 \pm 32.97}{2\pi} = \dfrac{92.97}{2\pi}$ or $\dfrac{27.03}{2\pi}$

$b = 14.80$ m or 4.30 m

Substituting in equation (1):

Either $200 = \pi a(14.80)$

$\qquad a = \dfrac{200}{\pi(14.80)} = 4.301$ m

or $200 = \pi a(4.30)$

$\qquad a = \dfrac{200}{\pi(4.30)} = 14.81$ m

Hence the semi-major and semi-minor axes are 14.81 m and 4.30 m, and **the maximum length and breadth of the plot of land are 29.62 m and 8.60 m**

Further problems on the area and perimeter of an ellipse may be found in Section 19.5 (Problems 1–10), page 359.

19.2 The frustum and zone of a sphere

For any sphere of radius r:

Surface area $= 4\pi r^2$ square units

Volume $= \frac{4}{3}\pi r^3$ cubic units

A frustum of a sphere is the portion intercepted between two parallel planes (for example, the portion ABCD shown in Fig. 19.7).

A zone is the curved surface of a frustum.

If h is the height, or thickness, of a frustum and r_1 and r_2 are the radii of the ends, then

Area of a zone $= 2\pi rh$ square units

Volume of frustum $= \dfrac{\pi h}{6}\,(h^2 + 3r_1^2 + 3r_2^2)$ cubic units

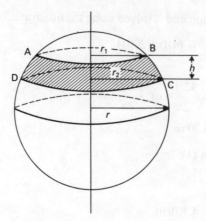

Fig. 19.7

Worked problems on frusta and zones of a sphere

Problem 1. Calculate (*a*) the volume and the surface area of a sphere of diameter 38.6 cm, (*b*) the volume of a frustum of the sphere if the radii of the ends of the frustum are 10.0 cm and 18.0 cm and the height of the frustum is 8.0 cm, and (*c*) the curved surface area of the frustum (i.e. the area of the zone).

(*a*) If diameter of sphere = 38.6 cm then the radius = 19.3 cm

Hence volume of sphere $= \frac{4}{3}\pi r^3 = \frac{4}{3}\pi(19.3)^3$

$$= \mathbf{30\,110 \ cm^3}$$

Surface area of sphere $= 4\pi r^2 = 4\pi(19.3)^2$

$$= \mathbf{4\,681 \ cm^2}$$

(*b*) Volume of a frustum of a sphere $= \dfrac{\pi h}{6}[h^2 + 3r_1^2 + 3r_2^2]$,

where $h = 8.0$ cm, $r_1 = 10.0$ cm and $r_2 = 18.0$ cm

Hence volume of the frustum

$$= \frac{\pi 8.0}{6}[(8.0)^2 + 3(10.0)^2 + 3(18.0)^2]$$

$$= \frac{\pi 8.0}{6}[64.0 + 300 + 972]$$

$$= \mathbf{5\,596 \ cm^3}$$

(c)   Area of a zone of a sphere $= 2\pi rh$, where $r = 19.3$ cm and $h = 8.00$ cm

Hence area of the zone of the sphere $= 2\pi(19.3)(8.00)$
$$= \textbf{970.1 cm}^2$$

Problem 2. Calculate the volume of the frustum of a sphere and the area of the zone formed, if the diameters of the ends of the frustum are 16.0 cm and 30.0 cm and the height (or thickness) of the frustum is 6.0 cm.

Volume of a frustum of a sphere $= \dfrac{\pi h}{6}[h^2 + 3r_1^2 + 3r_2^2]$, where $h = 6.0$ cm, $r_1 = \dfrac{16.0}{2} = 8.0$ cm and $r_2 = \dfrac{30.0}{2} = 15.0$ cm

Hence the volume of the frustum of the sphere
$$= \frac{\pi 6.0}{6}[(6.0)^2 + 3(8.0)^2 + 3(15.0)^2]$$
$$= \pi(36.0 + 192 + 675) = \textbf{2 837 cm}^3$$

The area of the zone of the sphere $= 2\pi rh$, where $r = $ radius of sphere and $h = 6.0$ cm

The radius of the sphere is not given in the question. It can, however, be calculated. Figure 19.8 shows a plane view through the centre of the sphere.

Using the theorem of Pythagoras:
$$OC^2 = DC^2 + OD^2$$
$$\text{i.e. } r^2 = (15.0)^2 + OD^2 \tag{1}$$
$$\text{Also } OB^2 = AB^2 + AO^2$$
$$\text{i.e. } r^2 = (8.0)^2 + (6.00 + OD)^2 \tag{2}$$

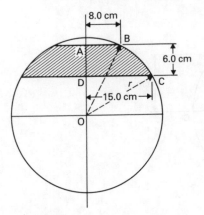

Fig. 19.8

Equating equations (1) and (2) gives:
$$(15.0)^2 + OD^2 = (8.0)^2 + (6.0 + OD)^2$$
Therefore $225 + OD^2 = 64.0 + 36.0 + 12.0(OD) + (OD)^2$
and $\qquad\qquad 225 = 100 + 12.0(OD)$
Hence $\qquad 12.0(OD) = 125$
$$OD = \frac{125}{12.0} = 10.42 \text{ cm}$$

Substituting in equation (1) gives:
$$r^2 = (15.0)^2 + (10.42)^2$$
$$r = \sqrt{[(15.0)^2 + (10.42)^2]} = 18.26 \text{ cm}$$

Hence the area of the zone of the sphere $= 2\pi(18.26)(6.0)$
$$= \textbf{688.4 cm}^2$$

Problem 3. A sphere has a diameter of 10.0 cm. Calculate its volume and surface area. A frustum of the sphere is formed by two parallel planes, one through the diameter, the other distance h from the diameter. If the curved surface area of the frustum (i.e. the area of the zone) is to be one-third of the total surface area of the sphere find the thickness and the volume of the spherical frustum. Express the volume of the frustum as a percentage of the volume of the sphere.

If the diameter of the sphere $= 10.0$ cm, then the radius $= 5.0$ cm

Volume of sphere $= \frac{4}{3}\pi(5.0)^3 = \textbf{523.6 cm}^3$
Surface area of sphere $= 4\pi(5.0)^2 = \textbf{314.2 cm}^2$

The frustum of the sphere is shown in Fig. 19.9

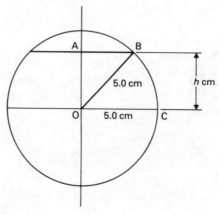

Fig. 19.9

Curved surface area of frustum $= 2\pi rh$

$$\text{Therefore } \frac{314.2}{3} = 2\pi(5.0)h$$

Thus the thickness $h = \dfrac{314.2}{(3)(2)(\pi)(5.0)} = \textbf{3.334 cm}$

The smaller radius of the frustum (i.e. AB in Fig. 19.9) is given by:

$AB = \sqrt{[(5.0)^2 - (3.334)^2]}$

$\quad = 3.726$ cm

The volume of a frustum $= \dfrac{\pi h}{6} [h^2 + 3r_1^2 + 3r_2^2]$, where

$h = 3.334$ cm, $r_1 = 3.726$ cm and $r_2 = 5.0$ cm

Hence the volume of the frustum

$$= \frac{\pi(3.334)}{6} [(3.334)^2 + 3(3.726)^2 + 3(5.0)^2]$$

$$= \frac{\pi(3.334)}{6} [11.12 + 41.65 + 75.0]$$

$$= \textbf{223.0 cm}^3$$

$\dfrac{\text{Volume of frustum}}{\text{Volume of sphere}} = \dfrac{223.0}{523.6} = 0.425\ 9$ or **42.59%**

Note that the curved surface area of the frustum is 33.33% of the whole spherical surface area but the volume of the frustum is 42.59% of the whole spherical volume (i.e. the curved surface area of a frustum is not proportional to its volume).

Problem 4. What percentage volume of a sphere 22.00 cm in diameter is contained between two parallel planes distance 3.00 cm and 5.00 cm from the centre and on opposite sides of it?

Figure 19.10 shows a view through the centre of the sphere, the parallel planes AC and FD forming a frustum ACDF.

To find the volume of a frustum it is necessary to know the radii of its ends, as well as its thickness.

Since the diameter of the sphere is 22.00 cm, the radius (i.e. OC or OD) is 11.00 cm.

$$BC = \sqrt{[OC^2 - OB^2]} = \sqrt{[(11.00)^2 - (3.00)^2]} = \sqrt{112}$$

∴ $\qquad$ BC $= 10.58$ cm $=$ radius of one end of frustum

Similarly, ED $= \sqrt{[OD^2 - OE^2]} = \sqrt{[(11.00)^2 - (5.00)^2]} = \sqrt{96.0}$

Therefore ED $= 9.798$ cm $=$ radius of other end of frustum.

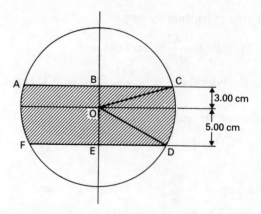

Fig. 19.10

Volume of frustum ACDF $= \dfrac{\pi h}{6}[h^2 + 3r_1^2 + 3r_2^2]$, where

$h = 3.00$ cm $+ 5.00$ cm $= 8.00$ cm, $r_1 = 10.58$ cm and $r_2 = 9.798$ cm

Hence volume of frustum ACDF

$$= \frac{\pi(8.00)}{6}[(8.00)^2 + 3(10.58)^2 + 3(9.798)^2]$$

$$= \frac{\pi(8.00)}{6}[64.0 + 335.8 + 288.0]$$

$$= 2\,881 \text{ cm}^3$$

Volume of sphere $= \frac{4}{3}\pi(11.00)^3 = 5\,575 \text{ cm}^3$

Hence the percentage volume contained between planes AC and FD is

$$\frac{2\,881}{5\,575} \times 100\%, \text{ i.e. } \mathbf{51.68\%}$$

Further problems on frusta and zones of spheres may be found in Section 19.5 (Problems 11–19), page 360.

19.3   Prismoidal rule for finding volumes

In Section 18.2 Simpson's rule for calculating the volume V of irregular solids was stated as

$$V = \frac{d}{3}[(1\text{st} + \text{last area}) + 2(\text{sum of odd ordinate areas})$$
$$+ 4(\text{sum of even ordinate areas})]$$

where d is the distance between areas.

If this rule is applied to a solid of length x divided by only three parallel equidistant planes then a formula known as the prismoidal rule is obtained. Referring to Fig. 19.11, the volume V of the solid is given by:

$$V = \frac{x/2}{3}[A_1 + A_3 + 4A_2]$$

i.e. $V = \frac{x}{6}[A_1 + 4A_2 + A_3]$ **cubic units**

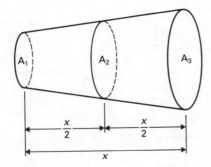

Fig. 19.11

The prismoidal formula gives an exact value of volume for pyramids, cones, spheres and prismoids (a prismoid being a solid approaching in form to a prism having two parallel plane similar figures for its ends).

Worked problems on the use of the prismoidal rule

Problem 1. Find (a) the volume of a sphere of radius r and (b) the volume of a cone of radius r and height h by using the prismoidal rule.

(a)   A sphere of radius r is shown in Fig. 19.12

Using the prismoidal rule: volume $V = \frac{h}{6}[A_1 + 4A_2 + A_3]$,

where $h = 2r$, $A_1 = A_3 = 0$ and $A_2 = \pi r^2$

Therefore volume $V = \frac{2r}{6}[0 + 4(\pi r^2) + 0]$

Hence the volume of a sphere $= \frac{4}{3}\pi r^3$ cubic units

(b)   A cone of radius r and height h is shown in Fig. 19.13

Using similar triangles: $\frac{h/2}{x} = \frac{h}{r}$

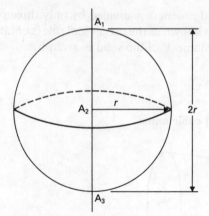

Fig. 19.12

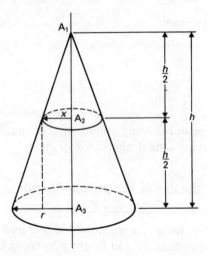

Fig. 19.13

$$\text{Hence } x = \frac{r}{2}$$

Using the prismoidal rule:

$$\text{volume } V = \frac{h}{6}[A_1 + 4A_2 + A_3]$$

$$\text{where } h = \text{height } h, \ A_1 = 0, \ A_2 = \pi x^2 = \pi\left(\frac{r}{2}\right)^2$$

$$\text{and } A_3 = \pi r^2$$

Hence volume $V = \dfrac{h}{6}\left[0 + 4\pi\left(\dfrac{r}{2}\right)^2 + \pi r^2\right]$

$= \dfrac{h}{6}[2\pi r^2] = \frac{1}{3}\pi r^2 h$ cubic units

Hence the volume of a cone $= \frac{1}{3}\pi r^2 h$ cubic units

---

Problem 2. A bucket is 20.0 cm internal diameter at the bottom and 36.0 cm internal diameter at the top. If the depth is 25.0 cm find the capacity of the bucket in cm^3, litres and dm^3.

The bucket is shown in Fig. 19.14

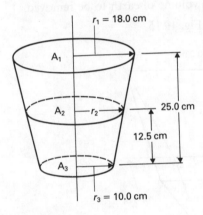

Fig. 19.14

At a distance of 12.5 cm from one end the radius r_2 will be $\dfrac{10.0 + 18.0}{2}$ or 14.0 cm since the sloping sides change uniformly.

Volume of bucket $= \dfrac{h}{6}[A_1 + 4A_2 + A_3]$, where $h = 25.0$ cm, $A_1 = \pi(18.0)^2$ cm^2, $A_2 = \pi(14.0)^2$ cm^2 and $A_3 = \pi(10.0)^2$ cm^2

Hence volume $= \dfrac{25.0}{6}[\pi(18.0)^2 + 4\pi(14.0)^2 + \pi(10.0)^2]$

$= \dfrac{25.0}{6}[324\pi + 784\pi + 100\pi]$

$= 15\,810$ cm^3

Capacity may be measured in litres (l) or in cubic decimetres (dm^3).

Strictly, $1\,l = 1.000\,028\ dm^3$. However, except in circumstances when results of high precision are required, 1 litre is assumed to be equal to 1 cubic decimetre, i.e. $1\,l \approx 1\ dm^3 = 1.000\,028\ dm^3$.

Hence the capacity of the bucket $= 15\,810\ cm^3$

$$= \mathbf{15.81\ litres\ or\ 15.81\ dm^3}$$

The prismoidal rule thus provides a shorter method than that used in Chapter 18, although none the less accurate, where the bucket would have been treated as a frustum of a cone.

Problem 3. A trench is to be dug in the form of a prismoid. The bottom is to be a rectangle 20.0 m long by 14.0 m wide; the top is also a rectangle, 30.0 m by 22.0 m wide. If the depth of the trench is to be 10.0 m, find the volume of earth to be removed.

The trench is shown in Fig. 19.15

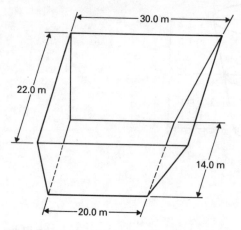

Fig. 19.15

If A_1 represents the area of the top of the trench then
$$A_1 = 30.0 \times 22.0 = 660\ m^2$$
If A_3 represents the area of the bottom of the trench then
$$A_3 = 20.0 \times 14.0 = 280\ m^2$$
If A_2 represents the rectangular area through the middle of the trench (the plane of A_2 being parallel to the planes of A_1 and A_3) then

$$\text{Length of rectangle} = \frac{30.0 + 20.0}{2} = 25.0\ m$$

$$\text{Width of rectangle} = \frac{22.0 + 14.0}{2} = 18.0\ m$$

$$\text{Then } A_2 = 25.0 \times 18.0 = 450\ m^2$$

Using the prismoidal rule the volume of the trench is given by:

Volume, $V = \dfrac{h}{6}[A_1 + 4A_2 + A_3]$, where $h = 10.0$ m, $A_1 = 660$ m^2,

$A_2 = 450$ m^2 and $A_3 = 280$ m^2

Volume, $V = \dfrac{10.0}{6}[660 + 4(450) + 280]$

$\qquad = 4\,567$ m^3

Hence the volume of earth to be removed is 4 567 m^3

Problem 4. The roof of a building is in the form of a frustum of a pyramid with a square base of side 6.00 m. The flat top is a square of side 1.50 m and all the sloping sides are pitched at the same angle. The vertical height of the flat top above the level of the eaves is 5.00 m. Calculate the volume enclosed by the roof and the total area of the sloping surface.

The roof is shown in Fig. 19.16

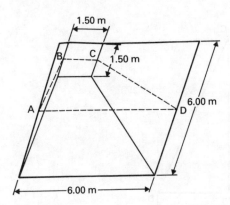

Fig. 19.16

If the area of the top of the frustum is A_1, then $A_1 = (1.50)^2 = 2.25$ m^2. If the area of the bottom of the frustum is A_3, then $A_3 = (6.00)^2 = 36.0$ m^2.

If the area of a section through the middle of the frustum is represented by A_2 (the plane of A_2 being parallel to the planes of A_1 and A_3) then since A_1 and A_3 are similar areas, i.e. squares, A_2 will also be a square, its sides being halfway between those of A_1 and A_3, that is $\left[\dfrac{6.00 + 1.50}{2}\right]$ m

Thus $A_2 = \left[\dfrac{6.00 + 1.50}{2}\right]^2 = (3.75)^2 = 14.06 \text{ m}^2$

Using the prismoidal rule:

volume of frustum, $V = \dfrac{h}{6}[A_1 + 4A_2 + A_3]$, where $h = 5.00$ m,
$A_1 = 2.25 \text{ m}^2$, $A_2 = 14.06 \text{ m}^2$ and $A_3 = 36.0 \text{ m}^2$

Volume, $V = \dfrac{5.00}{6}[2.25 + 4(14.06) + 36.0]$

$\qquad = \dfrac{5.00}{6}(94.49) = 78.74 \text{ m}^3$

Hence the volume enclosed by the roof is 78.74 m³

The area of a sloping side is the area of a trapezium of parallel sides 1.50 m and 6.00 m, with a perpendicular distance between the parallel sides given by $\sqrt{(5.00^2 + 2.25^2)}$ or 5.483 m from Fig. 19.17, which shows a vertical section of the roof (as shown by the broken lines in Fig. 19.16).

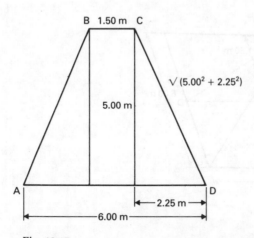

Fig. 19.17

Hence the area of a sloping side is $\left[\dfrac{6.00 + 1.50}{2}\right](5.483) = 20.56 \text{ m}^2$

Area of four sloping sides is 4(20.56) = 82.24 m²

Problem 5. A frustum of a sphere of radius 10.00 cm is formed by two parallel planes on opposite sides of the centre, each at a distance of 4.00 cm from the centre.

Calculate the volume of the frustum by (*a*) using the formula for the volume of a frustum of a sphere and (*b*) using the prismoidal rule.

(*a*) The frustum of the sphere is shown as ABFE in Fig. 19.18

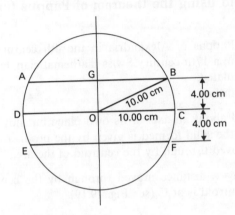

Fig. 19.18

Using the formula for the volume V of the frustum of a sphere, $V = \dfrac{\pi h}{6}[h^2 + 3r_1^2 + 3r_2^2]$, where $h = 8.00$ cm,

$r_1 = r_2 = $ length BG of Fig. 19.18

$$BG = \sqrt{(10.0^2 - 4.00^2)} = \sqrt{84.0} = 9.165$$

Hence the volume of frustum ABFE

$$= \frac{\pi 8.00}{6}[(8.00)^2 + 3(9.165)^2 + 3(9.165)^2]$$

$$= \frac{\pi(8.00)}{6}(568.0) = \mathbf{2\ 379\ m^3}$$

(*b*) Using the prismoidal rule, volume V of frustum ABFE is given by $V = \dfrac{h}{6}[A_1 + 4A_2 + A_3]$, where $h = 8.00$ cm,

$A_1 = \pi(9.165)^2$, $A_2 = \pi(10.00)^2$ and $A_3 = \pi(9.165)^2$

Hence the volume of frustum ABFE

$$= \frac{8.00}{6}[\pi(9.165)^2 + 4\pi(10.00)^2 + \pi(9.165)^2]$$

$$= \frac{8.00\pi}{6}(568) = \mathbf{2\ 379\ m^3}$$

Further problems on the use of the prismoidal rule may be found in Section 19.5 (Problems 20–26), page 361.

19.4 Volume of a solid using the theorem of Pappus (or Guldin)

A theorem discovered by Pappus of Alexandria in the 4th century and rediscovered by Paul Guldin, a 17th-century Swiss mathematician, enables volumes of solids to be calculated.

The theorem of Pappus states:

'If a plane area is rotated about an axis in its own plane but not intersecting it, the volume of the solid formed is given by the product of the area and the distance moved through by the centroid of the area.'

i.e. Volume generated = Area × distance moved through by the centroid. Let A be any area whose centroid is at C (see Fig. 19.19).

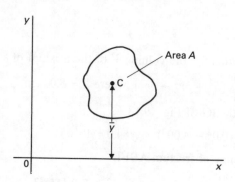

Fig. 19.19

If the distance of the centroid C from the axis of rotation is $\bar{y}$ then the distance moved through by the centroid when the area A is rotated 360° about the x-axis is $2\pi\bar{y}$ (i.e. the circumference of a circle of radius $\bar{y}$) and the volume V of the solid produced is given by

$$\text{Volume generated} = \text{Area} \times 2\pi\bar{y}$$
$$\text{i.e. } V = 2\pi A\bar{y} \text{ cubic units}$$

Worked problems using the theorem of Pappus

Problem 1. Using the theorem of Pappus, calculate the position of the centroid of a semicircle of radius r.

If the semicircular area shown in Fig. 19.20 is rotated about the axis XX then the volume of the solid generated will be that of a sphere, i.e. $\frac{4}{3}\pi r^3$. The area of the semicircular area is $\frac{\pi r^2}{2}$ (N.B. XX touches but does not intersect the semicircle.)

The centroid C will lie on the axis of symmetry OA shown in Fig. 19.20.

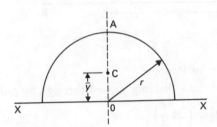

Fig. 19.20

Using the theorem of Pappus:
Volume generated = Area × distance moved through by centroid

$$\text{i.e. } \tfrac{4}{3}\pi r^3 = \left(\frac{\pi r^2}{2}\right)(2\pi\bar{y})$$

$$\bar{y} = \frac{4\pi r^3}{(3)\left(\dfrac{\pi r^2}{2}\right)(2\pi)}$$

$$\bar{y} = \frac{4r}{3\pi} \text{ units from the diameter}$$

Hence the centroid of a semicircle lies on the axis of symmetry at a distance of $\dfrac{4r}{3\pi}$ from the diameter

---

Problem 2. Derive the formula for the volume of a cone, using the theorem of Pappus, given that the centroid of a triangle lies at a distance of $\dfrac{h}{3}$ from the base, where h is the perpendicular height of the triangle.

Let the right-angled triangle PQR be as shown in Fig. 19.21 with its centroid C at a distance $\dfrac{r}{3}$ from PR. If the triangle is rotated 360° about axis XX the volume of a cone will result.

The area of triangle PQR $= \frac{1}{2}qr$

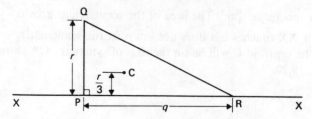

Fig. 19.21

Thus from the theorem of Pappus:

Volume generated = Area × distance moved through by centroid

$$= (\tfrac{1}{2}qr)\left(2\pi \frac{r}{3}\right)$$

$$= \tfrac{1}{3}\pi qr^2$$

Length r is the radius and length q the height of the cone

Hence the volume of a cone is given by $\dfrac{\pi}{3}$ × radius² × height

---

Problem 3. Calculate the volume of an anchor ring formed by rotating a circle of radius 5.0 cm about an axis at a distance of 20.0 cm from its centre.

The 5.0 cm radius circle is rotated about the axis XX as shown by the side elevation in Fig. 19.22(*a*), forming an anchor ring shown by a different elevation in Fig. 19.22(*b*).

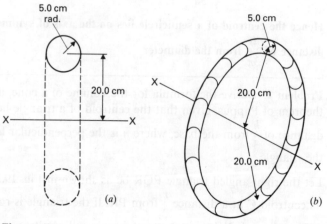

Fig. 19.22

Area of circle $= \pi(5.0)^2 = 25.0\pi$ cm^2

Distance moved by centroid C in one complete
revolution $= 2\pi(20.0) = 40.0\pi$ cm. Using the theorem of Pappus,
volume generated = area × distance moved through by the centroid.
i.e. Volume generated $= (25.0\pi)(40.0\pi) = 1\,000\pi^2 = 9\,870$ cm^3

Hence the volume of the anchor ring is 9 870 cm^3

Problem 4. A steel disc has a diameter of 12.00 cm and is of
thickness 2.50 cm. A semicircular groove of diameter 2.00 cm has
been machined centrally around the rim to form a pulley. Find
the volume of metal removed, the volume of the pulley and its
mass (in kg) if the density of steel is 7.80 g cm^{-3}.

The disc is shown in Fig. 19.23(a) with a side view of the rim in
Fig. 19.23(b)

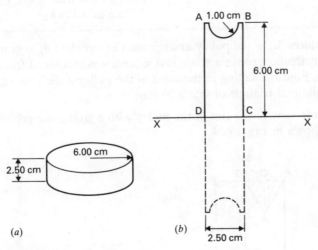

Fig. 19.23

If the area ABCD is rotated about axis XX the volume generated
will be that of the pulley.

The centroid of the semicircular area removed is at a distance
$\dfrac{4r}{3\pi}$ cm from its diameter, i.e. $\dfrac{4(1.00)}{3\pi}$ cm or 0.424 cm from AB

Hence the distance of the centroid of the semicircular area is
$(6.00 - 0.424)$ cm or 5.576 cm from axis XX. The distance moved
through by this centroid when rotated 360° about the axis XX is
$2\pi(5.576)$ cm.

The area of the semicircle $= \dfrac{\pi(1.00)^2}{2} = \dfrac{\pi}{2}$ cm

Hence the volume V generated by the semicircular area when rotated $360°$ about axis XX is given by the theorem of Pappus, i.e.

$V = $ (Area)(distance moved through by the centroid)

$= \left(\dfrac{\pi}{2}\right)(2\pi\ 5.576)$

$= 55.03$ cm^3

Hence the volume of metal removed is 55.03 cm³

The volume of the steel disc (i.e. before machining the groove) is $\pi(6.0)^2(2.50)$ or 282.7 cm^3

Hence the volume of the pulley $= (282.7 - 55.03)$ cm^3

$\qquad\qquad\qquad\qquad\quad = $ **227.7 cm³**

Mass of the pulley $= $ density $\times$ volume

$\qquad\qquad\qquad\quad = (7.80$ g cm$^{-3})(227.7$ cm$^3)$

$\qquad\qquad\qquad\quad = $ **1 776 g or 1.776 kg**

Problem 5. In the pulley arrangement of Problem 4, instead of a semicircular groove a triangular groove was machined from the rim. Find the saving in the mass of the pulley if the groove is an equilateral triangle of side 2.50 cm.

A side view of the rim of the pulley with a triangular groove PQR is shown in Fig. 19.24

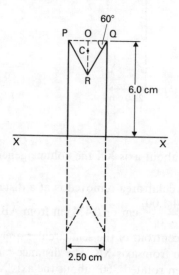

Fig. 19.24

The perpendicular height OR of triangle PQR

$$= 2.50 \sin 60° = 2.165 \text{ cm}$$

Area of triangle PQR $= \frac{1}{2}(2.50)(2.165) = 2.706$ cm

The centroid C is at a distance of $\frac{1}{3}(2.165)$ cm from O, i.e. 0.722 cm. Hence the distance of the centroid C from axis XX is $(6.00 - 0.722)$ cm or 5.278 cm. The distance moved through by the centroid when rotated 360° about the axis XX is $2\pi(5.278)$ cm.

Hence the volume V generated by triangle PQR when rotated 360° about axis XX is given by the theorem of Pappus, i.e.

$V =$ (Area)(distance moved through by centroid)

$\quad = (2.706)(2\pi\ 5.278) \text{ cm}^3$

$\quad = 89.74 \text{ cm}^3$

The volume of the pulley $= (282.7 - 89.74) \text{ cm}^3$

$$= 193.0 \text{ cm}^3$$

Mass of the pulley $= (7.80)(193.0)$ g $= 1\ 505$ g

Thus the saving in the mass by machining a triangular groove instead of a semicircular groove is $(1\ 776 - 1\ 505)$ g, i.e. 271 g

---

Further problems using the theorem of Pappus may be found in the following Section (Problems 27–35), page 362.

19.5   Further problems

The ellipse

1. The major axis of an ellipse is 125.0 mm and the minor axis 75.0 mm. Find the perimeter and the area of the ellipse. [314.2 mm, 7 363 mm²]

2. An ellipse is produced by a loop of string slipped around two points 15.0 cm apart and then sliding a pencil along the string in such a way as to keep it taut at all points. If the loop of string is 43.0 cm long find (a) the length of the major and minor axes, (b) the perimeter of the ellipse and (c) the area of the ellipse.
    (a) [28 cm, 23.64 cm]   (b) [81.12 cm]   (c) [519.9 cm²]

3. Find the total surface area and volume of a regular oval closed cylinder, the diameters being 320 mm (major axis) and 210 mm (minor axis) respectively and the height 125 mm. [2 096 cm², 6 597 cm³]

4. An elliptical fish pond has an area of 240.0 m². If its greatest length is 45.0 m, find the perimeter of the pond.   [81.35 m]

5. What are the circumference and area of an ellipse whose major and minor axes are 4.62 m and 2.86 m respectively? [11.75 m, 10.38 m²]

6. If fencing costs £5.00 per metre length find the cost (to the nearest pound) of enclosing an elliptical plot of land which has major and minor diameter lengths of 110.0 m and 70.0 m. [£1 414]

7. The axes of an ellipse are 5.00 cm and 3.00 cm respectively. Find the area, perimeter and the distance between foci. [11.78 cm², 12.57 cm, 4.00 cm]

8. A cycling track is in the form of an ellipse, the axes being 250 m and 150 m respectively for the inner boundary and 270 m and 170 m for the outer boundary. Calculate the area of the track.   [6 597 m²]

9. A rectangular metal plate measures 12.00 cm by 8.00 cm. If the maximum possible sized ellipse is cut from the plate find the area of metal wasted.   [20.60 cm²]

10. An ellipse has an area of 156.0 cm² and a circumference of 45.18 cm. Find the maximum length and the maximum breadth of the ellipse.   [17.24 cm, 11.52 cm]

The sphere

11. Calculate the volume and surface area of a sphere of diameter 42.0 cm.   [38 790 cm³, 5 542 cm²]

12. If the volume of a sphere is 2 360 cm³ find its diameter and surface area.   [16.52 cm, 857.4 cm²]

13. The surface area of a sphere is 426 cm². Find its radius and volume.   [5.82 cm, 825.8 cm³]

14. Calculate the volume and curved surface area of a frustum of a sphere if the diameters of the ends are 96.0 mm and 132.0 mm and the thickness is 24.0 mm. [258.3 cm³, 109.8 cm²]

15. A sphere has a diameter of 14.6 cm. Find its volume and surface area. A frustum of the sphere is formed by two parallel planes, one through the diameter and the other at a distance X from the diameter. If the curved surface area of the frustum (i.e. the zone) is to be $\frac{2}{5}$ of the surface area of the sphere find the height X of the frustum and its volume. [1 630 cm³, 669.7 cm², 5.84 cm, 769.1 cm³]

16. A sphere has a diameter of 46.0 mm. Calculate the volume of a frustum of the sphere contained between two parallel planes distances 10.0 mm and 14.0 mm from the centre and on opposite sides of it.   [35 960 mm³ or 35.96 cm³]

17. A frustum of a sphere is formed by two parallel planes on opposite sides of the centre. If the sphere is of diameters 18.6 cm and the diameter of the ends of the frustum are 11.4 cm and 5.8 cm find the ratio of the volume of the frustum to the volume of the sphere. Find also the ratio of the curved surface area of the frustum to the surface area of the sphere.   [0.967, 0.870]

18. A spherical storage vessel is filled with liquid to a depth of 40.0 cm. If the internal diameter of the vessel is 50.0 cm, find how many litres of liquid there are in the container (1 litre = 1 000 cm^3)   [58.64 litres]

19. A dome is in the form of a zone of a sphere and is 3.00 m deep between the horizontal circular sections which have radii of 4.00 m and 7.50 m respectively. Calculate the volume of the dome and the area of the spherical surface of the dome. [354.6 m^3, 172.1 m^2]

Prismoidal rule

20. Use the prismoidal rule to find the volume of the frustum of a sphere contained between two parallel planes on opposite sides of the centre, each of radius 9.00 cm and each 5.00 cm from the centre.   [3 068 cm^3]

21. A cone has a perpendicular height of 18.60 cm and base diameter 9.80 cm. Find its volume using the prismoidal rule. [467.7 cm^3]

22. A bucket is in the form of a frustum of a cone. The internal diameter of the bottom is 32.0 cm and that of the top 40.0 cm. If the depth is 30.0 cm find the capacity (in dm^3) of the bucket (1 dm^3 = 1 000 cm^3).   [30.66 dm^3]

23. A hole is to be dug in the form of a prismoid. The bottom is a rectangle 24 m long by 18 m wide; the top is a rectangle 36 m long by 28 m wide. If the depth of the hole is to be 15 m, find the volume of earth to be removed. [10 500 m^3]

24. Find the capacity (in litres) of a small water reservoir, the top being a rectangle measuring 25.0 m by 10.0 m, the bottom a rectangle measuring 18.0 m by 8.0 m and the depth 4.0 m (1 litre = 1 000 cm^3).   [779 000 litres]

25. The length of a trench is 10.0 m and the trench is of uniform depth 1.3 m. The trench is 5.0 m wide at the top and 4.0 m wide at the bottom. Find the volume of soil removed in excavating the trench.   [58.5 m^3]

26. The roof of a building is in the form of a frustum of a pyramid with a square base of 5.00 m. The flat top is a square of 1.20 m and all the sloping sides are pitched at the same angle. The vertical height of the flat top above the level of the eaves is 4.00 m. Calculate the volume enclosed by the roof and the total area of the sloping surfaces.
[43.25 m³, 54.91 m²]

The theorem of Pappus

27. A rectangle measuring 11.00 cm by 5.00 cm revolves 360° about one of its longest sides as axis. Find the volume of the resulting cylinder by using the theorem of Pappus.
[863.9 cm³]

28. A right-angled isosceles triangle of hypotenuse 7.75 cm is revolved 360° about one of its equal sides as axis. Find the volume of the solid generated.   [172.3 cm³]

29. A quadrant of a circle of radius r is rotated 360° about one of its radii to form a hemisphere. Using the theorem of Pappus find the distance of the centroid of the quadrant from the centre of the circle. $\left[\dfrac{4\sqrt{2}r}{3\pi} \text{ or } 0.6r\right]$

30. One edge of an equilateral triangle of side 9.00 cm is parallel to an axis YY lying at a distance of 15.00 cm from the edge, as shown in Fig. 19.25. If the triangle is rotated one complete revolution about this axis find the volume of the solid generated.   [3 878 cm³]

31. The shape shown in Fig. 19.26 is rotated 360° about axis YY. Calculate the volume of the solid generated.   [29 130 cm³]

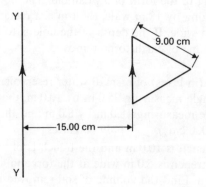

Fig. 19.25

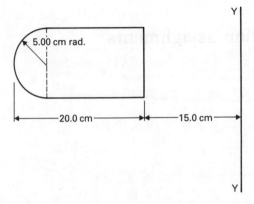

Fig. 19.26

32. Find the volume of an anchor ring formed by rotating a circle of radius 42.0 cm about an axis distance 15.3 cm from its centre. [5 327 cm^3]

33. Find the volume of a life buoy formed by rotating an ellipse, having major and minor axes of 12.00 cm and 8.00 cm, about an axis distance 36.00 cm from its centre. [17 060 cm^3]

34. A metal disc has a radius of 5.30 cm and is of thickness 2.00 cm. A semicircular groove of diameter 2.00 cm is machined centrally around the rim to form a pulley. Find the volume of metal removed and the volume of the pulley. [48.12 cm^3, 128.4 cm^3]

35. A steel disc has a diameter of 8.00 cm and is of thickness 1.80 cm. A triangular groove is machined centrally around the rim to form a pulley. If the triangular shape is isosceles, with base 1.80 cm (coinciding with the edges of the rim) and perpendicular height 1.20 cm, find the volume of metal removed and the mass of the pulley if steel has a density of 7.80 g cm^{-3}. [24.43 cm^3, 515.2 g]

Chapter 20

Mensuration assignments

Introduction

There are several reasons for carrying out an assignment at this stage of study but the main one is to enable you to **use** the mathematics recently learned in a **real** situation. In this case the assignment is related to the sections on mensuration (Chapters 17 to 19). It could be, for example, to measure the dimensions of a drilling machine or a lathe to determine the mass, and hence the cost of the materials in their construction. You are then required to assess which measurements to take and the best way to use these measurements to obtain the answer required.

This chapter consists of a number of exercises with chosen values to give some idea of the types of problems which might be considered as a basis for an assignment. As examples are given with an answer which can be checked, you will be able to get some practice in the methods chosen in solving a particular problem and ensure that you are using them correctly.

Assignment No. 1

A marquee is made from canvas in the shape of a cylinder surmounted by a cone. The total vertical height of the marquee is 15.0 m and the cylindrical portion has a height of 8.0 m with a diameter of 35.0 m.

(a) If the cost of canvas is £3.20 per square metre, and an extra 10% of material is needed for joins and waste, calculate the cost of making the marquee.

(b) Determine the volume of the marquee, correct to the nearest cubic metre.

(c) Determine the perpendicular height of a marquee in the shape of a conical pyramid if it has the same volume as that determined in part (b) and a base radius of 17.5 m.

(*d*) If only 5% extra material is needed for joins and waste with the conical marquee, calculate its cost if it is made of canvas costing £3.20 per square metre.

(*a*) [£6 744]   (*b*) [9 942 m³]   (*c*) [31.0 m]   (*d*) [£6 576]

Assignment No. 2

Figure 20.1 shows a plan of the floor of a building which is to be carpeted. The following measurements were taken: AB = 1.40 m, BC = DC = 2.00 m, DE = HI = 1.60 m, EF = GH = 0.90 m, FG = 0.60 m, IL = 3.40 m, JK = 1.80 m. Other measurements are shown in the plan in Fig. 20.1. Calculate the area of the floor in m². A carpet costing £14.90 per m² is chosen. Assuming 30% extra carpeting is required due to wastage in fitting, calculate the cost of carpeting the floor to the nearest pound.
[17.53 m²; £340]

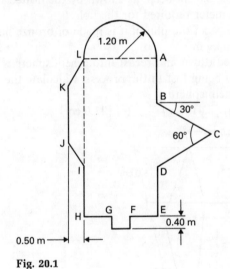

Fig. 20.1

Assignment No. 3

The circumference of a 6.0 m length of timber of varying circular cross-section is measured at intervals of 1.0 m along its length. The results obtained were:

Distance from one end (m)	0	1.0	2.0	3.0	4.0	5.0	6.0
Circumference (m)		6.55	6.20	5.63	4.95	4.18	3.65	3.30

(*a*) Determine the cross-sectional area of the timber at the 1.0 m intervals and hence estimate the volume of the timber in cubic metres.

(b) Calculate the mass of the timber if the wood has a density of 600 kg m^{-3}.

(c) What are the dimensions of the largest square cross-section of 6.0 m length which can be cut from the timber?

(d) What percentage of the timber does the square prism of part (c) represent?

(a) [12.13 m^3]   (b) [7 278 kg]   (c) [74.27 cm by 74.27 cm]
(d) [27.28%]

Assignment No. 4

Figure 20.2 shows a tapered plug in the form of a 7.0 cm thick frustum of a pyramid having square ends of side 10.0 cm and 5.0 cm respectively.

(a) The plug can be lightened by 5% by boring a cylindrical, flat-bottomed hole 3.0 cm deep, as shown by the dotted line. Calculate the diameter required for the hole.

(b) Determine the mass of the plug if it is made of bronze having a density of 8 800 kg m^{-3}.

(c) The plug is melted down and recast into a hemisphere with 10% of the metal being lost in the process. Calculate the diameter of the hemisphere.

(a) [2.94 cm]   (b) [3.414 kg]   (c) [11.0 cm]

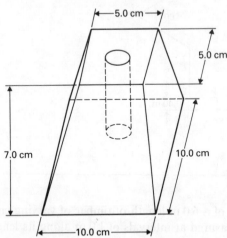

Fig. 20.2

Assignment No. 5

The cross-section of part of a circular ventilation system is shown in Fig. 20.3, ends AB and CD being open.

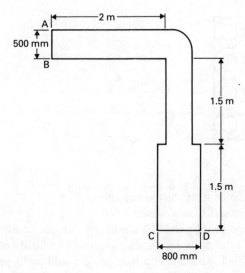

Fig. 20.3

Calculate:

(a) the volume of air contained in the part of the system shown, neglecting the sheet metal thickness;

(b) the cross-sectional area of the sheet metal used to make the system;

(c) the cost of the sheet metal used if the material costs £8 per square metre; and

(d) the mass of the material used if its density is $2\,800$ kg m^{-3} and its thickness is 1.5 mm.

State answers (a), (b) and (d) correct to 5 significant figures and answer (c) correct to the nearest penny.   (a) [1.457 6 m^3] (b) [9.770 4 m^2]   (c) [£78.16]   (d) [41.036 kg]

Assignment No. 6

A steel washer has external and internal diameters of 20 mm and 12 mm respectively, and is 1 mm thick. The washers are stamped from sheet steel, a sheet measuring 1 m by 0.5 m by 1 mm thick. The pattern of the stampings is shown in Fig. 20.4.

Calculate:

(a) the circular cross-sectional area of each washer;

(b) the maximum number of washers which can be produced from one sheet of steel;

(c) the mass of material wasted per sheet of steel, assuming a density of $7\,900$ kg m^{-3};

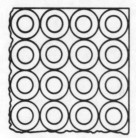

Fig. 20.4

(d) the percentage of steel wasted per sheet;
(e) the mass in grams of one washer; and
(f) the percentage profit per 1 000 washers produced, assuming
     one sheet of steel costs £6, production and overhead costsare
     £150 per 100 000 washers, and washers are sold at £1 per100.
Express the answers correct to 4 significant figures.
(a) [201.1 mm²]   (b) [1 250]   (c) [1.965 kg]
(d) [49.73%]   (e) [1.588 g]   (f) [58.73%]

Assignment No. 7

A template, made of brass, is to be of the shape shown in
Fig. 20.5, the circular area being removed.
(a) Determine the cross-sectional area of the template.
(b) The template is cut from a square sheet of brass of side
     200 mm. Determine the percentage waste for each template,
     correct to the nearest percent.

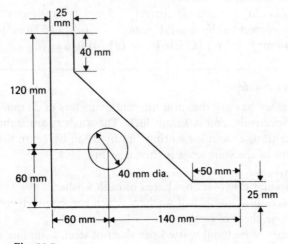

Fig. 20.5

(c) If a rectangular sheet of brass having dimensions 200 mm by 240 mm is used, two such templates can be cut from it. Determine the percentage waste per template.

(d) Determine the volume of the template if the thickness of the plate is 2 mm.

(e) Find the mass of the template if the density of brass is $8\,500\ \mathrm{kg\ m^{-3}}$.

(f) A dozen such templates cost £9.60 to produce. A profit margin of 35% is added when sold. Determine the cost of an order from a firm of 200 templates.

(a) [14 806 mm²]   (b) [63%]   (c) [38.3%]
(d) [29 612 mm³]   (e) [251.7 g]   (f) [£216]

Assignment No. 8

A container is in the shape of an inverted frustum of a cone of maximum diameter 1 m and minimum diameter 50 mm. The perpendicular height of the container is 1.5 m. Water enters the container at a constant rate of $6\,000\ \mathrm{cm^3\ s^{-1}}$ and leaves at a constant rate of $4\,000\ \mathrm{cm^3\ s^{-1}}$ (see Fig. 20.6).

Fig. 20.6

Calculate, correct to 4 significant figures:
(a) the volume of the container;
(b) the external surface area of the container, assuming the top and bottom are open;

(c) how long it will be after the water flow starts, before the container overflows, assuming it is initially empty.

(a) [0.413 3 m³]   (b) [2.595 m²]   (c) [3 min 27 s]

Assignment No. 9

Part of a hydraulic system is made up of the following components:

(i) a cylindrical reservoir of length 100 mm and internal and external diameters of 60 mm and 65 mm respectively, and closed at both ends by discs, each of diameter 65 mm and each of width 2.5 mm (i.e. giving an overall length of the reservoir of 105 mm);

(ii) feeder pipes having a total length of 1.5 m and internal and external diameters of 3 mm and 4 mm respectively; and

(iii) a cylindrical piston mechanism having a maximum length of travel of 90 mm and internal and external diameters of 30 mm and 32 mm respectively, and closed at both ends by discs, each of diameter 32 mm and width 1 mm (i.e. giving an overall length of 92 mm).

The reservoir is made of aluminium alloy of density $2\,800$ kg m^{-3}.

The feeder pipes are made of copper of density $8\,930$ kg m^{-3}.

The piston is made of mild steel of density $7\,860$ kg m^{-3}.

The system is filled with hydraulic oil of density 900 kg m^{-3}.

Calculate:

(a) the total volume of hydraulic oil required to completely fill the system;

(b) the mass of oil required to completely fill the system;

(c) the external surface area of the system, neglecting pipe entry areas;

(d) the volume of metal used in making each of the three parts of the system, neglecting the pipe entry volume; and

(e) the total mass of the system, excluding the oil.

Express the answers correct to 5 significant figure accuracy.

(a) [356 960 mm³]   (b) [0.32 127 kg]   (c) [57 785 mm²]

(d) [reservoir 65 679 mm³; pipes 8 246.7 mm³; piston 10 374 mm³]

(e) [0.33 918 kg]

Assignment No. 10

Figure 20.7 shows a plan of an oil container to be made from sheet steel. The shaded areas are flanges, each 5 mm wide, to be soldered after bending along the broken lines into a rectangular

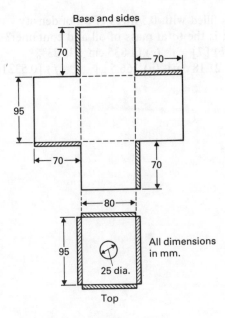

Base and sides

70

70

95

70

70

80

95

All dimensions
in mm.

25 dia.

Top

Fig. 20.7

prismoid. The circular area in the top is to be removed (to insert
a pouring funnel).

(a) Determine the area of sheet steel used for one container. Give
your answer in square centimetres, correct to 4 significant
figures.

(b) The shapes shown in Fig. 20.7 are cut from a sheet of steel
measuring 240 mm wide by 2 m long. Determine the
maximum number of complete containers that can be
produced from this sheet of steel.

(c) From your answer to part (b) determine the amount of
wasted material per sheet of steel and express this as a
percentage of the whole sheet.

(d) If the sheet steel used costs £10.50 per square metre, calculate
the cost of the material (including wastage) needed for one
container.

(e) If the thickness of the steel is 0.5 mm determine:
   (i) the volume of material used per container,
   and (ii) the mass of material per container, in grams, if the
       density of the sheet steel is $7\,860\ \mathrm{kg\,m^{-3}}$.

(f) Determine the capacity of the container in litres
   (1 litre = $1\,000\ \mathrm{cm^3}$).

(*g*) If the container is filled with 0.5 litres of oil of density
900 kg m^{-3}, what is the total mass of oil and container?

(*a*) [423.6 cm^2]   (*b*) [7]   (*c*) [1 835 cm^2; 38.23%]

(*d*) [72p]   (*e*) [(i) 21.18 cm^3, (ii) 166.5 g]   (*f*) [0.5321]

(*g*) [616.5 g]

Chapter 21

Introduction to differentiation

21.1 Introduction

Calculus is a branch of mathematics involving or leading to calculations dealing with continuously varying functions. The subject falls into two parts, namely **differential calculus** (usually abbreviated to **differentiation**) and **integral calculus** (usually abbreviated to **integration**).

The central problem of the differential calculus is the investigation of the rate of change of a function with respect to changes in the variables on which it depends.

The two main classes of integral calculus are firstly, finding such quantities as the length of a curve, the area enclosed by a curve, or the volume enclosed by a surface, and secondly, the problem of determining a variable quantity given its rate of change.

There is a close relationship between the processes of differentiation and integration, the latter being considered as the inverse of the former.

Calculus is a comparatively young branch of mathematics; its systematic development started in the middle of the 17th century. Since then there has been an enormous expansion in the scope of calculus and it is now used in every field of applied science as an instrument for the solution of problems of the most varied nature.

Before such uses can be investigated it is essential to grasp the basic concepts and to understand the notations used. The following text deals with this necessary preparatory work.

21.2 Functional notation

An expression such as $y = 4x^2 - 4x - 3$ contains two variables. For every value of x there is a corresponding value of y. The variable x is called the **independent variable** and y is called the **dependent variable**. The variable y is said to be a function of x and is written as $y = f(x)$. Hence from above
$$f(x) = 4x^2 - 4x - 3$$
The value of the function $f(x)$ when $x = 0$ is denoted by $f(0)$. Similarly when $x = 1$ the value of the function is denoted by $f(1)$ and so on.

If $f(x) = 4x^2 - 4x - 3$

then

$$f(0) = 4(0)^2 - 4(0) - 3 \quad = -3$$
$$f(1) = 4(1)^2 - 4(1) - 3 \quad = -3$$
$$f(2) = 4(2)^2 - 4(2) - 3 \quad = 5$$
$$f(3) = 4(3)^2 - 4(3) - 3 \quad = 21$$
$$f(-1) = 4(-1)^2 - 4(-1) - 3 = 5$$

and $$f(-2) = 4(-2)^2 - 4(-2) - 3 = 21$$

Figure 21.1 shows the curve $f(x) = 4x^2 - 4x - 3$ for values of x between $x = -2$ and $x = 3$. The lengths represented by $f(0)$, $f(1)$, $f(2)$, etc. are also shown.

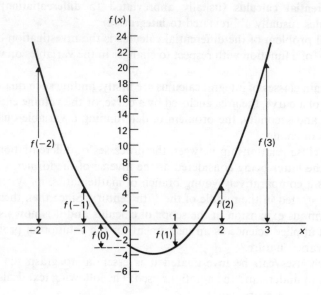

Fig. 21.1   Graph of $f(x) = 4x^2 - 4x - 3$

Worked problems on functional notation

Problem 1. If $f(x) = 5x^2 - 3x + 1$ find $f(0)$, $f(3)$, $f(-1)$, $f(-2)$ and $f(3) - f(-2)$

$$f(x) = 5x^2 - 3x + 1$$
$$f(0) = 5(0)^2 - 3(0) + 1 \quad = 1$$
$$f(3) = 5(3)^2 - 3(3) + 1 \quad = 37$$
$$f(-1) = 5(-1)^2 - 3(-1) + 1 = 9$$
$$f(-2) = 5(-2)^2 - 3(-2) + 1 = 27$$
$$f(3) - f(-2) = 37 - 27 \quad\quad = 10$$

Problem 2. For the curve $f(x) = 3x^2 + 2x - 9$ evaluate $f(2) \div f(1)$, $f(2 + a)$, $f(2 + a) - f(2)$ and $\dfrac{f(2 + a) - f(2)}{a}$

$$f(x) = 3x^2 + 2x - 9$$

$$f(1) = 3(1)^2 + 2(1) - 9 = -4$$

$$f(2) = 3(2)^2 + 2(2) - 9 = 7$$

$$f(2 + a) = 3(2 + a)^2 + 2(2 + a) - 9$$

$$= 3(4 + 4a + a^2) + 4 + 2a - 9$$

$$= 12 + 12a + 3a^2 + 4 + 2a - 9$$

Hence $f(2 + a) = 7 + 14a + 3a^2$

$$f(2) \div f(1) = \frac{f(2)}{f(1)} = \frac{7}{-4} = -1\tfrac{3}{4}$$

$$f(2 + a) - f(2) = 7 + 14a + 3a^2 - 7$$

$$= 14a + 3a^2$$

$$\frac{f(2 + a) - f(2)}{a} = \frac{14a + 3a^2}{a} = 14 + 3a$$

Further problems on functional notation may be found in Section 21.5 (*Problems 1–5*), page 387.

21.3   The gradient of a curve

If a tangent is drawn at a point A on a curve then the gradient of this tangent is said to be the gradient of the curve at A.

In Fig. 21.2 the gradient of the curve at A is equal to the gradient of the tangent AB.

Consider the graph of $f(x) = 2x^2$, part of which is shown in Fig. 21.3.

The gradient of the chord PQ is given by:

$$\frac{QR}{PR} = \frac{QS - RS}{PR} = \frac{QS - PT}{PR}$$

At point P, $x = 1$ and at point Q, $x = 3$

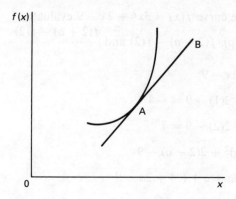

Fig. 21.2

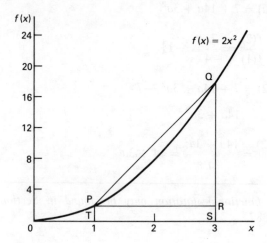

Fig. 21.3

Hence the gradient of the chord PQ $= \dfrac{f(3) - f(1)}{3 - 1}$

$$= \dfrac{18 - 2}{2}$$

$$= \dfrac{16}{2} = 8$$

More generally, for any curve (as shown in Fig. 21.4):

Gradient of PQ $= \dfrac{f(x_2) - f(x_1)}{x_2 - x_1}$

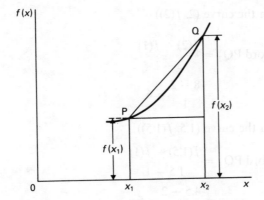

Fig. 21.4

For the part of the curve $f(x) = 2x^2$ shown in Fig. 21.5 let us consider what happens as the point Q, at present at $(3, f(3))$, moves closer and closer to point P, which is fixed at $(1, f(1))$. Note that if the chord PQ is extended to R, then PR is called a secant.

Let Q_1 be the point on the curve $(2.5, f(2.5))$.

$$\text{Gradient of chord PQ}_1 = \frac{f(2.5) - f(1)}{2.5 - 1}$$

$$= \frac{12.5 - 2}{1.5} = 7$$

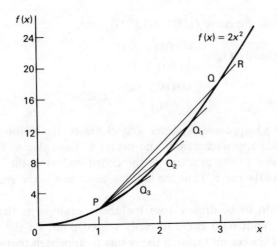

Fig. 21.5

Let Q_2 be the point on the curve $(2, f(2))$

$$\text{Gradient of chord } PQ_2 = \frac{f(2) - f(1)}{2 - 1}$$

$$= \frac{8 - 2}{1} = 6$$

Let Q_3 be the point on the curve $(1.5, f(1.5))$

$$\text{Gradient of chord } PQ_3 = \frac{f(1.5) - f(1)}{1.5 - 1}$$

$$= \frac{4.5 - 2}{0.5} = 5$$

The following points, i.e. Q_4, Q_5 and Q_6 are not shown on Fig. 21.5.
Let Q_4 be the point on the curve $(1.1, f(1.1))$

$$\text{Gradient of chord } PQ_4 = \frac{f(1.1) - f(1)}{1.1 - 1}$$

$$= \frac{2.42 - 2}{0.1} = 4.2$$

Let Q_5 be the point on the curve $(1.01, f(1.01))$

$$\text{Gradient of chord } PQ_5 = \frac{f(1.01) - f(1)}{1.01 - 1}$$

$$= \frac{2.040\,2 - 2}{0.01} = 4.02$$

Let Q_6 be the point on the curve $(1.001, f(1.001))$

$$\text{Gradient of chord } PQ_6 = \frac{f(1.001) - f(1)}{1.001 - 1}$$

$$= \frac{2.004\,002 - 2}{0.001} = 4.002$$

Thus as the point Q approaches closer and closer to the point P the gradients of the chords approach nearer and nearer to the value 4. This is called the **limiting value** of the gradient of the chord and **at P the secant becomes the tangent to the curve.** Thus the limiting value of 4 is the gradient of the tangent at P.

A general conclusion to be drawn from the above example is that the process of moving a point on a curve towards a fixed point on the curve causes the gradient of the secant through the points to approach that of the tangent to the curve at the fixed point.

It can be seen from the above example that deducing the gradient of the tangent to a curve at a given point by this method is a lengthy process. A much more convenient method is shown below.

21.4   Differentiation from first principles

Let P and Q be two points very close together on a curve as shown in Fig. 21.6.

Let the length PR be δx (pronounced delta x), representing a small increment in x, and the length QR, the corresponding increase in y, be δy (pronounced delta y). It is important to realise that δ and x are inseparable, i.e. δx does not mean δ times x. Let P be any point on the curve with coordinates (x, y). Then Q will have the coordinates $(x + \delta x, y + \delta y)$

$$\text{The gradient of the chord PQ} = \frac{\delta y}{\delta x}$$

But from Fig. 21.6, $\delta y = (y + \delta y) - y = f(x + \delta x) - f(x)$

Hence $\dfrac{\delta y}{\delta x} = \dfrac{f(x + \delta x) - f(x)}{\delta x}$

The smaller δx becomes, the closer the gradient of the chord PQ approaches the gradient of the tangent at P. That is, as $\delta x \to 0$, the gradient of the chord $\to$ the gradient of the tangent. (Note '$\to$' means 'approaches'.) As δx approaches zero, the value of $\dfrac{\delta y}{\delta x}$ approaches what is called a **limiting value**.

There are two notations commonly used when finding the gradient of a tangent drawn to a curve.

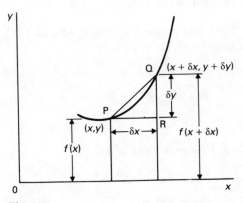

Fig. 21.6

1. The gradient of the curve at P is represented as $\underset{\delta x \to 0}{\text{limit}} \dfrac{\delta y}{\delta x}$

    This is written as $\dfrac{dy}{dx}$ (pronounced dee y by dee x), i.e.

    $$\dfrac{dy}{dx} = \underset{\delta x \to 0}{\text{limit}} \dfrac{\delta y}{\delta x}$$

    This way of stating the gradient of a curve is called **Leibniz notation**.

2. The gradient of the curve at P $= \underset{\delta x \to 0}{\text{limit}} \left\{ \dfrac{f(x + \delta x) - f(x)}{\delta x} \right\}$

    This is written as $f'(x)$ (pronounced f dash x)

    i.e. $f'(x) = \underset{\delta x \to 0}{\text{limit}} \left\{ \dfrac{f(x + \delta x) - f(x)}{\delta x} \right\}$

    This way of stating the gradient of a curve is called **functional notation**.

    $\dfrac{dy}{dx}$ equals $f'(x)$ and is called the **differential coefficient**, or simply the **derivative**.

    The process of finding the differential coefficient is called **differentiation**.

    In the following worked problems the expression for $f'(x)$, which is a definition of the differential coefficient, will be used as a starting point.

Worked problems on differentiation from first principles

Problem 1. Differentiate from first principles $f(x) = x^2$ and find the value of the gradient of the curve at $x = 3$.

To 'differentiate from first principles' means 'to find $f'(x)$' by using the expression

$$f'(x) = \underset{\delta x \to 0}{\text{limit}} \left\{ \dfrac{f(x + \delta x) - f(x)}{\delta x} \right\}$$

$$f(x) = x^2$$

$$f(x + \delta x) = (x + \delta x)^2 = x^2 + 2x\delta x + \delta x^2$$

$$f(x + \delta x) - f(x) = x^2 + 2x\delta x + \delta x^2 - x^2$$

$$= 2x\delta x + \delta x^2$$

$$\frac{f(x + \delta x) - f(x)}{\delta x} = \frac{2x\delta x + \delta x^2}{\delta x}$$

$$= 2x + \delta x$$

As $\delta x \to 0$, $\dfrac{f(x + \delta x) - f(x)}{\delta x} \to 2x + 0$

Therefore $f'(x) = \underset{\delta x \to 0}{\text{limit}} \left\{\dfrac{f(x + \delta x) - f(x)}{\delta x}\right\} = 2x$

At $x = 3$, the gradient of the curve, i.e. $f'(x) = 2(3) = 6$

Hence if $f(x) = x^2$, $f'(x) = 2x$. The gradient at $x = 3$ is 6

Problem 2. Find the differential coefficient of $f(x) = 3x^3$, from first principles.

To 'find the differential coefficient' means 'to find $f'(x)$' by using the expression:

$$f'(x) = \underset{\delta x \to 0}{\text{limit}} \left\{\frac{f(x + \delta x) - f(x)}{\delta x}\right\}$$

$$f(x) = 3x^3$$

$$\begin{aligned} f(x + \delta x) &= 3(x + \delta x)^3 \\ &= 3(x + \delta x)(x^2 + 2x\delta x + \delta x^2) \\ &= 3(x^3 + 3x^2\delta x + 3x\delta x^2 + \delta x^3) \\ &= 3x^3 + 9x^2\delta x + 9x\delta x^2 + 3\delta x^3 \end{aligned}$$

$$\begin{aligned} f(x + \delta x) - f(x) &= 3x^3 + 9x^2\delta x + 9x\delta x^2 + 3\delta x^3 - 3x^3 \\ &= 9x^2\delta x + 9x\delta x^2 + 3\delta x^3 \end{aligned}$$

$$\frac{f(x + \delta x) - f(x)}{\delta x} = \frac{9x^2\delta x + 9x\delta x^2 + 3\delta x^3}{\delta x}$$

$$= 9x^2 + 9x\delta x + 3\delta x^2$$

As $\delta x \to 0$, $\dfrac{f(x + \delta x) - f(x)}{\delta x} \to 9x^2 + 9x(0) + 3(0)^2$

i.e. $f'(x) = \underset{\delta x \to 0}{\text{limit}} \left\{\dfrac{f(x + \delta x) - f(x)}{\delta x}\right\} = 9x^2$

Problem 3. Differentiate from first principles $y = 3x$

The object is to find $\dfrac{dy}{dx}$

$$\frac{dy}{dx} = f'(x) = \lim_{\delta x \to 0} \left\{ \frac{f(x + \delta x) - f(x)}{\delta x} \right\}$$

$y = f(x) = 3x$

$f(x + \delta x) = 3(x + \delta x) = 3x + 3\delta x$

$f(x + \delta x) - f(x) = 3x + 3\delta x - 3x$

$$= 3\delta x$$

$$\frac{f(x + \delta x) - f(x)}{\delta x} = \frac{3\delta x}{\delta x} = 3$$

Hence $\dfrac{dy}{dx} = \lim_{\delta \to 0} \left\{ \dfrac{f(x + \delta x) - f(x)}{\delta x} \right\} = 3$

Another way of writing $\dfrac{dy}{dx} = 3$ is $f'(x) = 3$

or $\dfrac{d}{dx} (3x) = 3$ since $y = 3x$

Problem 4. Find the derivative of $y = \sqrt{x}$

Let $y = f(x) = \sqrt{x}$

To 'find the derivative' means 'to find $f'(x)$'

$$f'(x) = \lim_{\delta x \to 0} \left\{ \frac{f(x + \delta x) - f(x)}{\delta x} \right\}$$

$f(x) = \sqrt{x} = x^{\frac{1}{2}}$

$f(x + \delta x) = (x + \delta x)^{\frac{1}{2}}$

$f(x + \delta x) - f(x) = (x + \delta x)^{\frac{1}{2}} - x^{\frac{1}{2}}$

$$\frac{f(x + \delta x) - f(x)}{\delta x} = \frac{(x + \delta x)^{\frac{1}{2}} - x^{\frac{1}{2}}}{\delta x}$$

Now from algebra, $(a - b)(a + b) = a^2 - b^2$, i.e. the difference of two squares. Therefore, in this case, multiplying both the numerator and the denominator by $[(x + \delta x)^{\frac{1}{2}} + x^{\frac{1}{2}}]$, to make the numerator of the fraction of $(a + b)(a - b)$ form, gives:

$$\frac{f(x + \delta x) - f(x)}{\delta x} = \frac{[(x + \delta x)^{\frac{1}{2}} - x^{\frac{1}{2}}][(x + \delta x)^{\frac{1}{2}} + x^{\frac{1}{2}}]}{\delta x[(x + \delta x)^{\frac{1}{2}} + x^{\frac{1}{2}}]}$$

$$= \frac{[(x + \delta x)^{\frac{1}{2}}]^2 - [x^{\frac{1}{2}}]^2}{\delta x[(x + \delta x)^{\frac{1}{2}} + x^{\frac{1}{2}}]}$$

$$= \frac{(x + \delta x) - (x)}{\delta x[(x + \delta x)^{\frac{1}{2}} + x^{\frac{1}{2}}]}$$

$$= \frac{\delta x}{\delta x[(x + \delta x)^{\frac{1}{2}} + x^{\frac{1}{2}}]}$$

$$= \frac{1}{(x + \delta x)^{\frac{1}{2}} + x^{\frac{1}{2}}}$$

As $\delta x \to 0$, $\dfrac{f(x + \delta x) - f(x)}{\delta x} \to \dfrac{1}{(x + 0)^{\frac{1}{2}} + x^{\frac{1}{2}}}$

Therefore $f'(x) = \underset{\delta x \to 0}{\text{limit}} \left\{ \dfrac{f(x + \delta x) - f(x)}{\delta x} \right\} = \dfrac{1}{x^{\frac{1}{2}} + x^{\frac{1}{2}}} = \dfrac{1}{2x^{\frac{1}{2}}}$

Hence if $f(x) = \sqrt{x}$, $f'(x) = \dfrac{1}{2\sqrt{x}}$ or $\frac{1}{2}x^{-\frac{1}{2}}$

Another way of writing this is:

If $y = \sqrt{x}$, $\dfrac{dy}{dx} = \dfrac{1}{2\sqrt{x}}$

or $\dfrac{d}{dx}(\sqrt{x}) = \dfrac{1}{2\sqrt{x}}$

Problem 5. Differentiate from first principles $f(x) = \dfrac{1}{2x}$

$$f'(x) = \underset{\delta x \to 0}{\text{limit}} \left\{ \frac{f(x + \delta x) - f(x)}{\delta x} \right\}$$

$$f(x) = \frac{1}{2x}$$

$$f(x + \delta x) = \frac{1}{2(x + \delta x)}$$

$$f(x + \delta x) - f(x) = \frac{1}{2(x + \delta x)} - \frac{1}{2x}$$

$$= \frac{x - (x + \delta x)}{2x(x + \delta x)}$$

$$= \frac{-\delta x}{2x(x + \delta x)}$$

$$\frac{f(x + \delta x) - f(x)}{\delta x} = \frac{-\delta x}{2x(x + \delta x)\delta x}$$

$$= \frac{-1}{2x(x + \delta x)}$$

As $\delta x \to 0$, $\dfrac{f(x + \delta x) - f(x)}{\delta x} \to \dfrac{-1}{2x(x + 0)}$

Therefore $f'(x) = \underset{\delta \to 0}{\text{limit}} \left\{ \dfrac{f(x + \delta x) - f(x)}{\delta x} \right\} = \dfrac{-1}{2x(x)} = -\dfrac{1}{2x^2}$

Problem 6. Find the differential coefficient of $y = 5$

The differential coefficient of $y = 5$ may be deduced as follows:

If a graph is drawn of $y = 5$ a straight horizontal line results and the gradient or slope of a horizontal line is zero. Finding the differential coefficient is, in fact, finding the slope of a curve, or, as in this case, of a horizontal straight line.

Hence $\dfrac{dy}{dx} = 0$

This may also be shown by the conventional method since:

$$\frac{dy}{dx} = f'(x) = \underset{\delta x \to 0}{\text{limit}} \left\{ \frac{f(x + \delta x) - f(x)}{\delta x} \right\}$$

$y = f(x) = 5$

$\therefore f(x + \delta x) = 5$

$\therefore \dfrac{dy}{dx} = f'(x) = \underset{\delta x \to 0}{\text{limit}} \left\{ \dfrac{5 - 5}{\delta x} \right\}$

$= \dfrac{0}{\delta x} = 0$

More generally, if C is any constant, then if

$f(x) = C, f'(x) = \mathbf{0}$

i.e. **If** $y = C$ **then** $\dfrac{dy}{dx} = \mathbf{0}$

Problem 7. Differentiate from first principles $f(x) = 3x^2 + 6x - 3$ and find the gradient of the curve at $x = -2$

$$f'(x) = \text{limit} \atop{\delta x \to 0} \left\{ \frac{f(x + \delta x) - f(x)}{\delta x} \right\}$$

$$f(x) = 3x^2 + 6x - 3$$

$$\therefore f(x + \delta x) = 3(x + \delta x)^2 + 6(x + \delta x) - 3$$

$$= 3(x^2 + 2x\delta x + \delta x^2) + 6x + 6\delta x - 3$$

$$= 3x^2 + 6x\delta x + 3\delta x^2 + 6x + 6\delta x - 3$$

$$\therefore f(x + \delta x) - f(x) = (3x^2 + 6x\delta x + 3\delta x^2 + 6x + 6\delta x - 3)$$

$$- (3x^2 + 6x - 3)$$

$$= 6x\delta x + 3\delta x^2 + 6\delta x$$

$$\therefore \frac{f(x + \delta x) - f(x)}{\delta x} = \frac{6x\delta x + 3\delta x^2 + 6\delta x}{\delta x}$$

$$= 6x + 3\delta x + 6$$

As $\delta x \to 0$, $\dfrac{f(x + \delta x) - f(x)}{\delta x} \to 6x + 3(0) + 6$

Therefore $f'(x) = \text{limit} \atop{\delta x \to 0} \left\{ \dfrac{f(x + \delta x) - f(x)}{\delta x} \right\} = 6x + 6$

At $x = -2$ the gradient of the curve, i.e. $f'(x)$, is $6(-2) + 6$, i.e. -6

Hence if $f(x) = 3x^2 + 6x - 3$, $f'(x) = 6x + 6$ and the gradient of the curve at $x = -2$ is -6

Three basic rules of differentiation emerge from these results:

Rule 1. The differential coefficient of a constant is zero.

Rule 2. Constants associated with variables are carried forward.

$$\text{For example } \frac{d}{dx}(3x^2) = 3 \cdot \frac{d}{dx}(x^2)$$

Rule 3. $\dfrac{d}{dx}(ax^n) = anx^{n-1}$, where a and n are constants.

A summary of the results of Problems 1 to 6 is shown below where rule 3 is demonstrated. (See also Chapter 22.)

Summary

$y = ax^n$	$\dfrac{dy}{dx}$	a	n	$\dfrac{dy}{dx} = an\,x^{n-1}$		
x^2	$2x$	1	2	$(1)(2)x^{2-1}$	$= 2x$	
$3x^3$	$9x^2$	3	3	$(3)(3)x^{3-1}$	$= 9x^2$	
$3x$	3	3	1	$(3)(1)x^{1-1}$	$= 3x^0$	$= 3$
$x^{\frac{1}{2}}$	$\dfrac{1}{2}x^{-\frac{1}{2}}$	1	$\dfrac{1}{2}$	$(1)\left(\dfrac{1}{2}\right)x^{\frac{1}{2}-1}$	$= \dfrac{1}{2}x^{-\frac{1}{2}}$	$= \dfrac{1}{2\sqrt{x}}$
$\dfrac{1}{2x}\left(=\dfrac{1}{2}x^{-1}\right)$	$-\dfrac{1}{2x^2}$	$\dfrac{1}{2}$	-1	$\left(\dfrac{1}{2}\right)(-1)x^{-1-1}$	$= -\dfrac{1}{2}x^{-2}$	$= -\dfrac{1}{2x^2}$
$5\,(= 5x^0)$	0	5	0	$(5)(0)x^{0-1}$	$= 0$	

Problem 8. Differentiate from first principles $f(x) = \dfrac{1}{5x + 3}$

$$f'(x) = \underset{\delta x \to 0}{\text{limit}}\left\{\frac{f(x + \delta x) - f(x)}{\delta x}\right\}$$

$$f(x) = \frac{1}{5x + 3}$$

$$f(x + \delta x) = \frac{1}{5(x + \delta x) + 3}$$

$$f(x + \delta x) - f(x) = \frac{1}{5(x + \delta x) + 3} - \frac{1}{5x + 3}$$

$$= \frac{[5x + 3] - [5(x + \delta x) + 3]}{[5(x + \delta x) + 3][5x + 3]}$$

$$= \frac{5x + 3 - 5x - 5\delta x - 3}{[5(x + \delta x) + 3][5x + 3]}$$

$$= \frac{-5\delta x}{[5(x + \delta x) + 3][5x + 3]}$$

$$\therefore \frac{f(x + \delta x) - f(x)}{\delta x} = \frac{-5\delta x}{[5(x + \delta x) + 3][5x + 3]\delta x}$$

$$= \frac{-5}{[5(x + \delta x) + 3][5x + 3]}$$

As $\delta x \to 0$, $\dfrac{f(x + \delta x) - f(x)}{\delta x} \to \dfrac{-5}{[5(x + 0) + 3][5x + 3]}$

Therefore $f'(x) = \underset{\delta x \to 0}{\text{limit}}\left\{\dfrac{f(x + \delta x) - f(x)}{\delta x}\right\} = \dfrac{-5}{(5x + 3)(5x + 3)}$

Hence if $f(x) = \dfrac{1}{5x + 3}$, $f'(x) = \dfrac{-5}{(5x + 3)^2}$

In the above worked problems the questions have been worded in a variety of ways. The important thing to realise is that they all mean the same thing. For example, in worked problem 8, on differentiating from first principles $f(x) = \dfrac{1}{5x + 3}$ gives $\dfrac{-5}{(5x + 3)^2}$

This result can be expressed in a number of ways.

1. If $f(x) = \dfrac{1}{5x + 3}$ then $f'(x) = \dfrac{-5}{(5x + 3)^2}$

2. If $y = \dfrac{1}{5x + 3}$ then $\dfrac{dy}{dx} = \dfrac{-5}{(5x + 3)^2}$

3. The differential coefficient of $\dfrac{1}{5x + 3}$ is $\dfrac{-5}{(5x + 3)^2}$

4. The derivative of $\dfrac{1}{5x + 3}$ is $\dfrac{-5}{(5x + 3)^2}$

5. $\dfrac{d}{dx}\left(\dfrac{1}{5x + 3}\right) = \dfrac{-5}{(5x + 3)^2}$

Further problems on differentiating from first principles may be found in the following Section (21.5) (Problems 6–34).

21.5 Further problems

Functional notation

1. If $f(x) = 2x^2 - x + 3$ find $f(0)$, $f(1)$, $f(2)$, $f(-1)$ and $f(-2)$
    [3, 4, 9, 6, 13]
2. If $f(x) = 6x^2 - 4x + 7$ find $f(1)$, $f(2)$, $f(-2)$ and $f(1) - f(-2)$
    [9, 23, 39, -30]
3. If a curve is represented by $f(x) = 2x^3 + x^2 - x + 6$ prove
    that $f(1) = \dfrac{1}{3}f(2)$
4. If $f(x) = 3x^2 + 2x - 9$ find $f(3)$, $f(3 + a)$ and $\dfrac{f(3 + a) - f(3)}{a}$
    [24, $3a^2 + 20a + 24$, $3a + 20$]
5. If $f(x) = 4x^3 - 2x^2 - 3x + 1$ find $f(2)$, $f(-3)$ and
    $\dfrac{f(1 + b) - f(1)}{b}$ [19, -116, $4b^2 + 10b + 5$]

Differentiation from first principles

6. Sketch the curve $f(x) = 5x^2 - 6$ for values of x from $x = -2$ to $x = +4$. Label the coordinate $(3.5, f(3.5))$ as A. Label the coordinate $(1.5, f(1.5))$ as B. Join points A and B to form the chord AB. Find the gradient of the chord AB. By moving A nearer and nearer to B find the gradient of the tangent of the curve at B [25, 15]

In problems 7–27 differentiate from first principles:

7. $y = x$ [1]
8. $y = 5x$ [5]
9. $y = x^2$ [2x]
10. $y = 7x^2$ [14x]
11. $y = 4x^3$ [12x^2]
12. $y = 2x^2 - 3x + 2$ [4x - 3]
13. $y = 2\sqrt{x}$ $\left[\dfrac{1}{\sqrt{x}} \text{ or } x^{-\frac{1}{2}}\right]$
14. $y = \dfrac{1}{x}$ $\left[-\dfrac{1}{x^2}\right]$
15. $y = \dfrac{5}{6x^2}$ $\left[-\dfrac{5}{3x^3}\right]$
16. $y = 19$ [0]
17. $f(x) = 3x$ [3]
18. $f(x) = \dfrac{x}{4}$ $\left[\dfrac{1}{4}\right]$
19. $f(x) = 3x^2$ [6x]
20. $f(x) = 14x^3$ [42x^2]
21. $f(x) = x^2 + 16x - 4$ [2x + 16]
22. $f(x) = 4x^{\frac{1}{2}}$ $\left[2x^{-\frac{1}{2}} \text{ or } \dfrac{2}{\sqrt{x}}\right]$
23. $f(x) = \dfrac{16}{17x}$ $\left[-\dfrac{16}{17x^2}\right]$
24. $f(x) = \dfrac{1}{x^3}$ $\left[-\dfrac{3}{x^4}\right]$
25. $f(x) = 8$ [0]
26. $f(x) = \dfrac{1}{\sqrt{x}}$ $\left[-\dfrac{1}{2\sqrt{x^3}} \text{ or } -\dfrac{1}{2}x^{-3/2}\right]$
27. $f(x) = \dfrac{1}{3x - 2}$ $\left[\dfrac{-3}{(3x - 2)^2}\right]$

28. Find $\dfrac{d}{dx}(6x^3)$ $[18x^2]$

29. Find $\dfrac{d}{dx}(3\sqrt{x}+6)$ $\left[\dfrac{3}{2\sqrt{x}}\right]$

30. Find $\dfrac{d}{dx}(2x^{-2}+7x^2)$ $[-4x^{-3}+14x]$

31. Find $\dfrac{d}{dx}\left(13-\dfrac{3}{2x}\right)$ $\left[\dfrac{3}{2x^2}\right]$

32. If E, F and G are the points $(1, 2)$, $(2, 16)$ and $(3, 54)$ respectively on the graph of $y = 2x^3$, find the gradients of the tangents at the points E, F and G and the gradient of the chord EG $[6, 24, 54, 26]$

33. Differentiate from first principles $f(x) = 5x^2 - 6x + 2$ and find the gradient of the curve at $x = 2$ $[10x - 6, 14]$

34. If $y = \dfrac{7}{2}\sqrt{x} + \dfrac{3}{x^2} - 9$ find the differential coefficient of y with respect to x $\left[\dfrac{7}{4\sqrt{x}} - \dfrac{6}{x^3}\right]$

Chapter 22

Differentiation of algebraic and trigonometric functions

22.1 Differentiation of $y = ax^n$ by the general rule

It was shown in Chapter 21 that when differentiating a function of the form $y = ax^n$ from first principles the result is always:

$$\frac{dy}{dx} = an\ x^{n-1} \text{ (where } a \text{ and } n \text{ are constants).}$$

This is a general rule of differentiation. For an initial understanding of the meaning of the differential coefficient it is necessary to be able to differentiate from first principles. However, this is a time-consuming exercise and in all further work this statement of the general rule may be assumed to be correct in all cases.

When $y = 3x^2$, then by comparison with $y = ax^n$, $a = 3$ and $n = 2$

Using the general rule, $\dfrac{dy}{dx} = an\ x^{n-1} = (3)\ (2)\ x^{2-1} = \mathbf{6x}$

If the differential coefficient of $y = \dfrac{5}{x^2}$ is required then the function has to

be changed initially to the $y = ax^n$ form, i.e. $y = \dfrac{5}{x^2} = 5x^{-2}$

Hence $a = 5$ and $n = -2$

Thus, by the general rule, $\dfrac{dy}{dx} = (5)\ (-2)x^{-2-1} = -10x^{-3} = \dfrac{-10}{x^3}$

If the differential coefficient of $y = 6\sqrt{x^3}$ is required then again the function must be re-written in the $y = ax^n$ form, i.e. $y = 6\sqrt{x^3} = 6\ x^{\frac{3}{2}}$.

Hence $a = 6$ and $n = \dfrac{3}{2}$

Thus, by the general rule, $\dfrac{dy}{dx} = (6)\left(\dfrac{3}{2}\right)x^{\frac{3}{2}-1} = 9x^{\frac{1}{2}} = \mathbf{9\sqrt{x}}$

22.2   Graphical derivation of the differentiation of sin θ and cos θ

Figure 22.1(a) shows a graph of $y = \sin \theta$

The slope, or gradient, of a curve $y = f(\theta)$ is given by $\dfrac{dy}{d\theta}$. The gradient

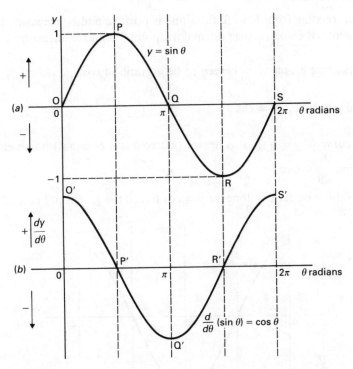

Fig. 22.1

of the graph $y = \sin \theta$ is continually changing as the curve proceeds from O to P to Q to R to S. If the values of the gradient $\left(\text{i.e. } \dfrac{dy}{d\theta}\right)$ at all points on the curve are plotted on a second set of axes in a corresponding position below $y = \sin \theta$, then the curve shown in Fig. 22.1(b) results. This curve is obtained as follows:

(i) At point O, the slope is positive and is at its steepest, making point O′ a maximum positive value.

(ii) Proceeding from O to P, the slope is positive but is decreasing in value until at point P, the slope is zero $\left(\text{shown as point P′ on the graph of } \dfrac{dy}{d\theta}\right)$.

(*iii*) Proceeding from P to Q, the slope is negative but is increasing numerically in value until at point Q the slope is at its steepest, making point Q′ a maximum negative value.

(*iv*) Proceeding from Q to R, the slope is still negative but is decreasing numerically until at point R the slope is zero $\left(\text{shown as point R' on the graph of } \dfrac{dy}{d\theta}\right)$

(*v*) Proceeding from R to S, the slope is positive and is increasing until at point S the slope is steepest, making point S′ a maximum positive value.

The resulting graph of $\dfrac{dy}{d\theta}$ is seen to be a graph of cos θ.

Thus if $y = \sin \theta$, $\dfrac{dy}{d\theta} = \cos \theta$

If the curve of $y = b \sin \theta$ is drawn (where b is a constant) it can easily be seen that $\dfrac{dy}{d\theta} = b \cos \theta$

If a similar exercise is done for $y = \cos \theta$ then the graphs of Fig. 22.2 result,

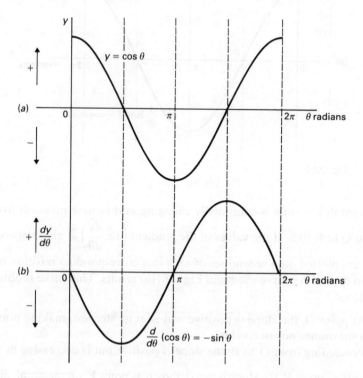

Fig. 22.2

showing $\dfrac{dy}{d\theta}$ to be a graph of sin θ, but displaced by π radians. If each point on the curve $y = \sin \theta$ (as shown in Fig. 22.1(a)) were to be made negative $\left(\text{i.e. } +\dfrac{\pi}{2} \text{ is made } -\dfrac{\pi}{2}, -\dfrac{3\pi}{2} \text{ is made } -\left(-\dfrac{3\pi}{2}\right) \text{ or } +\dfrac{3\pi}{2}, \text{ and so on}\right)$ then the graph shown in Fig. 22.2(b) would result. This latter graph thus represents the curve of $-\sin \theta$.

Thus, if $y = \cos \theta$, $\dfrac{dy}{d\theta} = -\sin \theta$

Also, if $y = b \cos \theta$, then $\dfrac{dy}{d\theta} = -b \sin \theta$

It may also be shown by completing similar constructions to those shown in Figs. 22.1 and 22.2 that:

If $y = \sin a\theta$, $\dfrac{dy}{d\theta} = a \cos a\theta$

and if $y = \cos a\theta$, $\dfrac{dy}{d\theta} = -a \sin a\theta$ where a is any constant.

Summary of standard differential coefficients

y or $f(x)$	$\dfrac{dy}{dx}$ or $f'(x)$
ax^n	$an\,x^{n-1}$
$\sin ax$	$a \cos ax$
$\cos ax$	$-a \sin ax$

Worked problems on standard differential coefficients

Problem 1. Using the general rule, differentiate the following with respect to the variable: (a) $y = 3x^8$; (b) $V = 4\sqrt{t}$; (c) $s = 7x(x-3)$

(a) Comparing $y = 3x^8$ with $y = ax^n$ shows that $a = 3$ and $n = 8$.

Using the general rule, $\dfrac{dy}{dx} = anx^{n-1} = (3)(8)x^{8-1} = 24x^7$

(b) $V = 4\sqrt{t} = 4t^{\frac{1}{2}}$

$\dfrac{dV}{dt} = (4)\left(\dfrac{1}{2}\right)t^{\frac{1}{2}-1} = 2t^{-\frac{1}{2}} = \dfrac{2}{t^{\frac{1}{2}}} = \dfrac{2}{\sqrt{t}}$

(c) $s = 7x(x - 3) = 7x^2 - 21x$

$$\frac{ds}{dx} = (7)(2)x^{2-1} - (21)x^{1-1} = 14x - 21 = 7(2x - 3)$$

Problem 2. Obtain the diffferential coefficient of

$$y = \frac{3}{5}x^3 - \frac{2}{x^2} + 5\sqrt{x^7} + 5$$

$$y = \frac{3}{5}x^3 - \frac{2}{x^2} + 5\sqrt{x^7} + 5 = \frac{3}{5}x^3 - 2x^{-2} + 5x^{\frac{7}{2}} + 5$$

$$\frac{dy}{dx} = \left(\frac{3}{5}\right)(3)x^{3-1} - (2)(-2)x^{-2-1} + (5)\left(\frac{7}{2}\right)x^{\frac{7}{2}-1} + 0$$

$$= \frac{9}{5}x^2 + 4x^{-3} + \frac{35}{2}x^{\frac{5}{2}}$$

Hence $\dfrac{dy}{dx} = \dfrac{9}{5}x^2 + \dfrac{4}{x^3} + \dfrac{35}{2}\sqrt{x^5}$

Problem 3. Differentiate the following with respect to the variable:
(a) $y = \sin 4\theta$; (b) $f(t) = \cos 2t$; (c) $y = 6 \sin 2x - 3 \cos x$.

(a) $y = \sin 4\theta$

$$\frac{dy}{d\theta} = 4 \cos 4\theta$$

(b) $f(t) = \cos 2t$
$f'(t) = -2 \sin 2t$
(c) $y = 6 \sin 2x - 3 \cos x$

$$\frac{dy}{dx} = (6)(2 \cos 2x) - (3)(-\sin x)$$

$$= 12 \cos 2x + 3 \sin x$$

Problem 4. An alternating voltage is given by $v = 80 \sin 100t$ volts, where t is the time in seconds. Calculate the rate of change of voltage when: (a) $t = 0.01$ s; and (b) $t = 0.02$ s

$v = 80 \sin 100t$

The rate of change of v, $\dfrac{dv}{dt} = (80)(100 \cos 100t)$

(a) When $t = 0.01$ s, $\dfrac{dv}{dt} = 8\,000 \cos [(100)(0.01)]$

$$= 8\,000 \cos 1$$

'cos 1' means 'the cosine of 1 radian', i.e. $\dfrac{180°}{\pi}$

Hence $\qquad \dfrac{dv}{dt} = 8\,000(0.540\,3) = \mathbf{4\,322}$ **volts per second**

(b) When $t = 0.02$ s, $\dfrac{dv}{dt} = 8\,000 \cos{[(100)(0.02)]}$

$$= 8\,000 \cos 2$$
$$= 8\,000(-0.416\,1)$$

Hence $\qquad \dfrac{dv}{dt} = \mathbf{-3\,329}$ **volts per second**

Problem 5. Differentiate, with respect to θ:

$$t = 3(\theta + 2)^3 - 2\left[4 \sin 6\theta + \frac{\cos 2\theta}{4}\right]$$

Expanding $t = 3(\theta + 2)^3 - 2\left[4 \sin 6\theta + \dfrac{\cos 2\theta}{4}\right]$ gives:

$$t = 3(\theta + 2)(\theta^2 + 4\theta + 4) - (2)(4) \sin 6\theta - \frac{2}{4} \cos 2\theta$$

$$= 3(\theta^3 + 6\theta^2 + 12\theta + 8) - 8 \sin 6\theta - \frac{1}{2} \cos 2\theta$$

i.e. $\qquad t = 3\theta^3 + 18\theta^2 + 36\theta + 24 - 8 \sin 6\theta - \dfrac{1}{2} \cos 2\theta$

$$\frac{dt}{d\theta} = (3)(3)\theta^2 + (18)(2)\theta + 36 + 0$$
$$- (8)(6) \cos 6\theta - \frac{1}{2}(-2 \sin 2\theta)$$

Hence $\dfrac{dt}{d\theta} = 9\theta^2 + 36\theta + 36 - 48 \cos 6\theta + \sin 2\theta$

Further problems on standard differential coefficients may be found in Section 22.4 (Problems 1–26), page 397.

22.3   Maximum and minimum values

Consider the curve shown in Fig. 22.3

The slope of the curve $\left(\text{i.e. } \dfrac{dy}{dx}\right)$ between points O and A is positive.

The slope of the curve between points A and B is negative and the slope between points B and C is again positive.

At point A the slope is zero and as x increases, the slope of the curve

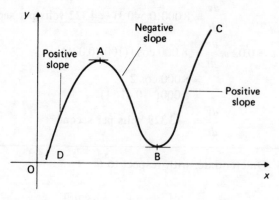

Fig. 22.3

changes from positive just before A to negative just after. Such a point is called a **maximum value**.

At point B the slope is also zero and, as x increases, the slope of the curve changes from negative just before B to positive just after. Such a point is called a **minimum value**.

Points such as A and B are given the general name of **turning-points**.

Maximum and minimum values can be confusing inasmuch as they suggest that they are the largest and smallest values of a curve. However, by their definition this is not so. A maximum value occurs at the 'crest of a wave' and the minimum value at the 'bottom of a valley'. In Fig. 22.3 the point C has a larger y-ordinate value than A and point D has a smaller y-ordinate than B. Points A and B are turning-points and are given the special names of maximum and minimum values respectively.

Summary

1. At a maximum point the slope $\dfrac{dy}{dx} = 0$ and changes from positive just before the maximum point to negative just after.

2. At a minimum point the slope $\dfrac{dy}{dx} = 0$ and changes from negative just before the minimum point to positive just after.

Thus for the sine wave shown in Fig. 22.1(a), point P $\left(\text{at } \theta = \dfrac{\pi}{2} \text{ and } y = 1\right)$ is a maximum point and point R $\left(\text{at } \theta = \dfrac{3\pi}{2} \text{ and } y = -1\right)$ is a minimum point.

22.4 Further problems

In Problems 1 to 21 write down the differential coefficients with respect to the variable.

1. $5x^7$ $[35x^6]$

2. $3x^4 - 2x + 1$ $[12x^3 - 2]$

3. $\dfrac{4}{x^2} - \dfrac{1}{x} + 2$ $\left[\dfrac{-8}{x^3} + \dfrac{1}{x^2} \right]$

4. $3\sqrt{V}$ $\left[\dfrac{3}{2\sqrt{V}} \right]$

5. $\dfrac{15}{7} \sqrt{t^7}$ $\left[\dfrac{15}{2} \sqrt{t^5} \right]$

6. $5y(y + 3)$ $[5(2y + 3)]$

7. $5x^{\frac{1}{2}} - 3x^{-3/2} + 6$ $\left[\dfrac{5}{2}x^{-\frac{1}{2}} + \dfrac{9}{2}x^{-5/2} \right]$

8. $\dfrac{(t + 2)^2}{t}$ $\left[1 - \dfrac{4}{t^2} \right]$

9. $\dfrac{(s - 1)^3}{\sqrt{s}} + 4$ $\left[\dfrac{5}{2}\sqrt{s^3} - \tfrac{9}{2}\sqrt{s} + \dfrac{3}{2\sqrt{s}} + \dfrac{1}{2\sqrt{s^3}} \right]$

10. $\dfrac{3}{5} \sqrt{x^3} - \dfrac{6}{\sqrt{x^5}}$ $\left[\dfrac{9}{10} \sqrt{x} + \dfrac{15}{\sqrt{x^7}} \right]$

11. $6 \sin 3y + \dfrac{3}{5} \sqrt{y} - \cos 2y$ $\left[18 \cos 3y + \dfrac{3}{10\sqrt{y}} + 2 \sin 2y \right]$

12. $3 + 2(u + 3)^3 - u$ $[6u^2 + 36u + 53]$

13. $\dfrac{f + 2}{\sqrt{f^3}}$ $\left[-\dfrac{1}{2\sqrt{f^3}} - \dfrac{3}{\sqrt{f^5}} \text{ or } \dfrac{-1}{\sqrt{f^3}}\left(\dfrac{1}{2} + \dfrac{3}{f} \right) \right]$

14. $\sin 2t$ $[2 \cos 2t]$

15. $-\cos \theta$ $[\sin \theta]$

16. $3 \sin 5x$ $[15 \cos 5x\}$

17. $2 \cos 4y$ $[-8 \sin 4y]$

18. $-5 \sin 3m$ $[-15 \cos 3m]$

19. $6 \sin 3t - 2 \cos 7t$ $[2(9 \cos 3t + 7 \sin 7t)]$

20. $24 \sin 100t$ $[2\,400 \cos 100t]$

21. $-35 \cos 50\theta$ $[1\,750 \sin 50\theta]$

22. An alternating voltage is given by $v = 125 \sin 80t$ volts where t is the time in seconds. Calculate the rate of change of voltage when $t = 0.02$ s $[-292$ volts per second$]$

23. An alternating current is given by $i = 50 \sin 120t$ amperes where t is in seconds. Calculate the rate of change of current when $t = 0.01$ s $[2\,174$ amperes per second$]$

24. $v = 24 \sin 50t$ represents an alternating voltage where t is the time in seconds. At a time of 32 ms find the rate of change of voltage.   [-35 volts per second]

25. An alternating current is given by $i = 28 \sin 60t$ where t is the time in seconds. Calculate the rate of change of current when $t = 150$ ms   [$-1\,531$ amperes per second]

26. Draw a graph of $v = 20 \sin 80t$ for values of t from 0 to 25 ms. Find the rate of change of voltage after (a) 10 ms and (b) 20 ms: (i) by drawing; and (ii) by calculation.
    (a) [1 115 volts per second]   (b) [-46.72 volts per second]

27. Sketch the following curves and state for each: (a) the value of and the position of the first maximum point, and (b) the value of and the position of the first minimum point. (i) $y = 3 \sin 2x$ (ii) $y = 5 \cos 3\theta$

$$(i) \ (a) \left[3 \text{ at } \left(\frac{\pi}{4}, 3 \right) \right] \quad (b) \left[-3 \text{ at } \left(\frac{3\pi}{4}, -3 \right) \right]$$

$$(ii) \ (a) \ [5 \text{ at } (0, 5)] \quad (b) \left[-5 \text{ at } \left(\frac{\pi}{3}, -5 \right) \right]$$

Chapter 23

Introduction to integration

23.1 Integration

The process of integration reverses the process of differentiation. In differentiation, if $f(x) = x^2$, then $f'(x) = 2x$. Since integration reverses the process of moving from $f(x)$ to $f'(x)$, it follows that the integral of $2x$ is x^2, i.e. it is the process of moving from $f'(x)$ to $f(x)$. Similarly, if $y = x^3$, $\dfrac{dy}{dx} = 3x^2$, and reversing this process shows that the integral of $3x^2$ is x^3.

Integration is also a process of summation or adding parts together and an elongated 'S', shown as $\int$, is used to replace the words 'the integral of'. Thus $\int 2x = x^2$ and $\int 3x^2 = x^3$

In differentiation, the differential coefficient $\dfrac{dy}{dx}$ or $\dfrac{d(f(x))}{dx}$ indicates that a function of x is being differentiated with respect to x, the dx indicating this. In integration, the variable of integration is shown by adding dx, or, more generally, d (the variable) after the function to be integrated. Thus, $\int 2x \, dx$ means 'the integral of $2x$ with respect to x' and $\int 3u^2 \, du$ means 'the integral of $3u^2$ with respect to u'. It follows that $\int y \, dx$ means 'the integral of y with respect to x', and since only functions of x can be integrated with respect to x, y must be expressed as a function of x before the process of integration can be performed.

The arbitrary constant of integration

The differential coefficient of x^2 is $2x$, hence $\int 2x \, dx = x^2$. Also, the differential coefficient of $x^2 + 3$ is $2x$, hence $\int 2x \, dx = x^2 + 3$. Since the differential coefficient of any constant is zero, it follows that the differential coefficient of $x^2 + c$ where c is any constant is $2x$. To allow for the possible presence of this constant, whenever the process of integration is performed, the constant must be added to the result. Hence the correct process of integration is:

$$\int 2x \, dx = x^2 + c$$

'c' is called the **arbitrary constant of integration** and it is important to include it in all work involving the process of determining integrals. Its omission will result in obtaining incorrect solutions in later work, such as in the solution of differential equations.

The general solution of integrals of the form ax^n

In Chapter 22, a general solution for finding the differential coefficient of x^n is used and is:

$$\frac{d(x^n)}{dx} = nx^{n-1}$$

Applying this to various function of x:

$$\text{e.g. } y = x, x^2, x^3, x^4, \ldots, x^n \tag{1}$$

$$\text{gives } \frac{dy}{dx} = 1, 2x, 3x^2, 4x^3, \ldots, nx^{n-1} \tag{2}$$

For integration, the positions of these two lines can be reversed, i.e.

$$\text{when } y = 1, 2x, 3x^2, 4x^3, \ldots, nx^{n-1} \tag{3}$$

$$\int y \, dx = x, x^2, x^3, x^4, \ldots, x^n \tag{4}$$

In order to express the functions of x in line (3) as their integrals in line (4) it is necessary to:

(a) increase the power of x by 1, i.e. the power of x^n is raised by 1 to x^{n+1} and

(b) the new power of x, i.e. x^{n+1} is divided by $(n + 1)$.

Thus to integrate x^2, the power of x is increased by 1 to $(2 + 1)$ or 3, giving x^3 and the term is divided by $(n + 1)$ or $(2 + 1)$, i.e. 3.

So the integral of x^2 is $\frac{1}{3}(x^3)$ or $\frac{x^3}{3}$

Applying this principle to the term x^n, gives

$$\int x^n \, dx = \frac{x^{n+1}}{n+1} + c$$

This is called a standard integral.

Although the development of this integral has been shown using integers only, the same rule applies for all values of n (fractions, decimal fractions, positive and negative values) except when $n = -1$. This special case will be developed in *Technician mathematics 3*. Thus applying the rule,

$$\int x^n \, dx = \frac{x^{n+1}}{n+1} + c$$

shows that: $\int x^5 \, dx = \dfrac{x^6}{6} + c$ (i.e. $n = 5$)

and $\qquad \int x^{\frac{3}{2}} \, dx = \dfrac{x^{(\frac{3}{2}+1)}}{\frac{3}{2}+1} + c = \dfrac{2}{5} x^{\frac{5}{2}} + c,$

and so on.

Rules of integration

Three of the basic rules of integration are:

1. The integral of a constant k is $kx + c$
    For example, $\int 3 \, dx = \int 3 \, x^0 \, dx$ since $x^0 = 1$

    Applying the standard integral, $\int x^n \, dx = \dfrac{x^{n+1}}{n+1} + c$ gives

    $$\int 3 \, dx = \frac{3(x^{0+1})}{0+1} + c = 3x + c$$

2. As in differentiation, constants associated with variables are carried forward, i.e. they are not involved in the integration. For example,

    $$\int 5x^4 \, dx = 5 \int x^4 \, dx$$
    $$= 5\frac{x^5}{5} + c$$
    $$= x^5 + c$$

3. As in differentiation, the rules of algebra apply where functions of a variable are added or substracted. For example:

    $$\int (x^2 + x^3) \, dx = \int x^2 \, dx + \int x^3 \, dx = \frac{x^3}{3} + \frac{x^4}{4} + c$$

    and $\quad \int (4x^5 - 3) \, dx = 4 \int x^5 \, dx - 3 \int dx = \dfrac{4x^6}{6} - 3x + c$

    $$= \frac{2x^6}{3} - 3x + c$$

    Including rule (2) with the standard integral for x^n gives

    $$\int ax^n \, dx = \frac{ax^{n+1}}{n+1} + c$$

where a and n are constants and n is *not* equal to -1. Integrals written in this form are called indefinite integrals, since their precise value cannot be found, i.e. 'c' cannot be calculated, unless additional information is provided.

Integrals of cos $a\theta$ and sin $a\theta$

From Chapter 22, $\dfrac{d}{d\theta}(\sin \theta) = \cos \theta$

Since integration is the reverse process of differentiation,

$$\int \cos \theta \; d\theta = \sin \theta + c$$

More generally, $\dfrac{d}{d\theta}(\sin a\theta) = a \cos a\theta$

Hence $\int a \cos a\theta \; d\theta = \sin a\theta + c$

and $\int \cos a\theta \; d\theta = \dfrac{1}{a} \sin a\theta + c$

Similarly, $\dfrac{d}{d\theta}(\cos \theta) = -\sin \theta$

Hence $\int -\sin \theta \; d\theta = \cos \theta + c$

and $\int \sin \theta \; d\theta = -\cos \theta + c$

More generally, $\dfrac{d}{d\theta}(\cos a\theta) = -a \sin a\theta + c$

Hence $\int -a \sin a\theta \; d\theta = \cos a\theta + c$

and $\int \sin a\theta \; d\theta = -\dfrac{1}{a} \cos a\theta + c$

Worked problems on indefinite integrals

Problem 1. Determine: (a) $\int x^7 \, dx$, (b) $\int 3y^{1.3} \, dy$ and (c) $\int 2z^{-\frac{3}{4}}$

Applying the standard integral, $\int ax^n = \dfrac{ax^{n+1}}{n+1} + c$ in each case

gives:

(a) $a = 1$ and $n = 7$, hence

$$\int x^7 \, dx = \frac{1(x)^{7+1}}{7+1} + c$$

$$= \frac{x^8}{8} + c$$

(b) $a = 3$ and $n = 1.3$, hence

$$\int 3y^{1.3}\, dy = \frac{3y^{1.3+1}}{1.3+1} + c$$

$$= \frac{3y^{2.3}}{2.3} + c$$

(c) $a = 2$ and $n = -\frac{3}{4}$, hence

$$\int 2z^{-\frac{3}{4}}\, dz = \frac{2z^{-\frac{3}{4}+1}}{-\frac{3}{4}+1} + c$$

$$= \frac{2z^{\frac{1}{4}}}{\frac{1}{4}} + c = 8z^{\frac{1}{4}} + c$$

Problem 2. Determine $\int (3x^3 - 4x^{\frac{1}{2}} + 5)\, dx$

Applying the standard integral for ax^n and the rules of integration gives:

$$\int (3x^3 - 4x^{\frac{1}{2}} + 5)\, dx = \int 3x^3\, dx - \int 4x^{\frac{1}{2}}\, dx + \int 5\, dx$$

$$= \frac{3(x)^{3+1}}{3+1} - \frac{4(x)^{\frac{1}{2}+1}}{\frac{1}{2}+1} + \frac{5(x)^{0+1}}{0+1} + c$$

$$= \frac{3x^4}{4} - \frac{4x^{\frac{3}{2}}}{\frac{3}{2}} + \frac{5x}{1} + c$$

$$= \frac{3x^4}{4} - \frac{8x^{\frac{3}{2}}}{3} + 5x + c$$

Although each of the integrals $3x^3$, $4x^{\frac{1}{2}}$ and 5 has its own arbitrary constant of integration, say, c_1, c_2 and c_3, these can be grouped together as a single arbitrary constant c, where $c = c_1 + c_2 + c_3$.

Problem 3. Determine the value of the arbitrary constant of integration if

$$m = \int \frac{t^2 - 4t}{t}\, dt \text{ and } t = 2, \text{ when } m = 3$$

It is only possible to integrate a standard integral. Expressions such as the one given must be simplified by algebraic or other means until they can be expressed in standard form. The expression given can be reduced to standard form as follows:

$$\frac{t^2 - 4t}{t} = \frac{t^2}{t} - \frac{4t}{t} = t - 4$$

Hence $\int \frac{t^2 - 4t}{t}\, dt = \int (t - 4)\, dt = \frac{t^2}{2} - 4t + c$

i.e. $$m = \frac{t^2}{2} - 4t + c \qquad (1)$$

But $m = 3$ when $t = 2$ and by substituting those values in equation (1) the value of c can be calculated. Thus:

$$3 = \frac{(2)^2}{2} - 4(2) + c$$

$$c = 3 - 2 + 8 = 9$$

Thus the value of the arbitrary constant, c is 9, i.e. $m = \dfrac{t^2}{2} - 4t + 9$

Problem 4. If $y = \displaystyle\int \left(q + \frac{1}{q} \right)^2 dq$, find the value of the arbitrary constant of integration if $y = \frac{1}{2}$ when $q = 1\frac{1}{2}$

Manipulating $\left(q + \dfrac{1}{q} \right)^2$ to obtain an expression in standard form gives:

$$\left(q + \frac{1}{q} \right)^2 = q^2 + 2(q)\left(\frac{1}{q} \right) + \left(\frac{1}{q} \right)^2$$

$$= q^2 + 2 + \frac{1}{q^2}$$

$$= q^2 + 2 + q^{-2}$$

Hence $y = \displaystyle\int \left(q + \frac{1}{q} \right)^2 dq = \int (q^2 + 2 + q^{-2}) \, dq$

$$= \frac{q^3}{3} + 2q + \frac{q^{-1}}{-1} + c$$

or $$y = \frac{q^3}{3} + 2q - \frac{1}{q} + c \qquad (1)$$

Now, $y = \frac{1}{2}$ when $q = 1\frac{1}{2}$, therefore, substituting in equation (1) gives:

$$\frac{1}{2} = \frac{(1\frac{1}{2})^3}{3} + 2(1\frac{1}{2}) - \frac{1}{1\frac{1}{2}} + c$$

or $$\frac{1}{2} = \frac{9}{8} + 3 - \frac{2}{3} + c$$

or $$c = -\frac{71}{24}$$

Hence the value of the arbitrary constant c, is $-\dfrac{71}{24}$

i.e. $y = \dfrac{q^3}{3} + 2q - \dfrac{1}{q} - \dfrac{71}{24}$

Problem 5. Determine (a) $\displaystyle\int 3 \cos 2x \, dx$ (b) $\displaystyle\int \frac{1}{2} \sin 5\theta \, d\theta$

(a) $\displaystyle\int \cos a\theta \, d\theta = \frac{1}{a} \sin a\theta + c$

Hence $\displaystyle\int 3 \cos 2x \, dx = 3 \int \cos 2x \, dx = 3\left(\frac{1}{2} \sin 2x\right) + c$

$$= \frac{3}{2} \sin 2x + c$$

(b) $\displaystyle\int \sin a\theta \, d\theta = -\frac{1}{a} \cos a\theta + c$

Hence $\displaystyle\int \frac{1}{2} \sin 5\theta \, d\theta = \frac{1}{2} \int \sin 5\theta \, d\theta = \frac{1}{2}\left(-\frac{1}{5} \cos 5\theta\right) + c$

$$= -\frac{1}{10} \cos 5\theta + c$$

Further problems on indefinite integrals may be found in Section 23.4 (Problems 1–10), page 417.

23.2   Limits and definite integrals

Limits

When a mathematical function is written in square brackets with numbers or letters at the top and bottom of the right-hand bracket, it is a mathematical operator and indicates that a certain sequence of operations should be performed. The numbers or letters outside of the bracket are called the **limits** and for $[x^2]_b^a$, 'a' is called the upper limit and 'b' the lower limit.

The sequence of operations is to substitute the upper limit value for the variable and then subtract the lower limit value substituted for the variable. Thus:

$$[x^2]_b^a = (a^2) - (b^2)$$

Similarly $[y^3]_3^4 = (4^3) - (3^3)$

$$= 64 - 27 = 37$$

and $[p + p^2]^2_{-2} = [2 + (2)^2] - [(-2) + (-2)^2]$
$$= 6 - 2 = 4$$

The process of applying limits is that of finding the increase in the value of the function as the variable increases from the lower limit value to the upper limit value. For $[x^2]^4_3$, the increase in the value of x^2 is found between $x = 3$ and $x = 4$.

Definite integrals

Limits can be applied to integrals and these integrals are then called definite integrals. The increase in the value of the integral $(x^2 + 4)$ as x increases from 1 to 3 can be written

$$[\int(x^2 + 4)\, dx]^3_1$$

However, this is invariably abbreviated by showing the value of the upper limit at the top of the integral sign and the value of the lower limit at the bottom, i.e.

$$[\int(x^2 + 4)\, dx]^3_1 = \int^3_1 (x^2 + 4)\, dx$$

The integral is evaluated as for an indefinite integral and then placed in the square brackets of the limits operator. Thus

$$\int^3_1 (x^2 + 4)\, dx = \left[\frac{x^3}{3} + 4x + c\right]^3_1$$

This can now be evaluated using the limits theory.
Thus:

$$\left(\frac{3^3}{3} + 4(3) + c\right) - \left(\frac{(1)^3}{3} + 4(1) + c\right) = (21 + c) - \left(4\frac{1}{3} + c\right) = 16\frac{2}{3}$$

The arbitrary constant of integration, c, always cancels out when limits are applied to an integral and is not usually shown when evaluating a definite integral.

Worked problems on limits and definite integrals

Problem 1. Evalulate (a) $[y^3]^3_2$, (b) $[y^3]^{-1}_4$ and (c) $[y^3]^{\frac{1}{2}}_{-\frac{3}{4}}$

(a) $[y^3]^3_2 = 3^3 - 2^3 = 27 - 8 = \mathbf{19}$
(b) $[y^3]^{-1}_4 = (-1)^3 - (4)^3 = -1 - 64 = \mathbf{-65}$
  Although limits are defined as the increase in the value of the function as it increases from its lower limit function value to the upper limit function value, the minus sign in this result shows that

the value of the function decreases when moving from the lower limit function to the upper limit function value. It is possible that in a practical problem the value of the upper limit is less than the value of the lower limit.

(c) $[y^3]^{\frac{1}{2}}_{-\frac{3}{4}} = (\frac{1}{2})^3 - (-\frac{3}{4})^3 = \frac{1}{8} + \frac{27}{64} = \frac{35}{64}$

Problem 2. Determine the value of $3x + 4 - \dfrac{5}{x}$ as x increases from

(a) 1 to 2, and (b) -1 to 3

(a) $\left[3x + 4 - \dfrac{5}{x}\right]^2_1 = \left(3(2) + 4 - \dfrac{5}{2}\right) - \left(3(1) + 4 - \dfrac{5}{1}\right)$

$\qquad = 7\frac{1}{2} - 2 = \mathbf{5\frac{1}{2}}$

(b) $\left[3x + 4 - \dfrac{5}{x}\right]^3_{-1} = \left(3(3) + 4 - \dfrac{5}{3}\right) - \left(3(-1) + 4 - \dfrac{5}{(-1)}\right)$

$\qquad = 11\frac{1}{3} - 6 = \mathbf{5\frac{1}{3}}$

Problem 3. Evaluate: $\displaystyle\int_2^3 (p - 1)^2 \, dp$

Since $(p - 1)^2$ is not in standard form, it must be altered until it is. Multiplying $(p - 1)$ by $(p - 1)$ gives $p^2 - 2p + 1$, which can now be integrated. Thus

$\displaystyle\int_2^3 (p - 1)^2 \, dp = \int_2^3 (p^2 - 2p + 1) \, dp$

$\qquad = \left[\dfrac{p^3}{3} - \dfrac{2p^2}{2} + p\right]^3_2$

$\qquad = \left(\dfrac{3^3}{3} - 3^2 + 3\right) - \left(\dfrac{2^3}{3} - 2^2 + 2\right)$

$\qquad = 3 - \frac{2}{3} = \mathbf{2\frac{1}{3}}$

Problem 4. Determine: $\displaystyle\int_0^a (ax - x^3) \, dx$ where a is a constant.

$\displaystyle\int_0^a (ax - x^3) \, dx = \left[\dfrac{ax^2}{2} - \dfrac{x^4}{4}\right]^a_0$

$\qquad = \left(\dfrac{a}{2}(a^2) - \dfrac{a^4}{4}\right) - \left(\dfrac{a}{2}(0)^2 - \dfrac{0^4}{4}\right)$

$$= \frac{a^3}{2} - \frac{a^4}{4} - 0$$

$$= \frac{a^3}{2}\left(1 - \frac{a}{2}\right)$$

Problem 5. Evaluate: (a) $\int_0^{\frac{\pi}{4}} 4 \sin 2x \, dx$, (b) $\int_0^1 3 \cos 3t \, dt$

(a) $\displaystyle\int_0^{\frac{\pi}{2}} 4 \sin 2x \, dx = \left[-\frac{4}{2} \cos 2x \right]_0^{\frac{\pi}{2}}$

$$= \left(-2 \cos 2\left(\frac{\pi}{2}\right) \right) - \left(-2 \cos 2(0) \right)$$

$$= (-2 \cos \pi) - (-2 \cos 0)$$

$$= (-2(-1)) - (-2(1))$$

$$= 2 + 2 = 4$$

(b) $\int_0^1 3 \cos 3t \, dt = \left[\frac{3}{3} \sin 3t \right]_0^1 = \left[\sin 3t \right]_0^1 = (\sin 3 - \sin 0)$

The limits in trigonometric functions are expressed in radians.

Thus 'sin 3' means 'the sine of 3 radians or $3\left(\dfrac{180}{\pi}\right)^\circ$', i.e. $171° \, 53'$

Hence $\sin 3 - \sin 0 = \sin 171° \, 53' - \sin 0°$
$$= 0.141\,2 - 0$$
Thus $\int_0^1 3 \cos 3t \, dt = \mathbf{0.141\,2}$

Further problems on limits and definite integrals may be found in Section 23.4 (Problems 11–20), page 418.

23.3   Determination of areas using integration

There are several instances in branches of engineering and science where the area beneath a curve is required to be accurately determined. For example, the area between limits of a:

> velocity/time graph gives distance travelled,
> force/distance graph gives work done,
> acceleration/time graph gives velocity,
> pressure/volume graph gives work done,
> normal distribution curve gives frequency,
> voltage/current graph gives power, and so on.

Provided there is a known relationship between the variables forming the axes of the above graphs, then the area may be calculated exactly using integration. If a relationship between variables is not known or if it is not possible to integrate, then the areas have to be approximately determined using such techniques as the trapezoidal rule, the mid-ordinate rule or Simpson's rule.

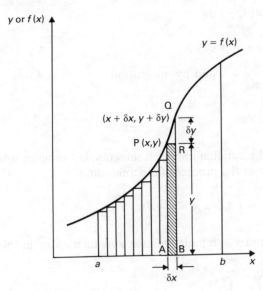

Fig. 23.1

Let A be the area enclosed between the curve $y = f(x)$, the x-axis and the ordinates $x = a$ and $x = b$, where $y = f(x)$ lies above the x-axis. Also, let A be subdivided into a number of elemental strips, each of width δx as shown in Fig. 23.1. One such strip is shown as PQRBA, with point P having coordinates (x, y) and point Q having coordinates $(x + \delta x)$, $(y + \delta y)$. Let the area of PQRBA be δA, which can be seen from Fig. 23.1 to consist of a rectangle PRBA of area $y\,\delta x$ and PQR which approximates to a triangle of area $\frac{1}{2}\delta x\,\delta y$, i.e. $\delta A \approx y\,\delta x + \frac{1}{2}\delta x\,\delta y$.
Dividing throughout by δx gives:

$$\frac{\delta A}{\delta x} \approx y + \frac{1}{2}\delta y$$

As δx is made smaller, the number of rectangles increases and all areas such as PQR become smaller. Also δy becomes smaller and in the limit, $\dfrac{\delta A}{\delta x}$ becomes the differential coefficient $\dfrac{dA}{dx}$ and δy can be considered as zero.

$$\lim_{\delta x \to 0}\left(\frac{\delta A}{\delta x}\right) = \frac{dA}{dx} = y + \tfrac{1}{2}(0) = y$$

Hence $\dfrac{dA}{dx} = y$ (1)

This shows that when a limiting value is taken,

$$\text{Area, } A = \lim_{\delta x \to 0} \sum_{x=a}^{x=b} y\,\delta x \qquad (2)$$

between the limits $x = a$ and $x = b$

From equation (1), $\dfrac{dA}{dx} = y$ and by integration,

$$\int \frac{dA}{dx}\,dx = \int y\,dx$$

When a function is differentiated and then integrated, it remains unaltered, since integration reverses the process of differentiation.

$$\text{Thus } \int \frac{dA}{dx}\,dx = \int dA = A = \int y\,dx$$

The ordinates $x = a$ and $x = b$ limit the area and such ordinate values are shown as limits. Thus

$$A = \left[\int y\,dx \right]_a^b = \int_a^b y\,dx \qquad (3)$$

Equations (2) and (3) show that:

$$\textbf{Area } A = \lim_{\delta x \to 0} \sum_{x=a}^{x=b} y\,\delta x = \int_a^b y\,dx$$

This statement, namely that the limiting value of a sum between limits is equal to the integral between the same limits, forms a **fundamental theorem of integration**. It can be illustrated by considering simple shapes of known areas. For example, Fig. 23.2 shows a rectangle bounded by the line $y = h$, ordinates $x = a$ and $x = b$ and the x-axis. Let the rectangle be divided into n equal vertical strips of width δx. The area of strip PQAB is $h \cdot \delta x$ and since there are n such strips making up the total area then total area $= nh\,\delta x$. The base length of the rectangle (i.e. $b - a$) is made up of n strips, each δx in width, hence $n\,\delta x = (b - a)$.

Hence, total area $= h(b - a)$. The total area is also obtained by adding the areas of all such strips as PQAB and is also independent of the value of n, that is, n can be infinitely large.

$$\text{Hence, total area} = \lim_{\delta x \to 0} \sum_{x=a}^{x=h} h\,\delta x = h(b - a) \qquad (4)$$

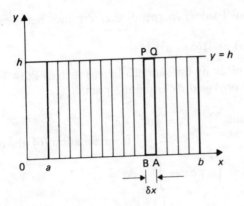

Fig. 23.2

Also, the total area is given by

$$\int_a^b y \, dx = \int_a^b h \, dx$$

$$= [hx]_a^b$$

$$= hb - ha$$

$$= h(b - a) \tag{5}$$

Hence from equations (4) and (5),

$$\lim_{\delta x \to 0} \sum_{x=a}^{x=b} h \, \delta x = \int_a^b h \, \delta x$$

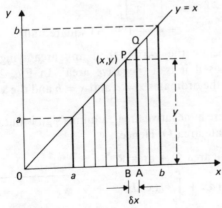

Fig. 23.3

Similarly, for say a trapezium bounded by the line $y = x$, the ordinates $x = a$ and $x = b$ and the x-axis (as shown in Fig. 23.3), the total area is given by:

(half the sum of the parallel sides) (perpendicular distance between these sides)

i.e. $\frac{1}{2}(a + b)(b - a)$ or $\frac{1}{2}(b^2 - a^2)$ (6)

Also the total area will be given by the sum of all areas such as PQAB which each have an area of $y \, \delta x$ provided δx is infinitely small.

i.e. total area $= \displaystyle\lim_{\delta x \to 0} \sum_{x=a}^{x=b} y \, \delta x = \frac{1}{2}(b^2 - a^2)$

from equation (6) above   (7)

Also the total area $= \displaystyle\int_a^b y \, dx = \int_a^b x \, \delta x$

$$= \left[\frac{x^2}{2}\right]_a^b = \frac{1}{2}(b^2 - a^2) \qquad (8)$$

Hence from equations (7) and (8), it has been shown that

$$\lim_{\delta x \to 0} \sum_{x=a}^{x=b} y \, \delta x = \int_a^b y \, \delta x \qquad (9)$$

The two simple illustrations used above show that equation (9) is valid and we shall therefore assume that the relationship is generally true, although a rigorous proof is beyond the scope of this book.

Thus finding the area beneath a curve is the same process as determining the value of a definite integral, previously discussed.

The area between the curve $y = 3x^2$, the ordinates $x = 1$ and $x = 4$ and the x-axis is found as follows:

$$\text{Required area} = \int_1^4 3x^2 \, dx = \left[3\frac{x^3}{3}\right]_1^4$$

$$= 4^3 - 1^3 = 63 \text{ square units}$$

Should the curve drop below the x-axis, then any area lying below the x-axis must be treated as if it is a 'negative area'. In Fig. 23.4 the total area between the curve, the ordinates $x = a$ and $x = b$ and the x-axis is: area P + area Q + area R.

Since Q is treated as if it were a 'negative area', a minus sign is put in front of the integral used to calculate area Q. Hence,

total area = area P + area Q + area R

$$\int_a^c f(x) \, dx - \int_c^d f(x) \, dx + \int_d^b f(x) \, dx$$

This is **not** the same as the area given by $\displaystyle\int_a^b f(x) \, dx$. For this reason, if there is any doubt about the shape of the graph of the function or any

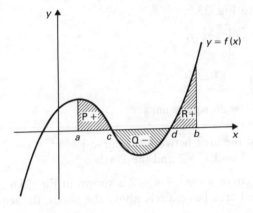

Fig. 23.4

possibility of all or part of it lying below the x-axis, a sketch should be made over the required limits to determine if any part of the curve lies below the x-axis.

Worked problems on determination of areas using integration

Problem 1. Verify by integration that the area of the triangle formed by the line $y = 2x$, the ordinates $x = 0$ and $x = 6$ and the x-axis is 36 square units.

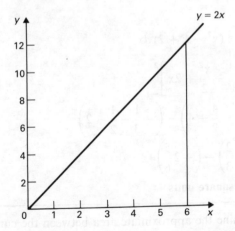

Fig. 23.5

With reference to Fig. 23.5:

$$\text{Area} = \int_0^6 y \, dx = \int_0^6 2x \, dx$$

$$= \left[2 \cdot \frac{x^2}{2} \right]_0^6 = 6^2 - 0^2$$

$$= \textbf{36 square units}$$

Problem 2. Find the area between the curve $y = x^2 - x + 2$, the ordinates $x = -1$ and $x = 2$ and the x-axis.

A sketch of the curve $y = x^2 - x + 2$ is shown in Fig. 23.6 and since the required area lies entirely above the x-axis, the required area is given by:

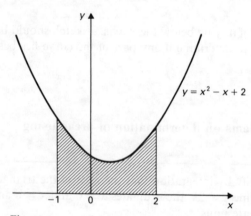

Fig. 23.6

$$\int_{-1}^{2} y \, dx = \int_{-1}^{2} (x^2 - x + 2) \, dx$$

$$= \left[\frac{x^3}{3} - \frac{x^2}{2} + 2x \right]_{-1}^{2}$$

$$= \left(\frac{8}{3} - \frac{4}{2} + 4 \right) - \left(-\frac{1}{3} - \frac{1}{2} - 2 \right)$$

$$= \left(4\frac{2}{3} \right) - \left(-2\frac{5}{6} \right)$$

Required area $= 7\frac{1}{2}$ **square units**

Problem 3. Determine the approximate area between the curve $y = x^3 + x^2 - 4x - 4$, the ordinates $x = -3$ and $x = 3$ and the *x*-axis by applying Simpson's rule. Compare the result obtained by this method with the true area obtained by integration.

Applying Simpson's rule for 12 equal spaces (i.e. 13 ordinates) for the curve shown in Fig. 23.7, taking all ordinate values as positive, since Simpson's rule relies on the lengths of the ordinates only, gives:

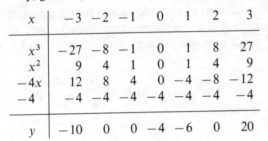

x	-3	-2	-1	0	1	2	3
x^3	-27	-8	-1	0	1	8	27
x^2	9	4	1	0	1	4	9
$-4x$	12	8	4	0	-4	-8	-12
-4	-4	-4	-4	-4	-4	-4	-4
y	-10	0	0	-4	-6	0	20

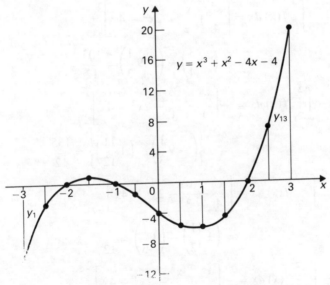

Fig. 23.7

Approximate area

$$= \frac{0.5}{3} \left[(y_1 + y_{13}) + 4(\text{sum of even ordinates}) + 2(\text{sum of odd ordinates}) \right]$$

$$= \frac{0.5}{3} \left[(10 + 20) + 4(3.4 + 0.9 + 1.9 + 5.6 + 4.4 + 7.9) + 2(0 + 0 + 4 + 6 + 0) \right]$$

$$= \frac{0.5}{3} (30 + 96.4 + 20) = 24.4 \text{ square units}$$

The exact area is achieved by integration. With reference to Fig. 23.7, since the curve lies both above and below the x-axis:

Required area

$$= \int_2^3 f(x)\,dx \;-\; \int_{-1}^2 f(x)\,dx \;+\; \int_{-2}^{-1} f(x)\,dx \;-\; \int_{-3}^{-2} f(x)\,dx$$

(the minus signs in front of the second and fourth integrals indicating areas treated as 'negative areas' because they are below the x-axis). Evaluating these integrals for $f(x) = x^3 + x^2 - 4x - 4$ gives:

$$\int f(x)\,dx = \int (x^3 + x^2 - 4x - 4)\,dx = \frac{x^4}{4} + \frac{x^3}{3} - 2x^2 - 4x + c$$

Hence

$$\int_2^3 f(x)\,dx = \left[\frac{x^4}{4} + \frac{x^3}{3} - 2x^2 - 4x \right]_2^3$$

$$= \left(-\frac{3}{4} \right) - \left(-9\frac{1}{3} \right) = \frac{103}{12}$$

$$-\int_{-1}^2 f(x)\,dx = -\left[\frac{x^4}{4} + \frac{x^3}{3} - 2x^2 - 4x \right]_{-1}^2$$

$$= -\left\{ \left(-9\frac{1}{3} \right) - \left(1\frac{11}{12} \right) \right\} = \frac{135}{12}$$

$$\int_{-2}^{-1} f(x)\,dx = \left[\frac{x^4}{4} + \frac{x^3}{3} - 2x^2 - 4x \right]_{-2}^{-1}$$

$$= \left(1\frac{11}{12} \right) - \left(1\frac{1}{3} \right) = \frac{7}{12}$$

$$-\int_{-3}^{-2} f(x)\,dx = -\left[\frac{x^4}{4} + \frac{x^3}{3} - 2x^2 - 4x \right]_{-3}^{-2}$$

$$= -\left\{ \left(1\frac{1}{3} \right) - \left(5\frac{1}{4} \right) \right\} = \frac{47}{12}$$

Adding these results gives,

$$\text{required area} = \frac{103 + 135 + 7 + 47}{12}$$

$$= 24\tfrac{1}{3} \text{ square units}$$

Thus the true area of $24\tfrac{1}{3}$ square units obtained by integration, compared with the approximate area obtained by applying Simpson's rule, which gave a result of 24.4, shows that Simpson's rule gives a result which compares very well with the actual result.

Further problems on determination of areas by integration may be found in the following Section (23.4) (Problems 21–34), page 420.

23.4 Further problems

Indefinite integrals

1. Determine: (a) $\int x^9 \, dx$, (b) $\int x^{-4} \, dx$ and (c) $\int 5 \, dx$

   (a) $\left[\dfrac{x^{10}}{10} + c \right]$ (b) $\left[\dfrac{x^{-3}}{-3} + c \text{ or } -\dfrac{1}{3x^3} + c \right]$

   (c) $[5x + c]$

2. Determine: (a) $\int y^{4.7} \, dy$, (b) $\int z^{-3.6} \, dz$ and (c) $\int p^{7.42} \, dp$

   (a) $\left[\dfrac{y^{5.7}}{5.7} + c \right]$ (b) $\left[\dfrac{z^{-2.6}}{-2.6} + c \text{ or } -\dfrac{1}{2.6z^{2.6}} + c \right]$

   (c) $\left[\dfrac{p^{8.42}}{8.42} + c \right]$

3. Determine: (a) $\int m^{\frac{5}{2}} \, dm$, (b) $\int x^{-\frac{3}{7}} \, dx$ and (c) $\int y^{-\frac{16}{9}} \, dy$

   (a) $\left[\dfrac{2m^{\frac{7}{2}}}{7} + c \right]$ (b) $\left[\dfrac{7x^{\frac{4}{7}}}{4} + c \right]$ (c) $\left[\dfrac{-9y^{-\frac{7}{9}}}{7} + c \right]$

4. Determine: (a) $\int (3r^2 + 5r - 7) \, dr$ $\left[r^3 + \dfrac{5r^2}{2} - 7r + c \right]$

   (b) $\int \left(x^2 - 6 + \dfrac{x}{2} \right) dx$ $\left[\dfrac{x^3}{3} - 6x + \dfrac{x^2}{4} + c \right]$

   and   (c) $\int \left(\dfrac{1}{4y^3} - \dfrac{1}{3y^2} + \dfrac{1}{5} \right) dy$ $\left[-\dfrac{1}{8y^2} + \dfrac{1}{3y} + \dfrac{y}{5} + c \right]$

5. Determine: (a) $\int \left(\dfrac{1}{4} t^{\frac{1}{2}} - 3t^{\frac{5}{2}} \right) dt$ $\left[\dfrac{1}{6} t^{\frac{3}{2}} - \dfrac{6}{7} t^{\frac{7}{2}} + c \right]$

   (b) $\int (3y^{-\frac{1}{2}} - 4y^{-3.7}) \, dy$ $\left[6y^{\frac{1}{2}} + \dfrac{4y^{-2.7}}{2.7} + c \right]$

   (c) $\int \left(\dfrac{3}{4} - 2p + \dfrac{4}{p^2} \right) dp$ $\left[\dfrac{3}{4} p - p^2 - \dfrac{4}{p} + c \right]$

6. Determine: (a) $\int \dfrac{a^4 + 4a + 3}{a^3}\, da$ $\left[\dfrac{a^2}{2} - \dfrac{4}{a} - \dfrac{3}{2a^2} + c\right]$

           (b) $\int p^{\frac{1}{2}}(1 + p^3)\, dp$ $\left[\dfrac{2p^{\frac{3}{2}}}{3} + \dfrac{2p^{\frac{9}{2}}}{9} + c\right]$

           (c) $\int\left(q^3 - \dfrac{\sqrt{q}}{2} + 4\right) dq$ $\left[\dfrac{q^4}{4} - \dfrac{\sqrt{q^3}}{3} + 4q + c\right]$

7. Determine: (a) $\int (2 - x)^2\, dx$ $\left[4x - 2x^2 + \dfrac{x^3}{3} + c\right]$

           (b) $\int \dfrac{3}{4} \sin 3\theta\, d\theta$ $\left[-\dfrac{1}{4} \cos 3\theta + c\right]$

           (c) $\int 7 \cos 7x\, dx$ $[\sin 7x + c]$

8. Determine the arbitrary constant of integration if

$$y = \int\left(4x - 5 + \dfrac{6}{x^2}\right) dx$$

and $y = 5$ when $x = -\frac{1}{2}$ $[-10]$

9. If $p = \int (q^2 + 5)^2\, dq$ and p is equal to $\frac{1}{4}$ when q is equal to 10, determine the arbitrary constant of integration.
[$-23\,583.08\dot{3}$]

10. Find the value of the arbitrary constant of integration if
$s = \int (u + at)\, dt$ and $s = 5$ when $t = 0.7$ and $u = 0$ and
$a = -\frac{3}{4}$ [$5.183\,75$]

Limits and the definite integral

11. Evaluate: (a) $[3x + 2]_1^4$ (b) $[3x + 2]_{-1}^2$ and   (c) $[3x + 2]_q^p$
    (a) [9]   (b) [9]   (c) $[3(p - q)]$

12. Evaluate: (a) $\left[\dfrac{p}{3} - \sqrt{(p)}\right]_4^9$ taking positive values of square

    roots only, (b) $\left[y^2 - y^{\frac{1}{3}}\right]_0^8$ and (c) $\left[\dfrac{q + q^3}{5}\right]_{-m}^0$

    (a) $\left[\frac{2}{3}\right]$ (b) [60]   (c) $\left[\dfrac{m}{5}(1 + m^2)\right]$

13. Evaluate: (a) $[m^2 + 3m - 4]_{1.3}^{\pi}$ correct to five significant

    figures,   (b) $[3\,ax - b^2x]_{-b}^a$ and   (c) $\left[r + \dfrac{3}{r^2} - 2.7r^2\right]_{-0.17}^{-1.3}$

    correct to three decimal places.   (a) [13.704]
    (b) $[(3a - b^2)(a + b)]$ (c) $[-107.646]$

14. Evaluate: (a) $[\theta^3 - \sqrt(\theta) + \frac{1}{2}\theta]_{\pi/3}^{\pi/2}$ correct to four significant figures, taking positive values of square roots only,

(b) $\left[b^2\left(1 - \frac{1}{b}\right)\right]_{-10}^{-4}$, and (c) $[\frac{1}{4}(x-1)(x+2)(x-3)]_0^4$

(a) [2.759]   (b) [−90]   (c) [3]

Evaluate the definite integrals shown in Problems 15 to 20.

15. (a) $\int_1^4 2x^2 \, dx$ [42]

(b) $\int_{-1}^2 4x \, dx$ [6]

(c) $\int_0^a 2x^3 \, dx$ $[\frac{1}{2}a^4]$

16. (a) $\int_0^1 (\sqrt(m) + 1)^2 \, dm$ $\left[2\frac{5}{6}\right]$

(b) $\int_{-2}^2 (8 - 2y^2) \, dy$ $[21\frac{1}{3}]$

17. (a) $\int_0^3 (4 - p)^2 \, dp$ [21]

(b) $\pi \int_1^2 (9x^2 - 6x^3 + x^4 - 4) \, dx$ $[0.5\,\pi]$

18. (a) $\int_1^2 \left(S^3 - \frac{1}{S^2}\right) dS$ [3.25]

(b) $2\pi \int_0^1 (x^2 - x^4) \, dx$ $\left[\frac{4\pi}{15}\right]$

19. (a) $\int_0^2 (x^2 - x - 6) \, dx$ $[-11\frac{1}{3}]$

(b) $\int_{-1}^1 (p^3 - p) \, dp$ [0]

20. (a) $\int_0^{\frac{\pi}{3}} 3 \sin 2x \, dx$ $[2\frac{1}{4}]$

(b) $\int_{\frac{\pi}{6}}^{\frac{\pi}{3}} 2 \cos t \, dt$ [0.7321]

Determination of areas by integration

21. Show by integration that the area of the rectangle formed by the line $y = 3$, the ordinates $x = 2$ and $x = 7$ and the x-axis is 15 square units.

22. By using integration, find the area of the triangle formed by the line $y = \dfrac{x}{2}$, the ordinate $x = 10$ and the x-axis.
    [25 square units]

23. Find the area of the triangular template formed by the line $y = x - 1$, the ordinate $x = 5$ and the x-axis, using integration methods.   [8 square units]

24. Determine the area bounded by the curve $y = 8x - 2x^2$, the ordinate $x = 2$ and the x-axis lying in the first quadrant.
    [$10\frac{2}{3}$ square units]

25. Find the area between the curve $y = (x - 1)(x - 2)$, the lines $x = 0$ and $x = 3$ and the x-axis. $\left[1\dfrac{5}{6} \text{ square units} \right]$

26. Determine the area between the curve $y = 8 + 2x - x^2$, the lines $x = 4$ and $x = -2$ and the x-axis.   [36 square units]

27. Find the area between the curve $y = 2(4 - x^2)$, the ordinates $x = -2$ and $x = 2$ and the x-axis.   [$21\frac{1}{3}$ square units]

28. Find the area between the curve $y = x(x - 1)(x + 3)$, the ordinates $x = -3$ and $x = 1$ and the x-axis.
    $\left[11\dfrac{5}{6} \text{ square units} \right]$

29. Determine the area of the metal plate bounded by the curve $y = 4x^3$, the ordinates $x = -2$ and $x = 2$ and the x-axis.
    [32 square units]

30. Use Simpson's rule to determine the approximate area enclosed by the curve $y = (x - 3)(x - 1)(x)$, the x-axis and $x = 0$, $x = 3$. Compare your result with the true area obtained by integration methods. $\left[3\dfrac{1}{12} \text{ square units} \right]$

31. A car has a velocity v of $(2 + 5t)$ metres per second after t seconds. How far does it move in the first three seconds? Find the distance travelled in the fourth second. (Distance travelled $= \displaystyle\int_{t_1}^{t_2} v \, dt$)   [$28\frac{1}{2}$ m; $19\frac{1}{2}$ m]

32. A vehicle has an acceleration a of $(25 + 3t)$ metres per second squared after t seconds. If the vehicle starts from rest find its velocity after two seconds.

$$\left(\text{Velocity} = \int_{t_1}^{t_2} a\, dt\right) \qquad [56\ \text{m s}^{-1}]$$

33. The force F newtons acting on a body at a distance x metres from a fixed point is given by $F = 3x + 2x^2$. Find the work done when the body moves from the position where

$$x = 2\ \text{m to that where}\ x = 5\ \text{m}\ \left(\text{work done} = \int_{x_1}^{x_2} F\, dx\right)$$

$[109\frac{1}{2}\ \text{N m}]$

34. The velocity v of a body t seconds after a certain instant is $(5t^2 + 4)\ \text{m s}^{-1}$. Find how far it moves in the interval from

$$t = 2\ \text{s to}\ t = 8\ \text{s.}\ \left(\text{Distance travelled} = \int_{t_1}^{t_2} v\, dt\right) \qquad [864\ \text{m}]$$

Chapter 24

Measures of central tendency

24.1   Some measures of central tendency

Mean

It is often convenient to have one value to represent a group of individual values to give a quick understanding of the general size of the group. The statement that the average family in a certain area buys three pints of milk a day should convey an impression that a milkman would expect to deliver about three hundred pints for every one hundred houses in this area. The number of houses having one or two pints would be balanced out by those having four or five pints, say. The word average is used frequently in many contexts. For example, batting and goal averages, average rainfall and temperature, average wages and so on. The mathematical term used for average is the **arithmetic mean** or just the **mean**. This is calculated by adding together the values of a set of items and dividing by the number of items. Thus, if five items are weighed and found to have masses of 33 kg, 37.5 kg, 38 kg, 40.3 kg and 42.7 kg, the total mass is $(33 + 37.5 + 38 + 40.3 + 42.7)$ kg and the number of items is five, giving an arithmetic mean of

$$\frac{33 + 37.5 + 38 + 40.3 + 42.7}{5} \text{ or } \frac{191.5}{5} \text{ or } 38.3 \text{ kg}$$

If there are n items in a set having values x_1, x_2, x_3 and so on to x_n, the arithmetic mean is

$$\frac{x_1 + x_2 + x_3 + \cdots + x_n}{n} = \frac{\sum x}{n}, \text{ which is defined as adding together}$$

all the x values and dividing by the total number of items n. The symbol used for the arithmetic mean value is $\bar{x}$, hence

$$\bar{x} = \frac{\sum x}{n}$$

For the set of numbers 7, 11, 13 and 17 the x-values are $x_1 = 7$, $x_2 = 11$, $x_3 = 13$ and $x_4 = 17$. There are four items so n is equal to 4 and the arithmetic mean is given by

$$\bar{x} = \frac{x_1 + x_2 + x_3 + x_4}{n} = \frac{\sum x}{n} = \frac{7 + 11 + 13 + 17}{4} = 12$$

For data grouped into a frequency distribution, the class mid-point is taken to be representative of that class. When this is multiplied by the frequency of the class, the total value of the class is obtained. By adding together all the classes making up the frequency distribution, the total value of all the items in the distribution is obtained. The symbol $\sum fx$ is used to indicate this total, where $\sum fx$ is the sum of all the (frequency × class mid-point) values. The arithmetic mean can be found by dividing this value by the total number of items obtained by adding together the frequencies of each of the classes, i.e. the $\sum f$ values. Thus for grouped data:

$$\bar{x} = \frac{f_1 x_1 + f_2 x_2 + \cdots + f_n x_n}{f_1 + f_2 + \cdots + f_n} = \frac{\sum (f \times \text{class midpoint})}{\sum f}$$

where f_1 is the frequency and x_1 the class mid-point of the first class and so on. A frequency distribution showing the masses of chickens kept in a store and the corresponding frequencies (the masses being correct to the nearest one-tenth of a kilogram) is as shown:

Mass in kilograms	1.3–1.5	1.6–1.8	1.9–2.1	2.2–2.4	2.5–2.7
Number of chickens	6	22	47	20	5

The value of the arithmetic mean can be determined as follows:

$$\sum (f \times \text{class midpoint}) = 6 \times 1.4 + 22 \times 1.7 + 47 \times 2.0$$
$$+ 20 \times 2.3 + 5 \times 2.6$$
$$= 198.8 \text{ kg}$$
$$\sum f = 6 + 22 + 47 + 20 + 5$$
$$= 100$$

Hence, mean value $\bar{x} = \dfrac{\sum (f \times \text{class midpoint})}{\sum f} = \dfrac{198.8}{100}$

That is, mean value $\bar{x} = 2.0$ kg correct to $\frac{1}{10}$ of a kilogram.

Median

One of the disadvantages of using an arithmetic mean value to represent a group of items occurs when one or two values of items in a set are largely different from the remainder of the set values. For example, the mean of the numbers 1, 2, 3, 4 and 100 is $\frac{110}{5}$ or 22, which is not representative of any of the numbers in the set. For a set of items having a few extreme values at the start or end, a measure of central tendency called the **median** value (middle item) can be used (3 for the above example). To obtain the median value of a set, the items are **ranked**, that is, placed in order of size. When this is done for sets containing an odd number of items, the middle term is selected as being representative of the set. For sets containing an even number of items, the arithmetic mean of the two middle items gives the value of the median. The set containing the numbers 3, 8, 8, 13, 4, 6, 7, 8, 15, 23, 2 and 4 is ranked as 2, 3, 4, 4, 6, 7, 8, 8, 8, 13, 15 and 23. Since there is an even number of items, and since 7, 8 are the middle items, then the median is

$$\frac{7 + 8}{2} \text{ or } 7.5$$

Mode

Another measure of central tendency used for sets containing extreme values is the **mode** or most commonly occurring value. For the set of numbers used above to illustrate the median, the most commonly occurring value is 8, which appears three times, giving the mode of this set as 8.

For grouped data, the modal value of a distribution may be determined from the histogram of the data. The modal value is obtained by using the rectangle of the histogram having the greatest height. The width of this rectangle is divided in the ratio of the different heights of the adjacent rectangles, as shown in Problem 6, and the variable value corresponding to this point of division is read from the horizontal scale of the histogram to give the modal value.

For some problems it is useful to be given more than one measure of central tendency. A builder who is about to build a housing estate may wish to esti-mate how many two-, three- or four-bedroomed houses to provide. The mean value of the number of children in a family does not give him very much help, but if he knows the mode, the mean and perhaps the median he can make a much better estimate. For example, in a survey, the mean value of the number of children is 3.1 due to there being one or two very large families. If in the same survey the median and mode is two, then the builder will know that the majority of his houses should have three bedrooms, to cater for the parents and two children (see Problem 4).

For distributions which are symmetrical about the mean value, the mean, median and modal values are all the same. However, for non-symmetrical distributions, particularly those having extreme values, the mean value is **not** the same as the median or modal values.

Worked problems on arithmetic mean, median and mode

Problem 1. Find the values of the mean, median and mode of the set of numbers: 9, 11, 15, 23, 17, 16, 15, 16, 14, 15 and 18.

The mean value of $\bar{x}$ is given by $\dfrac{\sum x}{n}$

Thus $\bar{x} = \dfrac{9 + 11 + 15 + 23 + 17 + 16 + 15 + 16 + 14 + 15 + 18}{11}$

i.e. **mean value** $= \dfrac{169}{11} = 15.\dot{3}\dot{6}$

Placing the set of numbers in rank order gives

9, 11, 14, 15, 15, 15, 16, 16, 17, 18 and 23

Selecting the middle item, since the set contains an odd number of items, gives **a median value of 15**

The most commonly occurring value is also 15, i.e. **the mode is 15**. Thus the mean is 15.36 and the median and mode are both 15. This result indicates that since the mean, median and mode are approximately the same value any small extreme values are counterbalanced by large extreme values if they exist.

Problem 2. Find the mean and modal class of the data given below:

Class	0–9	10–19	20–29	30–39	40–49	50–59
Frequency	1	3	8	12	9	2

The class mid-point of, say, the 10–19 class is obtained by adding the lower and upper class boundaries and dividing by two. This gives $\dfrac{9.5 + 19.5}{2}$ or 14.5. The other class mid-points are found similarly, giving 4.5, 14.5, 24.5, 34.5, 44.5 and 54.5

The arithmetic mean for grouped data is given by

$\dfrac{\sum f \times \text{class midpoint}}{\sum f}$, where f is the frequency.

$$\sum (f \times \text{class midpoints}) = 1 \times 4.5 + 3 \times 14.5 + 8 \times 24.5$$
$$+ 12 \times 34.5 + 9 \times 44.5 + 2 \times 54.5$$
$$= 1\,167.5$$
$$\sum f = 1 + 3 + 8 + 12 + 9 + 2$$
$$= 35$$

Thus the arithmetic mean, $\bar{x} = \dfrac{1\,167.5}{35} = 33.4$ correct to 3 significant figures. The modal class is the one having the largest frequency, namely the 30–39 class.

Thus **arithmetic mean = 33.4** correct to 3 significant figures and **the modal class is 30–39**

Problem 3. Find the arithmetic mean of the set of numbers:
2 374.8, 2 398.4, 2 414.5, 2 362.7 and 2 381.4

The calculation of an arithmetic mean can frequently be simplified. One way of simplifying sets containing data having a large number of significant figures is to assume a **working mean**. Any value can be chosen and deviations from the working mean can be either positive or negative. The nearer the value of the working mean is to the arithmetic mean, the smaller the numbers in the calculation. Inspection of the values of the numbers in the set above shows that the arithmetic mean is approximately 2 390.
The sum of the values can be written as:

$$(2\,390 - 15.2) + (2\,390 + 8.4) + (2\,390 + 24.5)$$
$$+ (2\,390 - 27.3) + (2\,390 - 8.6)$$

$$\text{and } \bar{x} = \frac{5 \times 2\,390 + (-15.2 + 8.4 + 24.5 - 27.3 - 8.6)}{5}$$

$$2\,390 - \frac{18.2}{5} = 2\,390 - 3.64 = \mathbf{2\,386.36}$$

This method of determining an arithmetic mean is useful when no calculator is available. The general rule when adopting this method is that the arithmetic mean is equal to the working mean plus the average value of the deviations from the working mean.

Problem 4. The number of houses in a street having 0, 1, 2, 3, ... children under 16 years of age is as shown:

Number of children	0	1	2	3	4	5	6
Number of houses	10	12	16	13	9	7	3

Find the mean, median and mode of this distribution.

The arithmetic mean will be: $\dfrac{\text{the total number of children}}{\text{the total number of houses}}$

$$\text{i.e. } \bar{x} = \frac{\begin{array}{c}(0 \times 10) + (1 \times 12) + (2 \times 16) + (3 \times 13) \\ + (4 \times 9) + (5 \times 7) + (6 \times 3)\end{array}}{10 + 12 + 16 + 13 + 9 + 7 + 3}$$

$$= \frac{12 + 32 + 39 + 36 + 35 + 18}{70} = \frac{172}{70}$$

$$= \mathbf{2.5} \text{ correct to two significant figures, i.e. 3 children.}$$

The median value can be obtained by writing all the items in the set in order of size and selecting the arithmetic mean of the two middle values, since the set contains an even number of items. Thus 0 would be written ten times, 1 twelve times, 2 sixteen times and so on, and the result would give **a median value of 2**. Alternatively, a line 70 units long, representing the total number of houses, can be drawn, and this line can be subdivided into 10 units representing no children, 12 units representing one child, 16 units representing two children and so on. The centre of this line indicates the median value. The most commonly occurring number of children is 2, hence **the mode is 2**.

Typical conclusions that a builder, say, can draw from these mean, median and modal values are as follows. The mean value of 3 children per house indicates that three- or four-bedroomed houses are likely to be most in demand (2 children + parents or 3 children + parents). The median and mode are both two, which indicates that more three-bedroomed houses are wanted than four-bedroomed houses. The actual distribution confirms these conclusions, since 51 of the 70 houses require 4 bedrooms or fewer, only 13 of which need have 4 bedrooms.

Problem 5. The frequency distribution shows the weekly income of 50 employees of a small company, correct to the nearest £1:

Class (£)	150–159	160–169	170–179	180–189	190–199	200–209
Number of employees	3	6	10	14	12	5

Find the mean wage.

As for Problem 3, a working mean x_w is assumed for the grouped data, in order to keep calculations small. In this case the mid-point of the 180–189 class is taken, namely 184.5

Class	Class mid-point (x)	Frequency (f)	Deviation from working mean ($x - x_w$)	$f(x - x_w)$
150–159	154.5	3	−30	−90
160–169	164.5	6	−20	−120
170–179	174.5	10	−10	−100
180–189	184.5	14	0	0
190–199	194.5	12	10	+120
200–209	204.5	5	20	+100
		Total: 50		**Total: −90**

The arithmetic mean = (the working mean)
+ (the mean of the deviations from the working mean)

$$x = 184.5 - \tfrac{90}{50} = 184.5 - 1.8 = 182.7$$

That is, **the mean wage of the frequency distribution given is £182.70**

Problem 6. Draw a histogram depicting the data given below. Determine the mean value by calculation and the modal value from the histogram.

Variable	4.6–5.4	5.6–6.4	6.6–7.4	7.6–8.4	8.6–9.4	9.6–10.4
Frequency	3	16	10	5	4	1

Mean value

Mean value, $x = \dfrac{\sum (f \times \text{class midpoint})}{f}$

$$= \frac{(5 \times 3) + (6 \times 16) + (7 \times 10)}{3 + 16 + 10 + 5 + 4 + 1}$$

$$= \frac{15 + 96 + 70 + 40 + 36 + 10}{39}$$

$$= \tfrac{267}{39} = \mathbf{6.9} \text{ correct to 2 significant figures}$$

Modal value

The histogram is drawn using the values of the variable for the horizontal scale and frequency for the vertical scale, as shown in Fig. 24.1.

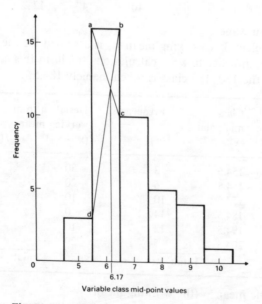

Fig. 24.1

Rectangles are drawn symmetrically, spanning the class mid-point values of the variable and touching one another at the vertical common line. The rectangle having the greatest height is the one having a frequency of 16 and by joining ac and bd, as shown in Fig. 24.1, the width of this rectangle is divided into the ratio of the difference in height of the rectangles adjacent to it, i.e. those having frequencies of 3 and 10 respectively. The vertical line drawn through the point of intersection of ac and bd cuts the horizontal scale at 6.17, giving a modal value for the distribution of **6.2**, correct to 2 significant figures.

Further problems on mean, median and modal values may be found in Section 24.3 (Problems 1–14), page 435.

24.2 Median, quartile, decile and percentile values

Median and quartile values

The **ogive** or **cumulative frequency** curve is introduced in *Technician mathematics 5* and this may be used to determine certain statistical parameters. For example, the median value for grouped data corresponds to the value of the variable where the 'less than' and 'or more' ogives cross, when drawn with respect to the same axes. The 'or more' ogive is a mirror image of the 'less than' ogive and the point where they intersect can be used to determine the median value if a vertical line is drawn to the horizontal axis (see worked Problems 1 and 2).

The median is defined as the value of the middle item in a set where the members have been placed in order of magnitude (ranked), such that it divides the set into two parts, so that the frequency of members above the median point is equal to the frequency of members below the median point. If a set of data is divided into four equal parts by values Q_1, Q_2 and Q_3, then these values are known as the first, second and third **quartiles** respectively. It follows that Q_2 is the median value.

For ungrouped data, consider a set containing, say, eleven members, marked 1 to 11 respectively, shown in Fig. 24.2. If the member marked 6 is

Fig. 24.2 A set having eleven members shown split into four equal parts by quartiles Q_1, Q_2 and Q_3

selected, then there are five members beneath it and five members above it, hence member 6 is the median member of the set. Similarly, members 3 and 9 divide the set (with member 6) into four equal parts and the values of these members give the quartile values. In general, if there are n members in a set, where n is an odd integer:

the median member, Q_2, is the $\left(\dfrac{n+1}{2}\right)$th member;

the first quartile member, Q_1, is the $\left(\dfrac{n+1}{4}\right)$th member; and

the third quartile member, Q_3, is the $\left(\dfrac{3(n+1)}{4}\right)$th member.

Applying these formulae to the set shown in Fig. 24.2 gives the median member, Q_2, as the $\left(\dfrac{11+1}{2}\right)$th member or the 6th member; the first quartile member, Q_1, as the $\left(\dfrac{11+1}{4}\right)$th member or the 3rd member; and the third quartile member, Q_3, as the $\left(\dfrac{3(11+1)}{4}\right)$th member or the 9th member. When n is an even integer:

the first quartile member, Q_1, is the $\left(\dfrac{(n+2)}{4}\right)$th member;

the median member, Q_2, is the $\left(\dfrac{2n+2}{4}\right)$th member; and

the third quartile member, Q_3, is the $\left(\dfrac{3n+2}{4}\right)$th member.

For a set having, say, 12 members, the first quartile member is the $\left(\dfrac{12+2}{4}\right)$th or the $3\frac{1}{2}$th member and the quartile value is found by taking the mean value of the 3rd and 4th members.

When using the median value to be representative of the values of members of a set, no indication is given of the range or deviation of the values of the set members from this median value. By determining the median and the quartile values of a set of data, some indication of deviation from the median is given. One or two extreme values do not generally affect the quartile values. Hence the median and two quartile values can be chosen to represent a set of data and to convey general information about the magnitude and deviation of the set. If the median wage earned by a group of employees is £160 a week and the quartile values are £150 and £180, then some of the information conveyed by this is:

(a) half the employees earn between £150 and £180,
(b) a quarter of the employees earn more than £180, and
(c) a quarter of the employees earn less than £150,

which begins to convey a much better understanding of the wages distribution than just the information that the median wage is £160.

The quartile values may be obtained for grouped data by using an ogive. Values of cumulative frequency are determined so that the total frequency is divided into four equal parts, and corresponding values of the variable are determined from the horizontal axis of the ogive. For example, if the total frequency is 60, cumulative frequency values of 15, 30 and 45 divide the total frequency into four equal parts and the variable values corresponding to these frequencies give Q_1, Q_2 and Q_3 respectively.

Deciles and percentiles

There are other schemes for dividing up a set of data or data presented as a frequency distribution. When data are divided into a number of groups, each of equal frequency, the general term used for each group is a **quantile**. When a set is split into ten groups, each group having the same frequency, the groups are given the special name of **deciles**, often signified by $D_1, D_2, D_3, \ldots$, D_{10}. A set split into 100 groups, each group containing the same number of members, is said to have its data grouped in **percentiles**, often signified by $P_1, P_2, P_3, \ldots, P_{100}$. In theory, a cumulative frequency curve can be used to determine percentiles in the same way as for quartiles, but because of the large numbers involved this is rarely feasible in practical situations.

Worked problems on quartiles, deciles and percentiles

Problem 1. Determine the median and quartile values for the set of numbers 55, 61, 57, 60, 57, 60, 58, 61, 59.

Ranking the numbers in ascending order of magnitude gives 55, 57, 57, 58, 59, 60, 60, 61 and 61. As there are 9 members in the set, $n = 9$.

The first quartile value, Q_1, will be that of the $\left(\dfrac{n+1}{4}\right)$th member since n is odd, that is $\dfrac{10}{4}$ or 2.5. This indicates that it lies between the second and third members and its value is the mean value of these two members. Hence $Q_1 = \dfrac{57+57}{2} = 57$

The median value is the $\dfrac{n+1}{2}$ or $\dfrac{10}{2}$, i.e. the 5th member value.

Hence **the median value is 59**

The third quartile value, Q_3, is that of the $\left(\dfrac{3(n+1)}{4}\right)$th member,

i.e. $\dfrac{3(9+1)}{4}$ or the 7.5th member. This indicates that it lies between

the 7th and 8th members and their mean value is taken.

Hence $Q_3 = \dfrac{60+61}{2} = 60.5$

Problem 2. Draw an ogive for the data given below and hence determine the median and first and third quartile values for this distribution.

Class intervals (mm)	Frequency	Cumulative frequency
1.24–1.26	2	2
1.27–1.29	4	6
1.30–1.32	4	10
1.33–1.35	10	20
1.36–1.38	11	31
1.39–1.41	5	36
1.42–1.44	3	39
1.45–1.47	1	40

Because the ogive is not specified as having a 'less than' or an 'or more' basis, the 'less than' basis is assumed.

Upper class boundary values are selected for the horizontal scale and cumulative frequency as the vertical scale. The ogive is shown in Fig. 24.3

The first quartile value, Q_1, occurs at a cumulative frequency value of $\dfrac{40}{4}$ or 10 and its value is **1.325 mm**

The second quartile or median value, Q_2, occurs at a cumulative frequency value of $\dfrac{40}{2}$ or 20 and its value is **1.355 mm**

The third quartile value, Q_3, occurs at a cumulative frequency value of $\dfrac{3 \times 40}{4}$ or 30 and its value is **1.382 mm**

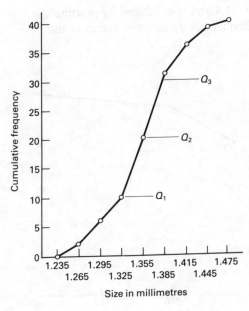

Fig. 24.3

Problem 3. Using the data given below, construct an ogive and determine the quartile and median values for this distribution.

Class interval	Frequency	Class interval	Frequency
1–4	2	20–49	24
5–9	10	50–99	14
10–19	15	100–199	5

A cumulative frequency distribution showing upper class boundary values and cumulative frequency is produced and is as shown.

Class interval	Frequency	Upper class boundary	Cumulative frequency
		Less than 0.5	0
1–4	2	Less than 4.5	2
5–9	10	Less than 9.5	12
10–19	15	Less than 19.5	27
20–49	24	Less than 49.5	51
50–99	14	Less than 99.5	65
100–199	5	Less than 199.5	70

The ogive is shown in Fig. 24.4 and is produced by plotting the values of the cumulative frequencies against the values of the upper class boundaries.

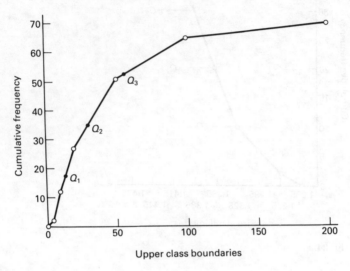

Fig. 24.4

The quartile values, Q_1, Q_2 and Q_3, correspond to frequencies of $\dfrac{70}{4}$, $\dfrac{2 \times 70}{4}$ and $\dfrac{3 \times 70}{4}$, that is $17\frac{1}{2}$, 35 and $52\frac{1}{2}$

From the ogive the first and third quartile values are 13 and 55 respectively and the median value is 29.5

Problem 4. Determine the members contained in the 30th to 39th percentile group and in the 9th decile group of the set of numbers shown below:

17 22 25 43 30 15 21 19 20 8 16 28
29 22 23 17 27 17 32 28

The set is ranked, giving

 8 15 16 17 17 17 19 20 21 22
22 23 25 27 28 28 29 30 32 43

There are 20 members in the set, hence the first 10% will be the two members 8 and 15, the second 10% will be 16 and 17 and so on.

Thus, **the 30th to 39th percentile group will be the members 19 and 20**

The first decile group is obtained by splitting the ranked set into 10 equal groups and selecting the first group, i.e. the members 8 and 15. The second decile group is the members 16 and 17, and so on.

Hence, **the 9th decile group contains the members 29 and 30**

Further problems on quartile, decile and percentile values may be found in Section 24.3 following (Problems 15–22), page 437.

24.3 Further problems

Arithmetic mean, median and mode

1. Find the values of the mean, median and mode of the set of numbers 52, 51, 52, 59, 60, 58, 54, 53 and 50   [54.$\dot{3}$, 53, 52]

2. Determine the arithmetic mean, the median and the mode of the lengths given below in metres:
    20, 28, 44, 32, 30, 28, 30, 26, 28 and 34   [30, 29, 28]

3. The average marks obtained by 20 students in an examination were: 51, 43, 30, 36, 42, 40, 27, 31, 61, 57,
          74, 64, 55, 42, 83, 52, 50, 37, 41, 71
    Determine the mean mark and the median mark.
    [49.35, 46.5]

4. The haemoglobin levels in a number of children are given in the following data:

Haemoglobin level	70–74	75–79	80–84
Number of children	40	91	165

Haemoglobin level	85–89	90–94	95–99
Number of children	248	200	156

    Using a working mean of 87, find the mean haemoglobin level.   [87.25]

5. The girth of 500 trees was measured, with the following results:

Girth (cm)	45–64	65–84	85–104	105–124
Number of trees	14	32	135	157

Girth (cm)	125–144	145–164	165–184
Number of trees	102	43	17

    Determine the mean girth of the trees.   [114.42 cm]

6. Calculate the arithmetic mean of the following data, correct to 4 significant figures:

881.7 882.3 884.6 879.8 882.3 847.9
885.3 906.4 882.9 885.3 876.7 882.9   [881.5]

7. Calculate the mean value correct to 4 significant figures of the set of numbers: 650 624 663 614 690 637 648 by

(a) taking 600 as a working mean,
(b) taking 650 as a working mean,
(c) taking 700 as a working mean.

Does the selection of a value for the working mean (a) alter the value of the result and (b) alter the magnitude of the calculations to be done?   [646.6, No, Yes]

8. The heights of a group of children were measured to the nearest centimetre and the results were as follows:

Height (cm)	158	159	160	161	162	163	164	165
Frequency	10	14	16	23	27	18	7	2

Find the mean height of the children, correct to 3 decimal places.   [161.154 cm]

9. The number of people occupying each house in a certain area is as shown. Determine the arithmetic mean correct to 3 decimal places.

Number of people per house	1	2	3	4	5	6	7	8
Frequency	26	64	83	79	37	13	4	2

[3.331]

10. The diameters of a sample of bolts produced by a company are as shown. Determine the mean diameter.

Diameter (cm)	2.247–2.249	2.250–2.252	2.253–2.255	2.256–2.258
Frequency	4	12	16	30

Diameter (cm)	2.259–2.261	2.262–2.264	2.265–2.267	2.268–2.270
Frequency	36	24	8	2

[2.258 5 cm]

11. The value of capacitance of a sample of capacitors is determined, correct to the nearest 0.1 of a microfarad.

The results are shown below:

Capacitance (µF)	7.7	7.8	7.9	8.0	8.1	8.2	8.3
Frequency	4	17	29	41	34	21	11

Determine the mean value of capacitance for this sample.
[8.02 µF]

12. The duration of telephone calls at an exchange was measured for a one-week period, between 6.00 p.m. and 10.00 p.m., and the results are as shown. Draw a histogram and hence determine the modal value of the distribution.

Duration in minutes	5	6	7	8	9	10	11	12	13	14
Frequency	20	72	110	173	246	357	296	74	13	9

[10.15 min]

13. The temperature rises (°C) in batteries being charged are recorded and the results are as shown. By drawing a histogram of this data, determine the modal value of the distribution.

Temperature rise (°C)	3.5	4.5	5.5	6.5	7.5	8.5
Number of batteries	4	14	24	34	18	6

[6.40°C]

14. The yield of 60 roots of potatoes had the following masses:

Mass (kg)	1.0–1.9	2.0–2.9	3.0–3.9
Number of roots	6	12	16
Mass (kg)	4.0–4.9	5.0–5.9	6.0–6.9
Number of roots	15	8	3

Construct a histogram and hence determine the modal value of the distribution.   [3.75 kg]

Quartiles, percentiles and deciles

15. The number of working days lost due to accidents for each of 12 one-monthly periods is as shown. Determine the median and first and third quartile values for this data.
15 10 14 16 12 13 19 17 19 16 11 15
[15, 12.5, 16.5 days]

16. The number of faults occurring on a production line in a nine-week period are as shown below. Determine the median and quartile values for these data.
    31 30 30 31 25 27 27 28   29
    [29, 27, 30.5 faults]

17. The number of emergency calls received per week during a 13-week period are as shown. Determine the quartile values for these data.
    17 20 26 21 23 27 16 27 18 25 20 27 19
    [18.5, 21, 26.5 calls]

18. The quantity of electricity used by an office over a 50-week period is shown below. Draw an ogive for these data and hence determine the median and quartile values.

Class intervals	71–79	80–88	89–97	98–106	107–115
Frequency	1	3	6	8	12

Class intervals	116–124	125–133	134–142	143–151
Frequency	8	5	4	3

$[Q_1 = 100.5$, median $= 111.5$, $Q_3 = 123.5]$

19. The length in millimetres of a sample of bolts is shown below. Produce a frequency distribution, an 'or more' ogive and a 'less than' ogive for these data and hence determine the median value.

Height (cm)	165	166	167	168	169	170	171
Frequency	5	14	18	28	36	29	29

Height (cm)	172	173	174	175	176	177
Frequency	24	19	15	6	3	2

[median $= 170$]

20. The data given below refer to the intelligence quotient of a group of children. Draw a histogram, frequency polygon and ogive for these data and hence determine the first and third quartile values and the median value.

I.Q.	60–79	80–89	90–94	95–99
Frequency	9	29	51	57

I.Q.	100–104	105–114	115–129	130–149
Frequency	61	53	31	9

$[Q_1 = 93.5$, median $= 99.8$, $Q_3 = 108]$

21.   Determine the members contained in the 6th decile group and in the 80th to 89th percentile groups for the set of numbers:

32 53 25 29 18 38 32 27 27 42
38 27 35 40 31 30 26 39 33 37

[32 and 33; 39 and 40]

22.   Determine the members in the 8th decile group and in the 20th to 29th percentile groups for the set of numbers:

23 32 50 31 49 54 48 50 56 50 56 58 55 59 63
57 64 72 62 74 76 71 77 79 75 80 23 78 25 32

[71, 72 and 74; 48, 49 and 50]

Chapter 25

Standard deviation

25.1 Standard deviation for ungrouped data

Having decided on an appropriate measure of central tendency (such as mean, median or mode), another important aspect of statistics is to measure how much the data in a set deviates from this measure of central tendency. The most important measure of dispersion or scatter of a set of data is the **standard deviation**. The Greek letter, small 'sigma' written σ, is frequently used to denote standard deviation. This is a quantity which will appear repeatedly as statistics is studied in more depth. It can be defined as the **root mean square deviation** and a small value of standard deviation indicates data which is closely packed about the measure of central tendency. For example, the standard deviation of a set of components, A, having an estimated life length of, say, 1 000 hours is found to be 100 hours. For another set, B having the same life length, it is found to be 10 hours. To a statistician, this information indicates that for set A, the large majority (95%) of failures will occur between 800 and 1 200 hours, but between 980 and 1 020 hours for set B. Thus a graph showing numbers of failures against hours will be loosely grouped about the mean value of 1 000 hours for set A, but tightly grouped for set B. These statements will become clearer when a study is made of the normal distribution curve. Unless otherwise stated, it is usual to take the mean value as the measure of central tendency.

The calculation of the standard deviation of a set of ungrouped data from the mean value is performed as follows:

(a) the arithmetic mean, $\bar{x}$ is calculated using $\bar{x} = \dfrac{\sum x}{n}$

(b) the differences of the values of the set of items from $\bar{x}$ (i.e. the deviation of each value from the mean) are calculated, i.e. $(x_1 - \bar{x})$, $(x_2 - \bar{x})$, ..., $(x_n - \bar{x})$

(c) the differences are squared, i.e.

$(x_1 - \bar{x})^2$, $(x_2 - \bar{x})^2$, ..., $(x_n - \bar{x})^2$ **'SQUARE'**

(d) all such squared terms are added together, i.e. $(x_1 - \bar{x})^2 + (x_2 - \bar{x})^2 + \cdots + (x_n - \bar{x})^2$

(e) the mean value of the sum of the squared terms is found by dividing by the number of items, i.e.

$$\frac{\sum (x - \bar{x})^2}{n}$$ **'MEAN'**

(f) the square root of the 'mean-squares' is calculated to reverse the squaring process in (d), i.e.

$$\sqrt{\left[\frac{\sum (x - \bar{x})^2}{n}\right]}$$ **'ROOT'**

Thus the standard deviation, $\sigma = \sqrt{\left[\dfrac{\sum (x - \bar{x})^2}{n}\right]}$

The reason for using an r.m.s. value, rather than the mean distance of values from the mean of the set is to give a more representative value when there are extreme values in a set. For example, the set 1, 2, 3 and 10 has a mean of $\frac{16}{4}$ or 4. The root-mean-square value is $\sqrt{\left[\dfrac{1^2 + 2^2 + 3^2 + 10^2}{4}\right]}$ or ± 5.34, and 5.34 is considered to be more representative of the set than 4

To find the standard deviation of the set of masses 4 kg, 6 kg, 4 kg, 4 kg, 7 kg and 5 kg correct to 3 decimal places, using this procedure gives:

(a) the mean $\bar{x} = \dfrac{4 + 6 + 4 + 4 + 7 + 5}{6} = \dfrac{30}{6} = 5$ kg

(b) the differences, $(x - \bar{x})$ are $(4 - 5)$, $(6 - 5)$, $(4 - 5)$, $(4 - 5)$, $(7 - 5)$ and $(5 - 5)$, or -1, 1, -1, -1, 2 and 0

(c) the **squares** of the differences are 1, 1, 1, 1, 4 and 0

(d) $\sum (x - \bar{x})^2$ are those squared values added, i.e. $1 + 1 + 1 + 1 + 4 + 0$ or 8

(e) the **mean** of the squares is $\dfrac{\sum (x - \bar{x})^2}{n}$ or $\dfrac{8}{6} = 1\frac{1}{3}$

(f) the **square root** of the mean of the squares is $\sqrt{1\frac{1}{3}}$ or ± 1.155 correct to 3 decimal places.

Thus the root mean square or standard deviation is 1.155 kg correct to 3 decimal places.
The value of the standard deviation affects the general shape of a distribution. The two sets of data shown below each refer to the heights of, say, 16 children selected at random from a class of 30 children, the heights being measured to the nearest 5 centimetres.

Height in centimetres	Frequency of set A	Frequency of set B
125	0	1
130	1	2
135	4	3
140	6	4
145	4	3
150	1	2
155	0	1

For set A, the mean $\bar{x}_A$ is given by:

$$\bar{x}_A = \frac{(1 \times 130) + (4 \times 135) + (6 \times 140) + (4 \times 145) + (1 \times 150)}{1 + 4 + 6 + 4 + 1}$$

$$\bar{x}_A = 140 \text{ cm}$$

The standard deviation, $\sigma = \sqrt{\left[\dfrac{\sum (x - \bar{x})^2}{n}\right]}$

Since, for example, 135 cm occurs four times, the $(x - \bar{x})^2$ term will be $4(135 - 140)^2$ and so on. Thus:

$$\sigma_A = \sqrt{\left[\frac{\begin{array}{c}(130 - 140)^2 + 4(135 - 140)^2 \\ + 6(140 - 140)^2 + 4(145 - 140)^2 + (150 - 140)^2\end{array}}{16}\right]}$$

$$= \sqrt{\frac{400}{16}} = \pm 5 \text{ cm}$$

For set B, the mean $\bar{x}_B$ is given by:

$$\bar{x}_B = \frac{\begin{array}{c}(1 \times 125) + (2 \times 130) + (3 \times 135) \\ + (4 \times 140) + (3 \times 145) + (2 \times 150) + (1 \times 155)\end{array}}{1 + 2 + 3 + 4 + 3 + 2 + 1}$$

$$\bar{x}_B = 140 \text{ cm}$$

The standard deviation, $\sigma = \sqrt{\left[\dfrac{\sum (x - \bar{x})^2}{n}\right]}$. Thus

$$\sigma_B = \sqrt{\left[\frac{\begin{array}{c}(125 - 140)^2 + 2(130 - 140)^2 \\ + 3(135 - 140)^2 + 4(140 - 140)^2 \\ + 3(145 - 140)^2 + 2(150 - 140)^2 + (155 - 140)^2\end{array}}{16}\right]}$$

$$= \sqrt{\frac{1\,000}{16}} = \pm 7.906 \text{ cm correct to 3 decimal places.}$$

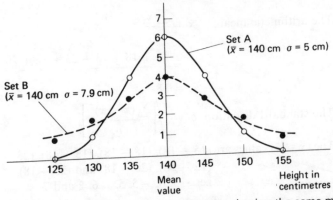

Fig. 25.1 Graph showing two sets of data having the same mean value but different standard deviations

By taking frequency as the vertical axis and heights as the horizontal axis, a graph can be drawn of these two sets of data and is shown in Fig. 25.1. This graph shows that although both sets have the same arithmetic mean value of 140 centimetres, set A, which has a smaller standard deviation, has its items sited more closely to the mean. The items in set B are more spread out relative to the mean, showing a greater variation in the heights of the children in this set.

A better estimate of the standard deviation of the population from which the sample is drawn is given by

$$s = \sqrt{\frac{\sum [f(x - \bar{x})^2]}{(\sum f) - 1}}$$

The reason for deducting 1 in the denominator is that in practical situations a better estimate of the true mean is obtained for small samples, $(\sum f < 30)$, by doing this.

In general, for large values of $\sum f$ (usually taken as $\sum f > 30$), there is very little difference between the standard deviation of a sample and that of the population from which it is drawn. As $\sum f$ gets smaller, the difference between the two values gets larger. Calculators with statistical functions usually have both these values of standard deviation programmed into them, and are marked 'σ_n' for use with non-grouped data and samples and 'σ_{n-1}' for the estimate of standard deviations of populations based on sample data.

Worked problems on standard deviation

Problem 1. Calculate the mean and standard deviation of the following set of numbers, correct to 3 decimal places:

15, 18, 13, 23, 12, 20 and 25

The arithmetic mean, $\bar{x} = \dfrac{\sum x}{n}$

$$= \frac{15 + 18 + 13 + 23 + 12 + 20 + 25}{7}$$

$$= 18$$

The standard deviation, $\sigma = \sqrt{\left[\dfrac{\sum (x - \bar{x})^2}{n}\right]}$

The $(x - \bar{x})$ values are $(15 - 18)$, $(18 - 18)$, $(13 - 18)$, $(23 - 18)$, $(12 - 18)$, $(20 - 18)$ and $(25 - 18)$,
i.e. $-3, 0, -5, 5, -6, 2$ and 7

Hence

$$\sum (x - \bar{x})^2 = (-3)^2 + (0)^2 + (-5)^2 + (5)^2 + (-6)^2 + (2)^2 + (7)^2$$
$$= 9 + 0 + 25 + 25 + 36 + 4 + 49$$
$$= 148$$

And

$$\frac{\sum (x - \bar{x})^2}{n} = \frac{148}{7} = 21.14$$

Hence the standard deviation,

$$\sigma = \sqrt{\left[\frac{\sum (x - \bar{x})^2}{n}\right]} = \sqrt{21.14}$$

$= \pm 4.598$ correct to 3 decimal places.

Problem 2. The monthly output of a coal pit in thousands of tonnes for twelve consecutive months is as shown. Determine the monthly mean output and the standard deviation from this mean.

5.3, 5.7, 5.6, 5.5, 5.4, 5.3, 5.2, 5.5, 5.7, 5.4, 5.7 and 5.4

The arithmetic mean,

$$\bar{x} = \frac{\sum x}{n}$$

$$= \frac{5.3 + 5.7 + 5.6 + 5.5 + 5.4 + 5.3 + 5.2 + 5.5 + 5.7 + 5.4 + 5.7 + 5.4}{12}$$

$= 5.475$ thousand tonnes.

The standard deviation, $\sigma = \sqrt{\left[\dfrac{\sum (x - \bar{x})^2}{n}\right]}$

The $(x - \bar{x})^2$ values are $(5.3 - 5.475)^2$, $(5.7 - 5.475)^2$, $(5.6 - 5.475)^2, \ldots (5.4 - 5.475)^2$

that is, $(-0.175)^2$, $(0.225)^2$, $(0.125)^2$, $(0.025)^2$, $(-0.075)^2$,
$(-0.175)^2$, $(-0.275)^2$, $(0.025)^2$, $(0.225)^2$, $(-0.075)^2$, $(0.225)^2$
and $(-0.075)^2$

Then, $\sum (x - \bar{x})^2 = 0.030\,625 + 0.050\,625 + 0.015\,625 + 0.000\,625$
$+ 0.005\,625 + 0.030\,625 + 0.075\,625$
$+ 0.000\,625 + 0.050\,625 + 0.005\,625$
$+ 0.050\,625 + 0.005\,625$

i.e. $\sum (x - \bar{x})^2 = 0.322\,5$

and $\dfrac{\sum (x - \bar{x})^2}{n} = \dfrac{0.322\,5}{12} = 0.026\,875$

Hence the standard deviation is: $\sqrt{(0.026\,875)} = \pm 0.164$ correct to 3 significant figures.

That is, **the standard deviation of the output is 164 tonnes from the mean, correct to 3 significant figures.**

Problem 3. Calculate the mean and standard deviation correct to 4 significant figures from the mean of the set of numbers:

1 450, 1 400, 1 455, 1 441, 1 448, 1 440, 1 452, 1 444, 1 446 and 1 459.

Assuming a working mean of 1 440, the mean value is

$$1\,440 + \frac{(10 - 40 + 15 + 1 + 8 + 0 + 12 + 4 + 6 + 19)}{10}$$

i.e. $\bar{x} = \mathbf{1\,440 + 3.5 = 1\,443.5}$

The $(x - \bar{x})$ values are $(1\,450 - 1\,443.5)$, $(1\,400 - 1\,443.5)$, ...,
$(1\,459 - 1\,443.5)$,

i.e. 6.5, -43.5, 11.5, -2.5, 4.5, -3.5, 8.5, 0.5, 2.5 and 15.5

$\sum (x - \bar{x})^2 = 42.25 + 1\,892.25 + 132.25 + 6.25 + 20.25 + 12.25$
$+ 72.25 + 0.25 + 6.25 + 240.25$

$= 2\,424.5$

Then, $\dfrac{\sum (x - \bar{x})^2}{n} = \dfrac{2\,424.5}{10} = 242.45$

The standard deviation $= \sqrt{\left[\dfrac{\sum (x - \bar{x})^2}{n}\right]}$

$= \sqrt{(242.45)}$

$= \pm \mathbf{15.57}$ correct to 4 significant figures.

Further problems on standard deviation of ungrouped data may be found in Section 25.3 (Problems 1–11), page 447.

25.2   Standard deviation for grouped data

For grouped data, the mean value of the distribution is determined using:

$\bar{x} = \dfrac{\sum (f \times \text{class mid-point values})}{\sum f}$, where the f-values are the class frequen-

cies. The standard deviation is found in a similar way to ungrouped data. The value of $(x - \bar{x})$ for each class are firstly determined, where the x-values are the class mid-point values. Each of the $(x - \bar{x})$ values is squared and multiplied by the appropriate class frequency to give $f(x - \bar{x})^2$ values for each class. The frequency values and $f(x - \bar{x})^2$ values are summed and the standard deviation is determined from

$$\sigma = \sqrt{\dfrac{\sum [f(x - \bar{x})^2]}{\sum f}}, \text{ as applied previously.}$$

For example, the heights of 100 people are measured, correct to the nearest centimetre and the frequency distribution is as shown.

Height (cm)	Class mid-point	Frequency
150–157	153.5	5
158–165	161.5	18
166–173	169.5	42
174–181	177.5	27
182–189	185.5	8

The mean value,

$$\bar{x} = \frac{(5 \times 153.5) + (18 \times 161.5) + (42 \times 169.5) + (27 \times 177.5) + (8 \times 185.5)}{5 + 18 + 42 + 27 + 8}$$

$$= \frac{17\,070}{100} = 170.7 \text{ cm}$$

A tabular method is used to determine the $\sum [f(x - \bar{x})^2]$ values.

Height (cm)	Class mid-point x	$(x - \bar{x}) = (x - 170.7)$	$(x - \bar{x})^2$	Frequency (f)	$f(x - \bar{x})^2$
150–157	153.5	−17.2	295.84	5	1 479.2
158–165	161.5	−9.2	84.64	18	1 523.52
166–173	169.5	−1.2	1.44	42	60.48
174–181	177.5	6.8	46.24	27	1 248.48
182–189	185.5	14.8	219.04	8	1 752.32

The column on the extreme right of the table is summed, giving $\sum [f(x - \bar{x})^2]$ as 6 064.

The standard deviation, $\sigma = \sqrt{\dfrac{\sum[f(x-\bar{x})^2]}{\sum f}}$

$$= \sqrt{\dfrac{6\,064}{100}} = \sqrt{60.64}$$

$$= \mathbf{7.79}, \text{ correct to 3 significant figures.}$$

Further problems on the standard deviation of grouped data may be found in the following Section (25.3) (Problems 12–15), page 449.

25.3   Further problems

Ungrouped data

1. The time to produce a component in a factory was measured ten times and the results are as shown. Calculate the mean and the standard deviation correct to 3 decimal places.

    Time in minutes: 10.1, 9.9, 10.6, 11.0, 10.2, 10.8, 9.9, 10.6, 10.2 and 10.7   [10.4, 0.369]

2. The percentages of defective items produced by an automatic process, sampled 10 times during a day are as shown:

    3.34, 2.96, 3.43, 2.87, 2.50, 3.27, 2.90, 3.04, 3.13 and 3.46

    Determine the standard deviation from the mean.
    [0.283 0]

3. Calculate the mean and standard deviation of the set of numbers:

    10, 8, 5, 9, 4, 10, 3, 6, 8 and 7   [7, 2.323 79]

4. Determine the standard deviation from the mean value for the numbers, shown below correct to 5 significant figures.

    3 461, 3 449, 3 452, 3 451, 3 455, 3 463, 3 458, 3 463, 3 453, 3 440, 3 444, 3 457, 3 448, 3 461, 3 440   [7.474 4]

5. The population of a village varied in a ten-year period as shown. Find the mean value and standard deviation of the population over this period correct to two decimal places.

    1 721, 1 689, 1 680, 1 677, 1 703, 1 724, 1 734, 1 743, 1 751 and 1 790   [1 721.2, 33.597]

6. The mass in kilograms of eight children at birth was:

    2.50, 4.09, 3.18, 4.31, 3.18, 3.63, 4.54 and 4.77

    Find the standard deviation of these weights correct to four significant figures.   [0.734 5 kg]

7. The mass in tonnes of steel supplied to a factory during a twenty-week period is as shown:

395, 395, 403, 431, 455, 471, 470, 474, 475, 480, 494, 469, 463, 511, 497, 513, 484, 517, 555 and 560

Calculate the mean and standard deviation correct to four significant figures.   [475.6 tonnes, 44.51 tonnes]

8. The molarity of a given solution of oxalic acid was determined by 20 workers with the following results.

0.100 2M   0.099 8M   0.099 9M   0.099 7M
0.100 4M   0.099 8M   0.100 1M   0.099 9M
0.099 9M   0.100 0M   0.100 3M   0.099 9M
0.100 2M   0.100 3M   0.100 2M   0.100 3M
0.100 3M   0.099 9M   0.100 4M   0.100 0M

Determine the mean of the results and the standard deviation from the mean. $[\bar{x} = 0.100\,08, \sigma = 0.000\,214]$

9. The percentage composition of the elements in a new product was determined. The results for nitrogen for 68 tests are given below.

Nitrogen (%)	3.40	3.50	3.60	3.70	3.80	3.90	4.00	4.10	4.20	4.30
No. of tests	2	3	8	12	17	13	8	4	0	1

Determine the mean value and the standard deviation of the distribution. $[\bar{x} = 3.80\%, \sigma = 0.174\%]$

10. The yields from a chemical process under reproducible conditions were

60%, 62%, 59%, 66%, 57%, 59%, 61%, 63%, 58%, 62%

Find the mean and standard deviation from this distribution.
$[\bar{x} = 60.7\%, \sigma = 2.53\%]$

11. The percentage fat content of 26 samples of milk analysed in one day gave the following results.

3.96   3.92   3.91   3.82   3.82
3.87   3.78   3.71   3.92   3.93
3.90   4.00   3.89   3.69
3.89   4.04   3.76   4.02
3.86   4.08   3.99   3.88
4.14   3.94   3.93   4.05

Determine the mean and standard deviation of the distribution.
$[\bar{x} = 3.91\%, \sigma = 0.107\%]$

Grouped data

12. The quantity of electricity used by an office over a 50-week period is shown below. Determine the standard deviation of the distribution, correct to 2 decimal places.

Consumption (kW h)	71–79	80–88	89–97	98–106	107–115
No. of weeks	1	3	6	8	12

Consumption (kW h)	116–124	125–133	134–142	143–151
No. of weeks	8	5	4	3

[17.33 kW h]

13. The length in millimetres of a sample of bolts is as shown below. Calculate the standard deviation, correct to 4 significant figures of the sample and estimate the standard deviation of the population from which the sample is drawn.

Length (mm)	165	166	167	168	169	170	171
Frequency	5	14	18	28	36	29	29

Length (mm)	172	173	174	175	176	177
Frequency	24	19	15	6	3	2

[$\sigma = 2.605$, $s = 2.611$]

14. The percentage of chlorine in a chloro-hydrocarbon is estimated 80 times with the following results:

21.83	21.84	21.85	21.86	21.87	21.88	21.89	21.90	21.91	21.92	21.93
1	1	2	6	8	10	28	13	6	3	2

Determine, correct to 3 significant figures, the standard deviation of the distribution. [0.018 7]

15. The resistance in kilohms of a sample of 40 similar resistors drawn from a large batch was measured and the results are as shown.

Resistance (kΩ)	1.24–1.26	1.27–1.29	1.30–1.32	1.33–1.35
Frequency	3	4	4	10

Resistance (kΩ)	1.36–1.38	1.39–1.41	1.42–1.44	1.45–1.47
Frequency	10	5	3	1

Calculate the standard deviation of the sample, and estimate the standard deviation of the batch from which the sample was drawn. [$\sigma = 0.051\,2$ kΩ, $s = 0.051\,8$ kΩ]

Index